KT-371-691

Beacon Hill

High House

Stone Ho.

WEST THURROCK

The Breach

MARSH

West Thur

LONG REACH

Marsh Street

Little Brook

Greenhithe

ST. CL

Cotton F^m

Stone

llege

Horn. Cros

18

ord

T.Pike

17

Knockho

16

Hedge Place

Stone Castle

ord rent

Roman

Road

wder lls

Black Dale

Lower Dale

Gore F^m

Darent Wood

Bean

MAP
of a
NATION

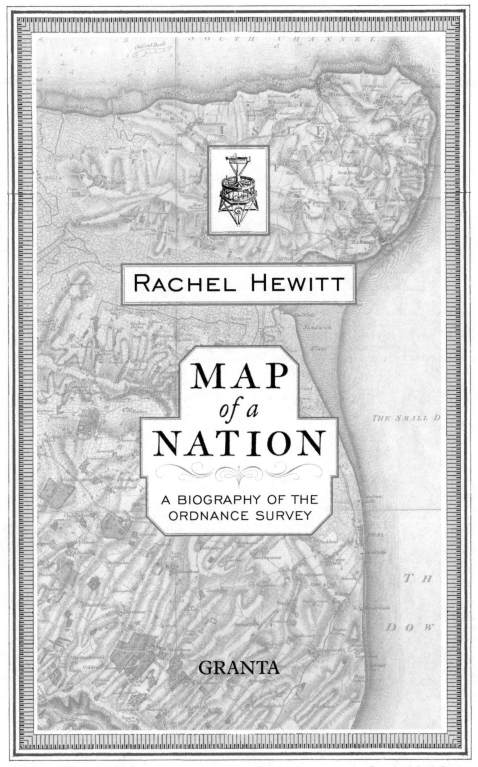

RACHEL HEWITT

MAP
of a
NATION

A BIOGRAPHY OF THE
ORDNANCE SURVEY

GRANTA

Scale of Statute Miles.

Furlongs

Miles.

Published 1st June 1810, by Lt. Colt. Mudge, Tower.

Engraved at the Drawing Room
in the Tower by Benjn. Baker & Assistants.
—— The Writing by Eben. Bourne.

Granta Publications, 12 Addison Avenue, London W11 4QR

First published in Great Britain by Granta Books 2010

A CIP catalogue record for this book
is available from the British Library.

5 7 9 10 8 6

ISBN 978 1 84708 098 1

Printed and bound in Great Britain by MPG Books Ltd, Bodmin, Cornwall

For Pete

Contents

List of Illustrations

ILLUSTRATIONS IN TEXT

Lost and Found

ON THE EVENING OF 16 April 1746 two men dined together in a castle on the south bank of the Beauly River, in Inverness-shire in Scotland. One was nearly eighty years old, stout but frail. His face was twisted by adversity and guile and dominated by an enormous yellowing periwig. Beneath it, heavily lidded eyes regarded his guest with concern. The second man was much younger, twenty-six years old. He had a naturally clear complexion, delicate lips and large eyes that were contorted with despair. In the months that followed, a tide of pamphlets, newspaper reports and biographies would speculate about the conversation that passed between the two men that evening. The younger, it was believed, had collapsed on the older, Simon Fraser, 11th Lord Lovat, in hysterical grief. 'My Lord,' he had cried, 'we are undone! My Army is routed! What will become of poor Scotland?' He was said to have fainted dead away on Lovat's bed.

Charles Edward Stuart had good reason for such misery. Since the so-called Glorious Revolution of 1688, which had seen the 'abdication' of his grandfather James II (VII of Scotland), there had been several armed rebellions aimed at reinstating the Stuart monarchy. Most Lowland Scots and the greater part of the English had not supported them: they had welcomed, or at least not strongly opposed, James II's departure. But the Scottish Highlands were different. The inhabitants of this often sublimely beautiful region were organised into clans to whose chiefs they looked for security and authority. For both religious and political reasons, the clans were drawn much more strongly to the absolutist and often Catholic Stuart kings than to the Protestant monarchs who reigned over Britain after 1688. In 1715 and 1719, serious Jacobite (from *Jacobus*, the Latin for James) uprisings had broken out in the Highlands, with the aim of ousting the Hanoverian British king, George I, and restoring the Stuarts. The situation then largely quietened for a quarter of a century until, in early 1744, whispers of renewed unrest started to circulate. After an initial attempt at rebellion failed, in July 1745 James II's grandson Charles Edward Stuart, known as the 'Young Pretender', had travelled from exile in Italy to board a frigate at Saint Nazaire on France's west coast. After a voyage fraught with danger, he had landed a month later on Eriksay in the Western Isles of Scotland.

Over the ensuing months, the Young Pretender had gathered a Jacobite army of several thousand, largely consisting of Highlanders. He had pronounced James II 'King of Britain', gained control of Edinburgh, and won a devastating victory against George II's army at Prestonpans, a few miles east of Scotland's capital. Stuart had then led his troops down through Scotland into England, arriving as far south as Derby on 4 December 1745, a mere 130 miles from the Hanoverian capital. North Britain's most popular newspaper, the *Caledonian Mercury*, described in fearful awe how 'the desperate Highlander's trusty broadsword and targe [is] headed by a person who can lie on straw, eat a dry crust, dine in five minutes and gain a battle in four'. But as the Hanoverian troops set out for Derby to face the Jacobite army, their opponents had suddenly, and quite unexpectedly, done an about-turn. The Jacobites' commander, Lord George Murray, had never much respected the Young Pretender, whom he considered something of a reckless adventurer. Should the rebels march on London, Murray had predicted a resounding defeat at the hands of the London militia. And to Charles Edward Stuart's utter dismay, his commander's motion for a retreat had won the vote.

Over the following four months, the Jacobite soldiers had travelled back into Scotland. George II's army was placed under the command of his fanatical, rotund son, William Augustus, the Duke of Cumberland, who led his men in pursuit of the withdrawing Jacobites. After four months of sidestepping one another in a martial *pas de deux*, on 16 April 1746 the Hanoverians and the Jacobites finally met on Culloden Moor, a few miles north-east of Inverness. Stuart's army was sadly dispersed, ravenously hungry and had not slept for two days. By contrast Cumberland's men were well rested, well fed and outnumbered their opponents by around three thousand. The Battle of Culloden began chaotically at around one o'clock in the afternoon. Forty-five minutes later it was over. In that short duration, the time it takes to enjoy a soak in the bath, between 1500 and 2000 Jacobites lost their lives. Only fifty of Cumberland's men had died. It was the last military battle to this day to be fought on the island of Great Britain. Stuart had not been able to watch its awful final throes. He had fled, ignominiously, with the cry 'Run, you cowardly Italian!' ringing in his ears, riding sixteen miles westward, towards Lovat's old stone fortress, Castle Dounie.

In the hours that followed the Battle of Culloden, the King's soldiers exercised little restraint in their retribution against the surviving rebels. One Hanoverian redcoat recalled that the 'moor was covered in blood; and our men, what with killing the enemy, dabbling their feet in the blood, and splashing one another, looked like so many butchers'. A decade after the event, the Scottish writer Tobias Smollett described the horrific aftermath of the battle:

> [The King's soldiers] laid waste the country with fire and sword . . . all the cattle and provision were carried off; the men were either shot upon the mountains, like wild beasts, or put to death in cold blood, without form of trial; the women, after having seen their husbands and fathers murdered, were subjected to brutal violation, and then turned out naked, with their children, to starve on the barren heaths . . . Those ministers of vengeance were so alert in the execution of their office, that in a few days there was neither house, cottage, man, nor beast, to be seen in the compass of fifty miles; all was ruin, silence, and desolation.

The bloodshed during and after Culloden earned Cumberland the nickname 'the Butcher' and the bitter quip that he should be made a member of the Worshipful Company of Butchers.

On Stuart's arrival at Castle Dounie, Lovat realised that neither man was safe for long. After a long, fraught career, Lovat had become infamous for his duplicity. Switching his allegiance back and forth between the Pretender's ancestors and those of the Hanoverian king had landed him with a conviction for high treason, a death sentence (later quashed through Lovat's notorious charm), outlawry, arrest in France as a double agent and a significant spell in the Bastille prison. During the rebellion of 1745–6 Lovat, who was then an elderly man, had tried to stay out of sight, placing his son in charge of the Fraser clan and commanding him to support the Young Pretender's cause, while he himself feigned illness and took to his bed. But as the rebellion had escalated and even seemed likely to succeed, Lovat's enthusiasm had got the better of him. He had left his bed to openly rally the rebels against the King with such zest that one journalist described him as 'the chief Author and Contriver of this wicked Scene'.

When Stuart told Lovat of the relish with which the King's forces had dispatched the rebels at Culloden, the old man realised how false was his hope that, should the rebellion fail, no government 'would be so cruel, as to endeavour to extirpate the whole Remains of the Highlanders'. An impassioned conversation took place between the two men in which Lovat tried to persuade the Young Pretender to make one further attempt, invoking the memory of Robert the Bruce's multiple defeats and his eventual success. But Stuart was disconsolate, and sought only a safe escape back to the Continent. Lovat concluded that, as an old sparring partner had recently put it to him, 'the double Game you have played for some Time past' was up.

On the morning of 17 April 1746, Lovat, Stuart and their retinues fled west into the Scottish Highlands. Numerous pamphlets published later that year liked to imagine them scaling the peak of Sgúrr a'Choir Ghlais, one of Scotland's most inaccessible mountains. From the hill's summit, over 3500 feet above sea level, the two men would have gazed over an extensive panorama of destruction, as George II's redcoated soldiers mercilessly pursued the fleeing rebels: 'Heaps of their Men lying in their Blood; others flying before their Enemies; Fire and Sword raging everywhere, and a great deal of it upon [Lovat's] own Estate, and among his Tenants.' The redcoats pursued a brutal strategy to subdue rebel clans to the King's authority in the months after the battle. They seized livestock, horses, gold and coin from the Highlanders, and burned houses and farms to the ground under a variety of pretexts, such as the failure to deliver up fugitive rebels, refusal to surrender weapons or sometimes for the sheer joy of pillaging. One soldier, who was posted at Fort Augustus in the summer of 1746, recalled: 'we had near twenty Thousand Head of Cattle brought in, such as Oxen, Horses, Sheep, and Goats, taken from the Rebels (whose Houses we also frequently plundered and burnt) by Parties sent out for them, and in Search of the Pretender; so that great Numbers of our Men grew rich by their Shares in the Spoil'. The redcoats were intent on driving out supporters of the Young Pretender from every corner of the Scottish Highlands. They sought to 'put an end' to the spirit of Jacobitism 'so effectually now, that it will never be able to break out again'.

Looking down on this scene of murder and horror, Lovat and Stuart decided to part company. It was the last time they would meet. The Young Pretender fled towards the Western Isles but Lovat made his way alone to Loch a'Mhuillidh, a small piece of water seventeen miles west of Castle Dounie in the midst of Glen Strathfarrar, where he turned a bothy on a small island midway down the loch into a temporary home. When the King's soldiers closed in on the glen, his hideaway became untenable. Evading the redcoats, the old man travelled south-west, swiftly disappearing into a seemingly impenetrable cluster of peaks which he had known since childhood. Detailed maps of the Scottish Highlands now allow anyone to name these towering mountains: Sgúrr na Lapaich, Carn nan Gobhar, Sgúrr nan Clachan Geala, Braigh a'Choire Bhig, An Riabhachan, Mullach a'Ghlas-thuill. We can trace their dramatic ascensions in closely bunched orange contour lines, quantify the immensity of their altitudes and pick out a safe path among teetering rocky outcrops, fields of boulders, vigorous streams and obscured lochs. But in 1746, the King's soldiers had no such assistance. They were beaten back by the formidable, seemingly unreadable scenery, and Lovat was able to make his way sixty miles south-west, unchallenged, around Loch Mullardoch, through the Kintail Forest, over the Five Sisters peaks, between Loch Hourn and Loch Quoich, through Glen Dessary, until, just short of the Scottish mainland's western extremity, he reached Loch Morar. Here Lovat secreted himself on an island towards the west end of the vast lake with his Roman Catholic bishop, his secretary and a band of followers, and for the time being considered himself safe.

THROUGHOUT THE COURSE of the rebellion, the redcoats had suffered from poor intelligence about the Highlands' geography. This was exacerbated by inadequate maps of the region. The surveying of strategic sites such as fortifications and barracks across Scotland had been an integral part of military practice since the seventeenth century, and the flatter, gentrified landscapes of the Lowlands possessed a number of estate and county maps. But there was no complete map of the whole Scottish nation that provided an accurate overview of the Highlands' mountains, rivers, lochs and paths. In March 1746, only a month before Culloden, Captain Frederick Scott had written to his commander from Castle Stalker, a fifteenth-century tower-house poised on a picturesque islet on Loch Laich, in Appin, on Scotland's west coast. Scott's letter betrayed his frustration that 'this Place is not marked on any of our Maps'. He had also noted the differences between the place names on his chart and those actually used by the locals.

The same deficiencies hampered the soldiers in their pursuit of the rebels in the battle's aftermath. Charged with policing the country that Scott had found so difficult to negotiate, Major General John Campbell had commented: 'by the Map of the country it appears very easy & a short cutt to cross over from Appin by the end of Lismore to Strontian'. In reality, this required island-hopping across Loch Linnhe, followed by a lengthy pathless scramble over the steep rocky overhangs and mountain streams that punctuate the scrubby cluster of hills south-east of Strontian. No wonder Campbell had concluded that, contrary to the map's advice, 'I have had [the route] viewed and it is impracticable in every respect.'

The problems caused by these poor maps were aggravated by the extreme inaccessibility of the country in which many of the rebels were hiding. The soldiers complained of 'the Want of Roads . . . the Want of Accommodation, the supposed Ferocity of the Inhabitants, and the

Difference of Language'. Official reports warned that other parts of the Highlands were 'extensive, full of rugged, rocky mountains, having a multiplicity of Cavities, & a great part of these cover'd with wood & brush. The places are of difficult access, thro' narrow passes, & most adapted of any to concealment of all kinds.' A newspaper gloomily confirmed that 'there are Hiding-places enough'. Furthermore, the Highlanders' superior intelligence networks always seemed to be one step ahead of the soldiers. These networks extended north, south, east and west, giving advance warning of the soldiers' approach to fugitive rebels. One of the King's officers noted that 'our Detachments have always been betrayed by People that the Rebels had on the top of the High Hills, who by some Signall agreed on could always convey any Intelligence from one to another in a short space of time'. Conversely the redcoats found that their own 'Intelligence [was] very difficult to obtain, notwithstanding Promises of reward & recommendation of Mercy'.

Because of the Hanoverian troops' lack of good geographical intelligence in the face of such complicated scenery, Lord Lovat, a partially crippled near-octogenarian, was able to evade them for almost two months. He was eventually caught in the second week of June, when a division of redcoats towed their own boat over the peninsula that separated Loch Morar from the sea and rowed to his island. There they found 'that wily old Villain' hiding in a hollow tree, betrayed only by a glimpse of tartan through a chink in the trunk. Most of Lovat's followers had already dispersed, melting into the surrounding mountains like ghosts. The redcoats' Commander-in-Chief, the Duke of Cumberland, was intensely relieved: 'The taking [of] Lord Lovat is a greater Humiliation & Vexation to the Highlanders than any thing that could have happened,' he wrote jubilantly. 'They thought it impossible for any one to be taken, who had these [mountainous] Recesses open, as well known to him to retire to.'

Greatly enfeebled by his flight, Lovat was carried on a litter to Fort William at the southern end of the Great Glen, then to Fort Augustus at its heart and finally to the Tower of London. On 18 March 1747 he was found guilty of high treason for the second time in his life. But this time there was no escape. On 9 April Simon Fraser, 11th Lord Lovat, chief of

Clan Fraser, was executed, earning the dubious accolade of being the last person in Britain to be publicly beheaded. Meanwhile, the 'burning scent' of the Young Pretender had gone stone cold. Despite the assertion of one Hanoverian commander that 'I should with infinite Pleasure walk barefoot from Pole to Pole' to find him, Charles Edward Stuart had been able to secrete himself away in Scotland's western islands, before finally securing a voyage home to Italy, and safety.

THE PERIOD THAT Lord Lovat spent in hiding from the King's troops marks a pivotal moment in the history of the nation's maps. It became unavoidably clear that the country lacked a good map of itself. Britain had national maps, but prior to the eighteenth century they had served a mostly symbolic function. When, in 1579, Christopher Saxton had published the first atlas of England and Wales, almost every sheet of his thirty-five maps was adorned with Elizabeth I's arms: Saxton's surveys were royalist insignia. Just over thirty years later, William Camden published the first English translation of his *Britannia*, in which his aim was overtly nationalistic. Camden wanted to celebrate 'the glory of the British name' by 'restor[ing] antiquity to Britaine, and Britaine to his antiquity', and his book's elaborately engraved frontispiece displayed a map surrounded by ancient icons of British patriotism, such as Stonehenge, to encapsulate this joined love of country and history. In 1610 John Speed had presented a county atlas to Britain's map-reading public, whose opening pages offered an exquisite, highly coloured map of 'Great Britaine and Ireland' beneath a coat of arms. An image of London in the top-right corner was balanced with one of Edinburgh in the top left, hinting perhaps at a convivial relationship between these capitals.

Although many early modern national maps were potent emblems of power, control, ownership and nationalism, their accuracy was often highly questionable. Surveys of nations were rarely made by measuring the ground itself, which was a time-consuming, skilled and expensive process. Instead,

1. The frontispiece to the 1600 edition of William Camden's Britannia.

map-makers tended to amalgamate information from a host of pre-existing maps to produce 'new' surveys, inevitably leading to the replication of old errors. And even when early surveyors did conduct their own measurements, many of their instruments and practices were the cause of further errors. The telescope was only invented at the beginning of the seventeenth century and it was not used at all in surveying instruments until 1670, which meant that map-makers' observations were restricted by the capabilities of their naked eyesight. After 1670 until around the middle of the eighteenth century, telescopic lenses were still liable to cause problems in the form of 'chromatic aberration', incongruous fringes of colour resulting from the lenses' failure to focus all colours to the same point. Accuracy in early map-making was further compromised by the susceptibility of many surveying instruments' materials to changes in temperature, which made them expand and contract. Chronometers and clocks had a frustrating tendency to speed up, slow down or stop altogether. And the measuring scales with which instruments were engraved were not sufficiently minute for a high level of accuracy, and mathematicians possessed no agreed method with which to reconcile conflicting results. Different maps of an identical piece of land could therefore look very dissimilar, and although estate maps were generally reliable enough for their owners' purposes, over larger areas errors proliferated. In 1734 a map-maker called John Cowley superimposed six existing maps of Scotland, and in doing so he created a striking image of radical divergence and disagreement. Some early map-makers did not even pretend to aspire to perfect truth and accuracy. The seventeenth-century Czech engraver Wenceslaus Hollar provided a telling admission of the limitations of early methods when he inscribed this motto onto one of his maps of London: 'The Scale's but small, Expect not truth in all.'

The 1745 Jacobite Rebellion vividly revealed the lack of, and need for, a complete and accurate map of the whole island of Britain. Over the next one hundred years, the myopia that had left Lovat's pursuers groping in the faintest of half-lights would lift agonisingly slowly until, at long last, Britons would be able to carry in their mind's eye, or in their pockets, a full and truthful image of the nation in which they lived.

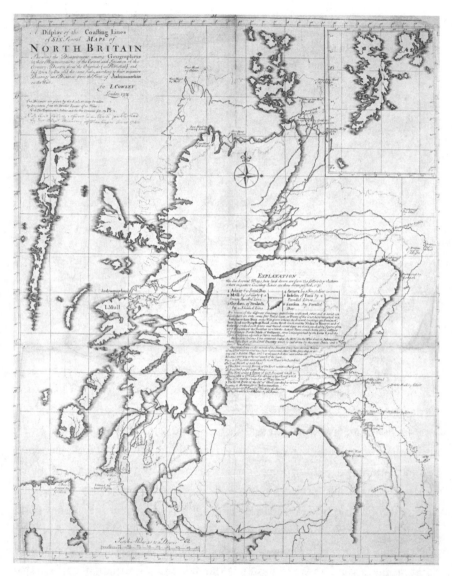

2. A Display of the Coasting Lines of Six Several Maps of North Britain, *compiled by John Cowley in 1734, and revealing the severe discrepancies between early-eighteenth-century maps of Scotland.*

The story that follows is an account of how, why and where Britain's national mapping agency came into being. Today the Ordnance Survey is easy to take for granted. From the Shetland Isles in the furthest north-easterly reaches of Britain,

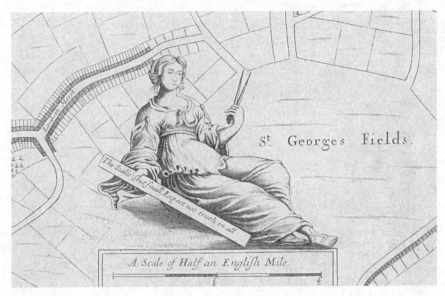

3. A detail from A New Map of the Cittyes of London and Westminster, *published around 1680 and attributed to Wenceslaus Hollar.*

to Penzance and the Scilly Isles in the south-west, it has surveyed every mile of the nation, dividing it into a cubist jigsaw of overlapping sheets. Today the Ordnance Survey's *Explorer* series covers Britain in 403 maps, and the *Landranger* series in 204. Each region, however inaccessible and forbidding, however urban and congested, possesses its own highly detailed map on a range of scales and with near-perfect accuracy. These surveys intimately trace the course of all the roads, national trails, minor footpaths, bridleways, power cables, becks, gills, streams, brooks and field boundaries that bisect a landscape of miniature woods, speckled scree slopes and orange contour lines. Black and blue icons represent sites of public interest, ranging from public toilets and telephones to National Trust properties and ancient ruins. In addition to its folded maps, the Ordnance Survey is also the source of the information contained in many A–Z street atlases, road maps, and walking and tourist guides published in the United Kingdom. Its depictions of the landscape fill the screens of 'satnavs' and similar hand-held navigational devices. The Ordnance Survey's massive digital database, the OS Master-Map, provides geographical data to a host of hungry consumers: national and local governments, transport organisations, the

Army and Navy, the emergency services, the National Health Service, housing associations, architects and insurance companies.

The national mapping agency has established a secure place in the affections of the modern British public. Ramblers lovingly fold dog-eared maps into plastic wallets to protect them from the country's soggy climate. Hikers become cheerily infuriated when the maps, like the nation's newspapers, refuse to crease cleanly at the required section and flap as capriciously as kites in sudden gusts of wind. At home, outspread across a carpet or table, OS maps display Britain's landscape in such vivid and exact detail that walkers can conjure in their imagination the course of eagerly anticipated or fondly remembered hikes, translating contours into visionary mountains and thin black lines into fantastical drystone walls. And the Ordnance Survey has left traces of its remarkable undertaking in the very landscape it mapped. Triangulation stations, or 'trig points', are those squat concrete pillars that hunker down amid rocks, heather or grass in the nation's most panoramic spots, precisely marking the points at which the OS's mapmakers placed their instruments. There are over five thousand in England alone, and trig points have become an obsession in their own right: members of the United Kingdom's 'Trigpointing Community' hunt after the pillars like prey. They have been described as one of the most beloved 'icons of England'.

This book describes the circumstances that led to the creation of the First Series of Ordnance Survey maps. It is usually said that the Ordnance Survey was born in 1791, when the Master-General of the Army's Board of Ordnance paid £373 14s for a high-quality surveying instrument and appointed the first three employees of a nationwide map. But in the following pages we will see how tricky it is to assign a decisive date to the start of that project. Arguably it really commenced in 1783, when a team of French astronomers approached the Royal Society to instigate a measurement between the observatories at Paris and Greenwich, a measurement that would form the backbone of the Ordnance Survey's early work. The year 1766 is also a key moment in its gestation, when a far-sighted Scottish surveyor petitioned the King to set up 'a General Military Map of England'. And it could even be suggested that the Ordnance Survey was first conceived twenty years earlier, from a map of Scotland that was conducted by that same surveyor in the wake of Culloden. Recognising all these landmarks that paved the way to the Ordnance Survey's completion of its First Series, this book reflects on what it was like to inhabit a nation that lacked a map of itself and how it felt to acquire that mirror.

At the same time as the Ordnance Survey was coming into being as a deeply loved and much-needed institution, the United Kingdom itself was also being formed through unions between England, Scotland and Ireland, and through the growth of national networks of travel, tourism and communications. Mile by mile, the Ordnance Survey painstakingly created an exquisite monochrome image of that new state, and the following pages describe how the changing face of the nation placed ever-increasing demands on the maps. But those requirements often led the map-makers astray from their principal focus, the creation of the First Series, so that – thanks to a wealth of fascinating digressions and distractions – it took seventy-nine years before those maps were completed in 1870. When that moment finally occurred, it was a landmark in Britain's history, similar to the publication of the first edition of the *Oxford English Dictionary* in 1928. Both events gave the reading public deeper and wider insights into the physical and intellectual landscapes in which we live our lives.

For its readers today, the name 'Ordnance Survey' may conjure up a Betjemanesque image of cycle-touring or hiking through idyllic pockets of the British countryside, with only a map or guidebook for company. But as we shall see, its surveys emerged largely from a long history of *military* map-making, in which that landscape was tramped over not by ramblers with a walking stick and a bag of sandwiches, but by redcoated soldiers.[1] During the Enlightenment, cartography became a symbol of the taxonomic intellect of the rational thinker, but philosophers were not the only ones to benefit.[2]

---

1 *Survey* has a much wider range of meanings than *map*, referring to sweeping representations of the world in words as well as in images. But to avoid excessive repetition of the word 'map', I will use the two terms largely interchangeably, clearly differentiating them where necessary.
2 The word *cartography* refers to 'the drawing of charts or maps' and derives from the French *carte*, card or chart, and Greek γρχφία, writing. Although it is a useful term that crops up frequently in this book, it is important to note that the word was only coined around 1859 and therefore was not used by the Ordnance Survey's earliest map-makers and readers.

Soldiers also became more 'map-minded' in this period, and the military was one of the principal crusaders in the improvement of British cartography over the course of the eighteenth century.

This book reveals a nation waking up, opening its eyes fully for the first time to the natural beauty, urban bustle and tantalising potential of its landscape. The ensuing chapters tell the stories of those who directed the Ordnance Survey's early course: scientists, mathematicians, leaders, soldiers, walkers, artists and dreamers. They relate the hostility and intermittent friendship that characterised the fraught relationship between England and France over a century in which a prolonged and violent series of wars repeatedly set one against the other: an awkward backdrop, as we can imagine, for Anglo-French map-making collaborations. They describe the inanimate celebrities of the Ordnance Survey's project – the most advanced, accurate, intricate measuring instruments that had ever existed; and the reactions of the maps' first consumers, which ranged from loving adulation to protest. But above all, this book recounts one of the great British adventure stories: the heroic tale of making from scratch the first complete and truly accurate maps of a nation.

CHAPTER ONE

'A Magnificent Military Sketch'

I

ON 29 JUNE 1704, in a north-western suburb of Edinburgh, Mary Baird, the well-to-do wife of a successful merchant called Robert Watson, gave birth to her eleventh child: a son they called David. David Watson's earliest years were spent in Muirhouse, which is now a sprawling housing estate to the north of Scotland's capital, but was then a prosperous area populated by traders, where his father had recently purchased an enviable house with surrounding land. David was the baby of a large family, whose siblings ranged in age from two to fourteen, and around whom a host of affluent aunts and uncles clustered. But at the age of eight the young boy's life underwent a dramatic upheaval: David's father suddenly died. And in that same year his eldest sister Elizabeth, whose portrait shows her to have had kind but nervous eyes, a hesitant smile and luxuriant auburn hair, married into one of Scotland's most influential families.

Elizabeth Watson's suitor was a 27-year-old lawyer called Robert Dundas, heir to a dynasty whose command of the Scottish legal system in the eighteenth century has led it to be termed the 'Dundas despotism'. His family played such significant roles in public life that their Victorian biographer concluded that 'to describe, in full detail, the various transactions in which they took the leading part would be to write the history of Scotland during the greater part of the eighteenth century'. Robert Dundas himself was a star in the legal firmament. He had been made Solicitor General of Scotland at a precociously young age, and rapidly attained the positions of Lord Advocate, Dean of the Faculty of Advocates and Lord President of the Court of Session. His reputation was immense: he was described as 'one of the ablest lawyers this country ever produced'.

Gruff, with a tendency to irritability, Dundas was hardly the smooth and urbane lawyer one might expect. Physically he was not prepossessing. A friend described him as 'ill-looking, with a large nose and small ferret eyes, round shoulders' and 'a harsh croaking voice' with a robust Lowland Scottish accent. He was a man of unpredictable extremes, whose temper was said to be characterised by 'heat' and 'impetuosity' but matched with 'abundance of tact'. He drank prodigiously – a bill for wine at his mansion over a nine-month period came to the equivalent of £11,000 – but he was still able to conduct his work with clarity and precision, even after 'honouring Bacchus for so many hours', as the novelist Walter Scott put it. And despite this erratic character, Dundas was highly respected. It was said that within three sentences his listener was invariably swayed by such 'a torrent of good sense and clear reasoning, that made one totally forget the first impression'. Taking into account his inheritance of the stunning and capacious estate of Arniston on the fertile banks of the river Esk in Midlothian, Robert Dundas was an entirely desirable prospect for Elizabeth Watson.

At the time of her marriage, Elizabeth had just lost her father, and it is possible that her mother was also deceased by this point, as she and her new husband became something of surrogate parents to her eight-year-old brother David. His contact with the Dundas family would change the course of David Watson's life. It was an exciting time: the atmosphere of Lowland Scotland was alive with optimism in the early eighteenth century. In 1707, an Act of Union had officially united Scotland and England into 'Great Britain' and, in the emerging 'Age of Enlightenment', the intellectual climate of this young nation was buzzing with a new intensity. The Dundases had played central roles in the Union and they considered themselves to be standard-bearers of the Scottish Enlightenment. His intimacy with this influential family would open up an array of opportunities to the young David Watson.

THE ONSET OF an Age of Enlightenment in Britain was enormously helped by two events that had occurred in 1687 and 1688. The publication of Isaac Newton's *Philosophiae Naturalis Principia Mathematica* (*Mathematical Principles of Natural Philosophy*), and the 'abdication' of King James II of England, followed by the arrival of his Protestant daughter Mary and her Dutch husband William as joint monarchs, were events that were both widely considered to demonstrate the potential of human powers of reason. Newton's *Principia Mathematica* revealed that, despite its semblance of arbitrariness and confusion, the cosmos was really a unified system. And the 'Glorious Revolution' that was marked by James II's departure set a new precedent for the relationship between the government and the Crown, founded on the rational principles that Britons deserved 'the right to choose our own governors, to cashier them for misconduct, and to frame a government for ourselves'. Following these momentous occurrences, the philosophers of the British Enlightenment emphasised that science, politics, geography, art and literature should be guided by reason above all else. They were confident that powers of rationality could uncover the truth about the world. One of the key aspirations of Enlightenment thinkers was the creation of a vibrant 'public sphere' in which every member of the populace would feel free to '*Sapere aude!*' – to dare to think for themselves.

The Enlightenment had important consequences for maps and mapmaking in Britain. A new ideal was dangled before surveyors: that cartography could be a language of reason, capable of creating an accurate image of the natural world. Enlightenment thinkers invested maps with the hope that the repeated observation and measurement of the landscape would build up an archive of knowledge that approached to perfect truth. The French *philosophes* Diderot and d'Alembert saw maps as images of the ordered, rational mind, and they compared their own *Encyclopédie* to 'a kind of world map' whose articles functioned like 'individual, highly detailed maps'. The natural philosopher Bernard de Fontenelle described the zeitgeist of the Age of Enlightenment as an '*esprit géométrique*'. The twentieth-century Argentine writer Jorge Luis Borges has encapsulated the desire of thinkers in this period for ultimate 'Exactitude in Science' by liken-ing it to the ultimately doomed objective to make 'a Map of the Empire' on the scale of one-to-one, whose 'size was that of the Empire, and which coin-cided point for point with it.'

Accuracy, or rather 'the quantifying spirit', thus became a new priority for map-makers in the eighteenth century, inspiring such dramatic advances in instruments and methods that, by the second half of the century, Britain was home to some of the most precise map-making and astronomical instru-ments in the world and the most diligent, rational surveyors. The emergence of relatively trustworthy maps had profound effects on the way they were used by the general populace. As we shall see later, the new maps assisted the process by which Britain's component regions were integrated into a unified nation. Progress in cartography occurred in parallel with the improvement of the nation's road networks, the innovations in coach design that made travel cheaper and less uncomfortable, and historical and cultural events that heralded a new dawn in the British tourist industry. Maps became hall-marks of an 'enlightened' mind and nation. And in 1720 a surveyor called George Mark issued a call to arms to the principal players of the Scottish Enlightenment, begging them to further the state of cartography: ''Tis truely strange why our Scotish [*sic*] Nobility and Gentry, who are so universally esteemed for their Learning, Curiosity and Affection for their Country, should suffer an Omission of this Nature . . . in what so much concerns the Honours of the Nation!'

DAVID WATSON GREW up in the early decades of the Scottish Enlightenment among a family who were enthusiastic sponsors of its values. In spite of a certain degree of anti-intellectual bluster on Robert Dundas's part, and his reputation for never having been 'known to read a book', Arniston's library was impressively stocked with travel-narratives, topo-graphical writings, atlases, maps and expensive globes. A theoretical knowledge of surveying was considered integral to the education of an enlightened gentleman, who was expected to be able to commission and judge maps of his own estate. Arniston House accordingly boasted an inspiring collection of surveying instruments and a series of cartographic depictions of the large and varied surrounding estate. We can imagine that David Watson, and Elizabeth and Robert Dundas's own children, looked on in fascination as well-known estate surveyors, and famous architects such as William Adam, laid measuring chains along the lengths of the youngsters' favourite avenues of trees, translating the familiar Midlothian landscape into numbers, angles and lines on a map.

The Dundases' enthusiasm for geography was such that they even attended the prestigious lectures on surveying that were delivered by the Edinburgh mathematician Colin Maclaurin. A child prodigy who was elected Professor of Mathematics at Aberdeen University at the age of nineteen, Maclaurin had so impressed Isaac Newton with his work that Newton had even offered to pay his salary. At Edinburgh, Maclaurin devised a rigorous course of mathematical education that emphasised the discipline's practical applications, especially to map-making. The *Scots Magazine* described how, in his lectures, Maclaurin 'begins with demonstrating the grounds of vulgar and decimal arithmetic; then proceeds to Euclid; and after . . . insists on surveying, fortification and other practical parts'. Maclaurin's lecture-theatre was an intellectual hothouse that produced a brood of illustrious surveyors, architects and mathematical instrument-makers such as Alexander Bryce, Murdoch Mackenzie, Robert Adam and James Short. Sitting in Maclaurin's audience, the Dundases, and perhaps Watson too, were among inspiring cartographical company.

As David Watson approached his mid teens, his sister and her husband applied themselves to furthering his career. Characteristically of younger relatives of the gentry, Watson expressed an interest in joining the Army. Robert Dundas used his influence to obtain a commission for his younger brother-in-law, and Watson duly spent much of his twenties and thirties on the Continent. He endured a 'long banishment at Gib[raltar]', where the British were building fortifications, and he enjoyed a martial 'Scuffle' during the War of the Polish Succession. But while Watson was fighting abroad, tragedy struck at Arniston. In November 1733 Elizabeth and Robert's young son came home from the local school in Dalkeith with signs of sickness. The symptoms rapidly worsened and revealed themselves as smallpox, and the boy was dead within a week. The couple's three other little children were infected and one by one between November 1733 and January 1734 they all died. By early December 1733 Elizabeth herself was 'confind to her Chamber and pretty much to her bed'. When her two small daughters died at the beginning of the new year, their mother was too weak to be told. By 6 February 1734 this poor woman had finally succumbed.

Robert Dundas retreated to London to mourn 'the best of mothers' and 'an incomparable wife'. But as one door closed for him, another was about to open. A couple of weeks before the arrival of the smallpox at Arniston, Dundas had paid a visit to a client in the Upper Ward of Lanarkshire, a region stretching south from the town of Carluke, about nineteen miles south-east of Glasgow. Sir William Gordon was the owner of two conjoined estates, Hallcraig and Milton, which extended west from Carluke along the dank northern edge of a small brook called Jock's Gill, then reached up onto wide fertile plains, before dropping down into the lush crook of the river Clyde, where the road stopped for want of a bridge. Sir William was a canny operator, one of the very few to have made money from the South Sea Bubble – the stockmarket crash that had devastated Britain's economy in 1720 – and he had been seeking legal advice from Dundas for over a decade. No doubt one of the chief attractions of Gordon's case for his lawyer was his young daughter Ann: a spirited, flirtatious and enormously beautiful woman. Her portrait, painted by the famous Joseph Highmore and now hanging on a staircase in Arniston House, shows large dark eyes slanted in readiness to laugh, fashionably alabaster skin and teak-coloured hair from whose arrangements a few unruly curls escaped. Elizabeth Dundas, to whom writing did not come naturally, had not been able to accompany her husband to Lanarkshire on this occasion and she laboured over a formal apology to Ann, hoping to 'have the happness [*sic*] to see' the eighteen-year-old woman in Edinburgh soon.

Robert Dundas's visit to Milton in October 1733 lasted only a few days. His carriage overturned on the bad roads that led out of Carluke on his way home, and he suffered what Elizabeth termed 'a truble in his bake' for many weeks. Upon returning to Arniston, Dundas too wrote to Ann. 'I won't omitt ane opportunity of writing to you however litle I may have to say,' he assured her. He was positive that, 'when I speak of affection[,] good oppinion and every good wish I am capable of, these will be no news to you'. Dundas designated Ann his 'rival wife', and he reflected ponderously that 'this is quite new for a man and two wifes to be all one'. He signed off by exhorting her to 'believe me Dear Anne' that he was 'with the greatest esteem and pleasure, as much yours as I can be . . . my Dear Girl'.

In the wake of Elizabeth's death only a few months later, Ann Gordon wrote back to Robert Dundas. She sympathised with 'the Loss You have . . . Made' but she firmly reminded him that 'You once gave me Reason to Pretend some Tittle to Your Heart'. With a mixture of sincerity and flirtatiousness, Ann openly informed this man, who was almost thirty years her senior, that 'I Intend to Pursue You with so Much Friendship & Contempt[,] Love and Indifference as Must Convince You'. He was easily convinced and the 'rival wife' soon became his lawful spouse. Ann and Robert's wedding took place four months after Elizabeth's death. There were few guests. In the circumstances of the recent devastation of Dundas's first family, a discreet wedding seemed appropriate. Dundas did not even inform his surviving son Robert, or 'Robin' as he was affectionately known, who was studying law at university in Utrecht, of his remarriage until three months after the fact. 'I did not incline to owne my mariage to any body,' he wrote.

Robert Dundas's second marriage to Ann Gordon did not mean that he excluded David Watson, his first wife's younger brother, from his affections. On the contrary, Robert and Ann too continued to take an active interest in Watson's career. In 1742 the Dundases were behind an alteration in his profession that capitalised on Watson's childhood love of maps. Dundas supplicated the Secretary of State for Scotland to help move his brother-in-law from the regular army to the Board of Ordnance. He successfully persuaded that minister to intercede with the Master-General of the Ordnance, and arranged for Watson to enter 'upon the Establishment of an Engineer'. David was instructed to 'immediately come here', to London, to clinch the deal, and the Secretary of State reflected with satisfaction that this 'would be a good beginning' for him. By 1743, thanks to Dundas's endeavours, Watson had taken up his new post as a military engineer under the employment of the Board of Ordnance. He became adept at reconnaissance, plotting march routes that transported armies between destinations. And one of his principal concerns was map-making.

A DEPARTMENT KNOWN as the 'Office of Ordnance' had existed in Britain since the late fourteenth century. The Ordnance was an adjunct to the Royal Arsenal, which was housed at the Tower of London, a large complex of buildings on the north side of the river Thames. Responsible for the administration of the monarch's armaments, arsenals and fortified castles, the body was renamed in 1518 as the Board of Ordnance, a name that stuck for almost 350 years. Ordnance is abbreviated from *ordinance*, a complex word that covers a host of meanings, all united by a general sense of 'that which is ordered or ordained', but the word's foremost association was military. Ordinance, or the French *ordonnance*, denoted an army's arrangement in ranks or lines and most importantly the 'ordinantia ad bellum', the military equipment, guns, cannons and explosives, for whose management the Board was responsible.

The Board of Ordnance was an independent military body, separate from the Army. Where regular soldiers answered to the monarch, employees of the Ordnance received their instructions from the Board. This was similar to the Navy, who operated under the thumb of the Board of Admiralty. By the end of the seventeenth century, the Ordnance's responsibilities were beginning to separate into two principal areas: artillery and engineering. On 16 May 1717 this polarisation was made official. Two companies were raised in the Board of Ordnance: the Royal Artillery and the Corps of Engineers. Guns and cannons were the business of the first; fortifications and harbours that of the second, whose rank and file were known as the Sappers. The intellectual sophistication of the Ordnance's staff meant that the Engineers and the Artillery were later jointly known as 'the scientific corps'.

Map-making was a responsibility, albeit a minor one, of both Ordnance companies. And it was also the concern of another separate body within the Board of Ordnance, which was tucked away in the White Tower at the heart of the Tower of London's complex. Here a Drawing Room could be found, in which a host of map draughtsman busied themselves. Comprised of civilians, not military men, from 1777 onwards young draughtsmen would train here from the age of eleven or twelve, receiving instruction in the conventions of military surveying and in mathematics, especially in trigonometry and geometry. To the Engineers, which David Watson entered, map-making was especially pertinent as a way of describing sites for fortifications. A Royal Warrant of 1683 stipulated that a military engineer 'ought to be well skilled in all parts of the mathematicks, more particularly in Stereometry, Altimetry, and Geodesia, to take Distances, Heights, Depths, Surveys of Land, Measure solid bodies' and to 'keep perfect draughts of every fortifications, forts and fortresses of our Kingdom'. Over the course of the eighteenth century the Board of Ordnance's employees produced a collection of large-scale surveys of specific spots in the country that were demarcated for the erection of forts and harbours.

For the senior officers involved in computation and advanced scientific techniques, its day-to-day work was mentally stimulating and could offer opportunities for travel across Great Britain, Europe, America, Canada and sometimes even further afield. Engineers had no choice but to roll up their sleeves and 'get their hands dirty', so the Corps often attracted those to whom practical work and fresh air were more important than dignity and reputation. It was a small operation and places were extremely limited, so prospective engineers with influential personal connections like David Watson exploited their advantage to the full.

AFTER ATTAINING a sought-after place in the Corps of Engineers, David Watson was delighted to find himself posted to Flanders in June 1743 to support King George II's troops in the War of the Austrian Succession.

Now in middle age, nearly forty years old, Watson was buoyant. He boasted a tall, athletic frame, large watchful eyes, a geometrically pleasing profile and a neat, determined mouth. Confident of his own abilities, and at ease in the presence of power, Watson achieved a degree of fame in Flanders by helping the King and his 25-year-old son William Augustus, the Duke of Cumberland, achieve a resounding victory against the French at the Battle of Dettingen. The Secretary of State for Scotland's private secretary reported excitedly that, although 'we do not live in an age, or in a land of Heros', nonetheless 'Mr Watson has gained universal character in Flanders'. But Watson's sojourn did not last long.

When the 1745 Jacobite Rebellion broke out, military commanders were initially reluctant to move soldiers deployed on the Continent back to Britain. But as Charles Edward Stuart's army penetrated deeper into England, one panicky diplomat emphasised the severity of the situation. 'The Pretender's son [has] near 3,000 rebels with new arms and French louis d'ors,' he angrily pointed out. 'We have not above 1,500 men [and] I cannot but wish for the return [of the army] without which I am certain we do not sleep in whole skins. Our danger is near and immediate, all our defence at a distance . . . I bewilder myself in scenes of misery to come, unless providentially prevented.' So in autumn 1745 the British troops were finally recalled to fight the Jacobites on home territory and David Watson was among them.

As the Jacobites retreated north, the King's soldiers followed in pursuit. Watson's commander, the Duke of Cumberland, offered him the role of Quartermaster General, which placed Watson in charge of army provisions. The winter of 1745 was bitterly cold and he was responsible for locating sufficient quantities of 'warm Jackets', adequate blankets, tents and new shoes for the rapidly moving troops. The Jacobite Rebellion convinced Watson that the rebels were aberrations, whose actions jeopardised 'the happyness this poor Country enjoy'd before the late irruption of these Barbarians'. And in the wake of what he termed 'our Interview with the Rebels upon Coloden Moor', he became something of a Hanoverian hero, due to the ardour with which he rooted out Charles Edward Stuart's surviving followers. Watson was a superficially charming man, at ease in his tall

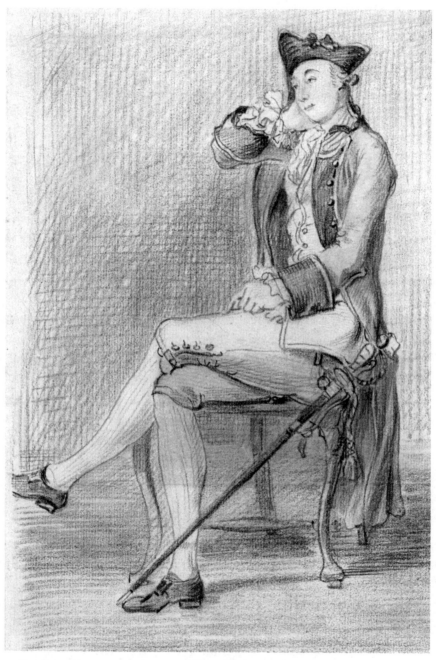

4. Col. David Watson on the Survey of Scotland, *1748, by Paul Sandby.*

soldier's physique and possessed of a biting wit. His letters openly admitted the 'temptations of pleasure or game' that he found in the 'hunt' after the Jacobites, especially in pursuit of the prize prey of Lord Lovat and the Young Pretender. Watson engineered 'the destruction of the Ancient Seat of the Glengarry' clan and was sardonically confident that 'the rest of the good people of Lochaber' who witnessed that violence will 'remember the Rebellion in the Year 46'. He was adamant that 'the Highlanders [are] the most despicable enemy that ever Troops mett with'.

Once the immediate brutality was over, Watson dedicated himself over the months that followed to 'the sole motive of restoring quiet, by convincing the World that Rebellion and its Authors are to be come at in the most inaccessable parts of the Highlands'. He was among a number of government sympathisers who applied themselves to a long-term programme of Highland reform. Together with his brother-in-law Robert Dundas, and Dundas's son Robin, Watson formulated strategies to systematically suppress Jacobite support, to destroy the loyalty that bound the clans together, to undermine Highland identity and to rebuild the region's infrastructure, driven by the ambition of unifying a hitherto divided Scotland and eradicating Jacobitism once and for all. Like many, the three considered the Highlands to be 'lawless' and 'barbaric' regions, whose clan system kept individual citizens in thrall to their chiefs, cowed beneath robberies, assaults and threats from neighbouring clans. In 1707 the Anglo-Scottish Union had constitutionally united England and Scotland into 'Great Britain', but it had only really succeeded in cementing sympathies between Lowland Scotland and England. The Highlands, with its autonomous legal system, language and culture, continued to be almost a separate nation.

Watson and the Dundases were not by any means the only ones who proposed practical strategies to 'pacify' the Highlands in the wake of the rebellion. Suggestions from others included the disarming of the clans, the establishment of soldiers to police the region, the building of better roads and the confiscation of Jacobite estates 'for the purposes of civilizing, and promoting the happiness of the Inhabitants upon the said Estate . . . by promoting amongst them the Protestant Religion, good Government, good Husbandry, Industry and Manufactures'. Many considered that the

Highland landscape was responsible for its inhabitants' rebellious behaviour. Mountains and lochs formed natural barriers between clans, fostering a profound sense of community and dividing Highlanders from Lowland influence. One man felt that Highland chiefs were mirrors of their surroundings in more metaphorical ways too: both were useless, obstructive and terrifying. 'Such *Noble-men* as these are like *Barren Mountains,* that bear neither Plants nor Grass for Publick Use,' he wrote. 'They touch the Skie, but are unprofitable to the Earth.' Based on this assumption, Watson's voice was heard, over and over again, passionately and articulately asserting 'the Benefit [which] must arise from protecting the Highlands by the Regular Troops' and the need to 'acquir[e] a perfect knowledge of the Country'. Watson particularly sought to remedy the deficiency of geographical information about the Highlands under which the King's army had suffered before, during and after the 'Forty-Five'. Watson pointed out that his colleagues had 'found themselves greatly embarassed for want of a proper Survey of the Country' and he emphasised the vital need for a good map of Scotland to facilitate the opening up of hitherto inaccessible areas of the Highlands.

Watson voiced his concerns persuasively to his commander, the Duke of Cumberland, who took the matter to his father, King George II. It was later reported that 'the Inconvenience was perceived and the Resolution taken, for making a Compleat and accurate Survey of Scotland'. But Watson had no intention of trying to execute such a mission himself. Instead, he turned to a man over twenty years younger: a man with no evidence of formal training in map-making and apparently no history of military involvement. It was a curious choice, but it would prove inspired.

THE UPPER WARD of Lanarkshire is a quiet region, composed of tightly wooded dells and lush fields. In the seventeenth century these tranquil slopes were the scene of violent clashes over the doctrine and organisation of the Scottish Church. Walter Scott's novel *Old Mortality* depicted a Lanarkshire whose 'broken glades' and 'bare hills of dark heath'

had been as dreadfully rent by those civil wars as by the ancient geological upheavals that, millennia earlier, had created the 'windings of the majestic Clyde'. In the wake of this century of turmoil, on 4 May 1726, a boy called William was born to Mary Stewart and her husband John Roy. Since the mid seventeenth century William Roy's grandfather and then his father had been employed at the estates of Hallcraig and Milton in the Upper Ward. These lands were owned first by Sir William Gordon, the father of Robert Dundas's wife Ann Gordon, and then by Ann's brother, a rather obstreperous lawyer called Charles Hamilton Gordon. The Roys lived in a small cottage on these estates' southern border called Miltonhead. This building and the estates' two mansions have since disappeared and the land now houses the Carluke Golf Club and, until fairly recently, an intricate Gothic mansion called Milton-Lockhart, which was the sometime home of Walter Scott's son-in-law and biographer, John Gibson Lockhart.[1]

The Roys were factors – land-stewards – and they managed the practical upkeep of the estates. An early-eighteenth-century guide to 'the Duty and Office of a Land Steward' described the many responsibilities of this position, which ranged from the relatively menial, such as trapping pests and catching poachers, to more onerous duties such as collecting tenants' rents, preventing subletting and ensuring the land was adequately farmed. Surveying was a vital part of the jobs of Roy's father and grandfather. 'It is necessary,' wrote one contemporary land-steward, 'to know the quantity and quality of every parcel of land belonging to his lord's estate', and to do so he advised that the factor 'should have a correct map of the whole . . . in which map, should be expressed every bend, corner, and irregular turn in the several hedges; all rivers, bridges, highways, gates, and stiles'. Other manuals to the land-steward's role emphasised that 'it is not only necessary that a Steward should be a good Accomptant, but also that he should have a tolerable Degree of Skill in *Mathematicks*, Surveying, Mechanicks, and Architecture'.

---

1 In 1987, after being bought by a Japanese actor, Milton-Lockhart was dismantled stone by stone. Once Mikhail Gorbachev had granted special permission, it was shipped in thirty containers on the Trans-Siberian Railway to Japan, where the mansion was reconstructed in woodland about a hundred miles from Tokyo. Renamed 'Lockheart Castle', it now hosts luxury boutiques and weddings.

Although little documentation has survived from John Roy's employment at Hallcraig and Milton, the estate maps on which he relied were almost certainly constructed with two instruments, called a 'plane table' and an 'alidade'. An alidade was a mechanism that allowed an object to be brought within a straight line of eyesight, and developed from the historical astronomical instrument known as the astrolabe, which was used to determine latitude. Deriving from an Arabic word meaning 'ruler', early alidades consisted of two vanes, each of which contained a thin hole or slot (without lenses) which was placed at either end of a small bar. A plane table comprised a level surface mounted on a robust base, and it had been used in surveying since the sixteenth century, perhaps even earlier, and generally with an alidade. We can imagine William's father teaching the young boy to make a map of his employer's estate using these two instruments. John Roy would have first measured a small baseline in a field on the Hallcraig estate, a straight, flat distance whose length depended on the size of the land to be surveyed. He drew this base on a sheet of paper that was fastened onto the surface of the plane table, and positioned the table directly over one end of the actual baseline so that it was in perfect alignment with the marking of

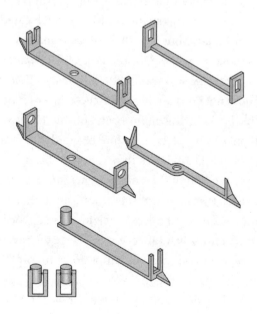

5. *Alidades.*

that same spot on the paper. Crouching down, and bringing his eye to the level of the table, John showed William how one of the alidade's vanes should be placed at the mark indicating the end of the base. He then looked along its length, gently shifting the instrument's position until a landmark on the estate, such as a fountain, was brought into view through both vanes. John now traced onto the paper the straight line of eyesight that passed from the mark of the baseline's end, along the length of the alidade, to this landmark. Then he moved his plane table to the other end of the baseline, again perfectly aligning the actual spot with that drawn on the paper. John Roy now traced a line from *this* end of the baseline to the fountain in the same way as before. Two lines were thus etched onto the embryonic map, radiating out from either end of the baseline to the fountain. The spot where these lines intersected produced the position of the landmark for the map. John showed William how this technique could be replicated over and over again, until an outline of the complex landscape of the estate was plotted onto paper by means of the simplest form of geometry.

It is likely that William Roy was proficient from a young age in the skills of estate surveying and plane-tabling, and it seems that he demonstrated a remarkable early aptitude for map-making. But records regarding the young boy's formal education are inconclusive: Roy was educated at the nearby grammar school in Lanark, but accounts of his teenage years are silent. It has been suggested that he worked for the Post Office, that he was employed in road-building, or that he was trained by the Army's Board of Ordnance, but Roy's name does not crop up on any of those institutions' records in this period. When the Jacobite Rebellion broke out in 1745, he was nineteen years old. The Upper Ward of Lanarkshire erupted into frenzied activity to halt the progress of the Jacobites as they retreated from Derby back to the Highlands. Initially the Young Pretender's army intended to march straight through the Ward, as it was 'the only communication that is open, from the north to England', as one of the Roys' neighbours explained. Around 700 local men collected into a resistance force 'to act for the Defence of His Majesty King George and our present happy constitution'. But in the event, heavy rains meant that the river Clyde was too swollen to be crossed, and the Jacobites were forced to divert their path further east.

He seems to have avoided direct confrontation with the rebels, but the 1745 uprising would change William Roy's life in an unexpected fashion. The Roys' employers were relations of Robert Dundas's wife, Ann Gordon, and at some point they were introduced to Dundas's brother-in-law from his first marriage, David Watson. In the months that followed Culloden, it may have been that this extended family was brought together for a large cele-bratory meal at Hallcraig House. We can imagine the conversation turning to Highland reform and Watson describing the need for better maps, and Ann leaving the room to fetch the twenty-year-old son of her family's land-steward who had shown a prodigious talent in surveying. A seemingly quiet, rather shy young man, Roy may initially have been nervous before David Watson's confident personality. But the meeting between the two men, how-ever it occurred, was evidently a success. Watson made the decision to appoint Roy to direct the first 'proper Survey of the Country' of Scotland. It would prove a far-sighted selection. Roy's birthplace, the probable site of his first meeting with Watson, is commemorated with a memorial in the shape of a trig point that describes how the young man would grow up to become the most illustrious map-maker of his day, 'from whose Military Survey of Scotland (Made in 1747–1755) Grew the Ordnance Survey of Great Britain'.

II

AS 1747 REACHED its summer solstice, David Watson was to be found deep in the Scottish Highlands. He was at Fort Augustus, in the middle of the geological fault known as the Great Glen, on the southern tip of Loch Ness, surrounded by a wall of mountains. The garrison had been built on a site called Kiliwhimin in the wake of a Jacobite uprising in 1715, and it was later renamed after William Augustus, the Duke of Cumberland. Charles Edward Stuart's army had seized control of Fort Augustus during the 1745 rebellion and it seems likely that Watson, zealous in his loathing of Jacobitism, would have enjoyed particular satisfaction when the King's army reclaimed the set-tlement after the Battle of Culloden. From the fort, this venerated, fanatical engineer was helping to coordinate a military occupation of the Highlands.

The Highlands are now defined as the region that lies both north and west of the Southern Upland Fault (a line running diagonally between the Firth of Clyde and Stonehaven on the east coast, south of Aberdeen), and west of another natural boundary between Perth and the Moray Firth. To many early-eighteenth-century English travellers and Lowland Scots, Watson included, the area resembled the ends of the earth. A rare London tourist who visited Scotland in the 1720s commented that 'the Highlands are but little known even to the inhabitants of the low country of Scotland' but that 'to the People of England' the 'Highlands are hardly known at all, for there has been less, that I know of, written upon the subject, than on either of the Indies'. In the early eighteenth century the region was very healthily populated, certainly in comparison to today when, in the aftermath of a long history of clearances (among other reasons), it is one of the most sparsely peopled areas in Europe. But this abundance of population did not make the Highlands less strange. Numerous visitors were struck by the region's acute social difference, and one traveller described how 'that nervous expressive tongue', Scots Gaelic, was spoken everywhere. For centuries, writers had wondered if the Highlands were, in fact, the mystical lands of Ultima Thule, 'those regions in which there was no longer any proper land nor sea nor air, but a sort of mixture of all three of the consistency of a jellyfish in which one can neither walk nor sail, holding everything together'.

From Fort Augustus, Watson was charged with overseeing the repair of fortifications and barracks that had been damaged or destroyed during the rebellion, and implementing the construction of new ones. He was also supervising the dramatic extension of a network of military roads throughout the Highlands. In the mid eighteenth century Britain's roads were notoriously bad. Many had been churned over the last few centuries to 'mere beds of torrents and systems of ruts' under the large, heavy wheels of carts or ploughs. Although in 1555 a Highway Act had attempted to combat this decline by placing responsibility for the upkeep of roads on local parishes and setting aside certain days for their repair by unpaid local inhabitants, this system was uncentralised. As many labourers unsurprisingly resented working on Britain's roads for free, their quality remained notoriously unpredictable. It was not until the seventeenth century that a combination of the introduction of the stagecoach and the legislative innovation of the Turnpike Act had begun to revolutionise journeys made by road. Referring to the spiked barriers that controlled access to major highways, Turnpike Acts subjected road-users to a toll whose revenue was used for maintenance and repair. But the number of turnpike trusts multiplied agonisingly slowly and they were initially focused solely on major routes leading into London. In any case, prior to the 1720s these innovations to Britain's roads only applied to networks that ran across England and Lowland Scotland. All major thoroughfares stopped at the border with the Highlands, after which mostly drovers' routes and similar tracks criss-crossed the landscape.

After two serious Jacobite uprisings in 1715 and 1719, the Hanoverian administration had tried to take the Highlands' roads in hand. George I sent a senior military officer called George Wade, an Irishman descended from a staunchly anti-Jacobite family, into the region 'to inspect the present situation of the Highlanders, their customs, manners, and the state of the country'. Wade's conclusions were devastating. He estimated that 'the number of men able to bear arms in the Highlands (including the inhabitants of the Isles) are by the nearest computation about 22,000 men, of which number . . . 12,000 have been engaged in rebellions against your Majesty, and are ready, when ever encouraged by their Superiors, or Heads of Clans, to create new troubles, and rise in arms to favour the Pretender'. In vivid detail, Wade described the clans' practices of robbery and assault, their abundant possession of weapons, and 'the little regard they ever paid to the Laws of the Kingdom, both before and since the Union'. To defend against this terrifying foe, he proposed that a militia be created from Highlanders loyal to the Crown, and that permanent barracks of English and Scottish soldiers should be established at Fort William and Inverness.

Seriously alarmed, George I agreed to Wade's suggestions and by January 1725 the latter had 'caused an exact Survey to be taken of the several Lakes and that part of the Country lying between Inverness and Fort William, which extends from the East to the West Sea'. But Wade soon realised that his vision of an extensive arrangement of forts across the Highlands was of little use without the means to travel swiftly between them. At that time no major roads extended further than Perthshire, so Wade duly dedicated his energies to the project for which he is most vividly remembered: a network of military thoroughfares across Scotland. Wade's roads were the first real attempt to 'open up' the Highlands to outsiders. They allowed a steady train of soldiers to travel up from Edinburgh and Glasgow with supplies and artillery, spilling out into barracks, garrisons and forts at Inversnaid, Ruthven, Fort Augustus and Fort William, among other places. Wade's successor, a man called William Caulfeild, reverentially acknowledged his forebear's achievement, and he and his subordinates were occasionally overheard singing the mantra, 'Had you seen these roads before they were made/ You would lift up your hands and bless General Wade.'

It is probable that when David Watson first mooted the idea of making a 'Military Survey of Scotland', he was partly motivated by the thought that such a map would be extremely useful to the extension of Wade's roads. So in 1747, Watson summoned the son of his extended family's land-steward to join him in the Highlands, and he commanded William Roy to begin the map with a survey of the region's existing military highways. (Thanks to a break-in at a flat in London in the late 1920s, no confirmed image of Roy survives, but a possible posthumous caricature shows him in middle age to have sported deep-set dark eyes above a thin Roman nose, cheeks that blushed easily and a gently amused smile.) A single, tattered sheet of measurements in the archives of Blair Castle, in Blair Atholl in Perthshire, indicates that one of this young man's first tasks was to 'Measure the roads from Inverness[,] Fort Augustus[,] Fort William by Blair & Dunkeld to Perth.' At some point in the 1750s, Watson compiled a set of 'Orders and Instructions' that outlined just how such a survey should be made, and it is likely that he similarly directed Roy to 'carefully follow the line of the principal Roads . . . every half-mile's distance minutely expressing every variation or Change that happens'. Watson added that this map-maker should also compile notes about the terrain that immediately surrounded each thoroughfare and emphasised that 'a distinct and useful idea of the nature of any River or Water [is] always of the greatest Consequence'. Both roads and rivers were potential means of passage for troops.

William Roy began the daunting task of mapping Highland Scotland's roads in the summer of 1747. We can imagine this young man placing a staff

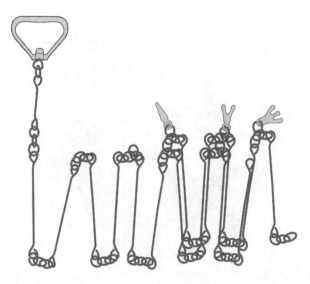

6. Gunter's chain.

on the perimeter of Fort Augustus, and marking the first bend or kink in the road to Inverness with another. He measured the distance between the two by simple pacing or with a perambulator (otherwise known as a surveyor's wheel, or 'waywiser'), an instrument consisting of a handle attached to a wheel, which was pushed along the ground. Each revolution marked out a specific length, usually a yard, and by counting the revolutions with a dial or 'click' mechanism, Roy could roughly deduce the length in question. For more accurate but slower measurements of the ground, he used a chain, probably based on the model made by Edmund Gunter in the early seventeenth century. Gunter had divided a basic measuring chain of sixty-six feet into one hundred links of equal sizes, which were marked off into groups of ten by brass rings that allowed the map-maker to quickly determine distance.

Roy needed to measure the angles as well as the lengths of the roads' twists and turns, and for this he cradled in his hands a simple version of a theodolite. Invented in England by a Kentish mathematician called Digges in the second half of the sixteenth century, the theodolite didn't make its way onto the Continent until the nineteenth century. Its birthplace was important to British surveyors; many thought of it as their national instrument. The

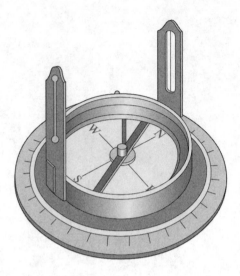

7. A circumferentor.

word's etymology is hazy, but it may have derived from two Greek words meaning 'I view' and 'clearly visible'. Digges had described his *theodelitus* as 'a circle divided into 360 degrees', which was attached to 'sights'. Roy's theodolite was a type called a *circumferentor* and consisted of an alidade fixed to a compass; like most theodolites in the first half of the eighteenth century, it did not contain a telescope. The instrument exploited the principle that light travels in a straight line in order to measure the angle of a surveyor's sight line from landmark to landmark.

Standing at the first staff that he had thrust into the ground, Roy placed his circumferentor on a tripod by its side. He then crouched down to bring his head to the instrument's level, closed one eye and placed the open one before the nearest vane of the alidade. Rotating the circumferentor's compass slowly with his left hand (it has been suggested that Roy was left-handed), he brought into view the second surveying staff, which was positioned at the nearest bend in the road, until its width filled the tiny slit of the alidade's furthest vane. Roy fixed the circumferentor at this position, straightened up and read the angle of his sight line off the instrument's compass. Jotting the angle and distance that separated the two staffs in a notebook, accompanied by a quick sketch, Roy then repeated the whole process. He pulled the first staff out of the ground and walked to the second bend in the road, where he plunged it into a new spot. Then, with the chain and circumferentor, he measured the distance and the angle of this second segment. Roy followed this method over and over, until he had surveyed the entire thirty-mile stretch between Fort Augustus and Inverness. But there was no rest. He also turned his attention to the slightly shorter distance that separated Fort Augustus from Fort William, and to the precipitous roads that fingered south-east towards Blair Atholl, Dunkeld and Perth. When Roy had finished mapping the Highlands' military roads, he applied the same technique to its major rivers. His task was to piece together a dense skeleton of the multiple networks that traversed the Highland landscape, as the basis for a new national map.

Roy's attention was not only trained on the roads and rivers that stretched before him. An array of landmarks clustered on either side of both, such as drovers' paths, crofts and the outer walls of large private estates. He selected the most prominent of those that were visible from the road, and measured compass bearings to them with his circumferentor. But instead of measuring with the chain, which would have been excrutiatingly time-consuming and difficult, due to the rugged terrain, he largely guessed at their proximity and duly positioned a smattering of features around the networks of waterways and highways that were marked on his sketch map. He also located visible mountains and lochs with his circumferentor and sketched in their rough outline by sight. The technique that Roy was using throughout the Military Survey was widely known as 'traverse' or 'route' surveying. It had been famously employed by the publisher and cosmographer John Ogilby in 1675 in his *Britannia Depicta*, which was 'an illustration of the kingdom of England and dominion of Wales: by a geographical and historical description of the roads thereof'. After measuring the roads of England and Wales using a 'great wheel' or perambulator, Ogilby had drawn a hundred 'strip maps' or traverse surveys, delineating the distances and nearby attractions of 2519 miles of road. But maps like Ogilby's risked giving their users tunnel vision, training readers' eyes solely on the road ahead and its closest landmarks. Roy's traverse surveys for the Military Survey were also focused primarily on the course of Scotland's major networks of rivers and thoroughfares, but he took in much of the intervening land too, even if only in a rather hazy

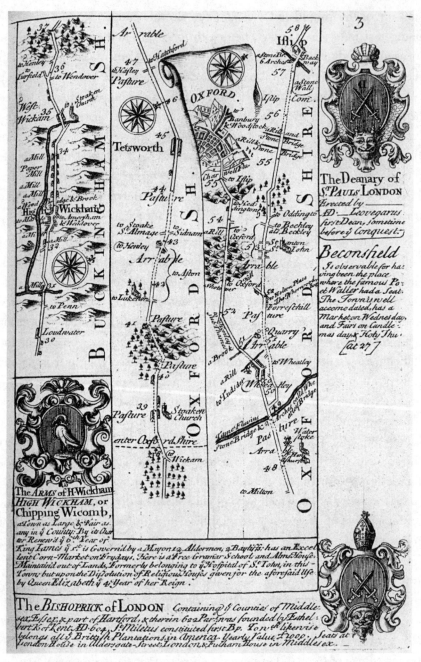

8. An extract from John Ogilby's Britannia Depicta, *1720, shows a 'strip map' or 'route survey' of the main road and surrounding territory between High Wycombe and Oxford.*

fashion, and his resulting surveys offered a far more expansive and coherent map of a nation than Ogilby's strip maps.

Ideally, traverse surveys would be conducted by groups: one man would carry the theodolite, two measured with the chain, two held the staffs, one made the observation and another acted as attendant. It is nearly impossible to do all this alone. William Roy must have had assistants, but it seems that initially he was the only trained map-maker on this mission to measure some of the most exposed and unforgiving corners of the Scottish landscape. The rugged scenery over which the young man trudged made his task extremely taxing. Wade's roads were breathtaking feats of engineering: they leapt up and over some of the highest passes and sharpest descents in the country and penetrated far into the Jacobites' heartlands. At times, Roy must have felt perilously vulnerable. He was likely to have been the victim of battering rain-storms, fierce winds, exhausting summits and a niggling terror that, behind his back, a host of hostile Jacobites might be gathering. Twenty-six years after Roy had first laid his measuring chain along the road from Fort Augustus to Inverness, the lexicographer and polymath Samuel Johnson followed in his footsteps on 'a journey to the Western Islands of Scotland'. Struggling on horseback up a precipitous series of hairpin bends in the road, Johnson cried out that 'to make this way . . . might have broken the perseverance of a Roman legion'.

Hostility from the locals may have added to Roy's woes. Map-makers were frequently an unwelcome presence in the Highlands, as many of the indigenous population assumed, often with reason, that they were tools of the state, surveying the land for purposes of taxation or surveillance. In the early 1740s, the Perthshire hydrographer Alexander Bryce had been engaged in mapping the north-west coast of Scotland between Assynt and Caithness for his *Map of the north coast of Britain*. But some smugglers were suspicious of his close attention to the region's harbours and bays, worried that he was engaged in state policing, and they confronted him with menace; finally he was forced to defend himself with firearms. The military engineers who descended on the Highlands after Culloden were also *personae non gratae*. In July 1747 a man called John Russell, who was charged with overseeing repairs at Inversnaid Barracks, in the stomping ground of Rob Roy

MacGregor's clan, reported: 'as I was sitting in my hutt at Innersnaid, two fellows who were Strangers to me came in without asking any Questions. On my Enquiring what they wanted, one of them drew a durk or rather a small Cutlass, and said he would let me know that presently.' The engineer only escaped death by the unexpected intervention of two Highland women who 'got between me and the Man who had the Durk, and saved me from receiving the Intended Blow'. Many Hanoverian officers and soldiers turned to drink while working in north and west Scotland, and the Chief Engineer there was forced to apologise to the Board of Ordnance for the 'scandalous Scrawl and form' of one of his underlings' reports, commenting with wry displeasure that 'I fancy he has consulted the Dram bottle.' The 'highwaymen', as the military road-builders were known, were notorious drunkards. Indeed, virtually the only consolation for what Wade admitted to be 'the remote situation' and 'unwholesomeness' of their working lives was the promise of unlimited home brew. After taking pity on a small team of men with a donation of a few shillings, Samuel Johnson recounted that he came across them again further down the road. They 'had marched at least six miles to find the first place where liquor could be bought', he recorded.

Despite the suspicion with which map-makers and Hanoverian military men were regarded in the Highlands, William Roy often had no choice but to confront passers-by or to knock at the doors of Highland crofts. He needed to discover the names of nearby mountains, rivers or settlements for his map; Watson was also keen on encouraging his employees 'to be particularly attentive to the produce of each part of a Country, and how Inhabited, if abounding in Grass and Hay'. All of these 'particulars you may be easily informed' about 'after some Acquaintance with a Judicious Countryman', Watson declared confidently. The first years of the Military Survey must have taken their toll on Roy and when on Christmas Day 1748 his father died, we can imagine that the prospect of returning to work in the spring to begin another season of mapping the Highlands mile by mile, alone, was daunting.

There was some compensation for Roy's graft, however. It is easy to romanticise the life of the map-maker: to imagine the exhilarating panoramas, the plentiful moments of peace and the loving familiarity engendered with every fold of the landscape. But although this was not the whole truth, Roy did rhapsodise over 'a scene the most wild and romantic that can be imagined' at Coigach, a peninsula in Wester Ross in the north-west Highlands. And the intense solitude and silence did not trouble him. Where David Watson revelled in easy anecdote among the higher reaches of Hanoverian society, Roy by his own admission did not 'love much talking'. 'I am for coming to the point,' he admitted ruefully. He would later write that 'a much truer notion may be formed' of the nation in maps 'than what could possibly be conveyed in many words'.

DAVID WATSON WAS all too aware that William Roy was faced with the mammoth task of constructing a map of the Scottish Highlands on his own. Since the earliest days of the endeavour Watson had repeatedly petitioned the Board of Ordnance for assistance for his young employee, but nothing had materialised. In 1748, almost a year after the Military Survey's commencement, Watson received the displeasing information that two engineers who had been expressly promised for the Military Survey were to be employed elsewhere. 'The Surveying Scheme has given me Infinite Pain,' he complained to the Chief of Engineers in Scotland, William Skinner. In March 1749 the Board finally relented. Almost two years into the project, George II consented to Roy 'having three more Assistants in the Survey he is making of Scotland'. Very little documentation from the Military Survey has survived, so it is hard to pinpoint how much of the Highlands Roy had managed to cover on his own, but we can surmise from Watson's continual appeals for assistance that his progress had been rather slow.

Watson travelled down to London to choose three new map-makers from among the cadets at the Royal Military Academy in Woolwich. Only founded eight years before, Woolwich had been designed to provide instruction for 'the people of its Military Branch to form good Officers of Artillery and perfect Engineers'. One of eight military academies established in Europe in the mid eighteenth century, the institution offered a thorough education in the mathematical theory and military science required by cadets who were destined for one of the two corps of the Board of Ordnance. Woolwich taught its cadets to survey according to different methods, to employ a variety of instruments, to use mathematical formulae to manipulate calculations of angles and lengths, and to reconcile a number of divergent measurements. The institution also instructed its cadets in landscape painting and the traditions of map draughtsmanship deriving from European military schools. It taught the technique of 'hachuring' to represent hills and mountains, a form of shading that consisted of small strokes drawn in the direction of the steepest slope. The teachers of the Woolwich Academy hoped to instil a uniformity of skill into the Corps of Engineers.

When Watson applied to the Royal Military Academy in 1749 for assistants for the Military Survey, that institution housed forty-seven cadets. Its exacting 'Deputy Head Master' John Muller was despairing of the quality of many of his students, who he dismissed as 'Idle', 'Lunatic' and lost to 'Debauchery, and a thorough neglect of Duty'. There were only a few of whose prospects he was hopeful: a sixteen-year-old Practitioner Engineer already working for the Corps called Hugh Debbieg, a young man with a determined jaw and petulent downturned mouth, and his friend, a cadet called John Williams. William Dundas, the twelve-year-old son of Watson's brother-in-law Robert Dundas and his second wife Ann Gordon, also happened to be studying at Woolwich. Watson duly made his choice, and Debbieg, Williams and Dundas were packed off to Scotland to join Roy on the Military Survey. Shortly afterwards, the Board of Ordnance found funds for two more assistants, and another Woolwich cadet called Thomas Howse and a slightly more experienced Practitioner Engineer called John Manson were also selected to work on the survey. Roy was thus presented with five assistant map-makers and even a few horses, plus 'additional Servants, Guides, Interpreters & otherwise'. This was really the bare minimum necessary to make such a survey, and the Board's provisions were long overdue. Nevertheless, Roy must have been delighted.

Among this band of enthusiastic but inexpert young engineers, Roy was also sent a draughtsman. After the failure of the 1745 Jacobite Rebellion, a sixteen-year-old Nottinghamshire boy with a sharp, intelligent face had sent samples of his landscape sketches to the Board of Ordnance. Its committee did not consider Paul Sandby worthy of permanent employment, and a satirical sketch he later made of an Ordnance meeting, portraying its squabbles and prejudices, was probably fuelled by enduring hurt at this rejection. But the Board did offer Sandby ad hoc work in Scotland, and he began by assisting David Watson in the repair and construction of forts by producing large-scale military surveys of their immediate environs. These maps were clearly the product of a nascent landscape painter: Sandby surrounded his plans with meticulously detailed watercolours of the scenes that rendered the redcoats' presence in the Highlands as natural as foliage.

In 1749 Watson sent Sandby to work with Roy on the Military Survey. In his first year of this employment, the small team of map-makers spent some time surveying the area around Kinloch Rannoch, especially the road that Wade and his successors had built from Stirling to Crieff and Dalnacardoch. From Sandby's surviving sketches, it seems that the scenery here appeared remarkably monochrome to the young draughtsman. And so it is, especially in cloudy weather: by the sides of Loch Rannoch, black mountains with snowy toupees stand as erect as pints of Irish stout. Fog clings to their sides like wisps of grey candyfloss, and the treacly loch is wrinkled by reflections of the trees that serrate its edges. Sandby spent some of his time there producing drawings of his teammates amid this muted Perthshire landscape.

Paul Sandby made one particularly fascinating pen-and-ink and watercolour painting of a scene a mile or so east of the loch's foot, on the banks of the River Tummel. Sandby's *View Near Loch Rannoch* showed the ragtag bunch of young surveyors in action. On a flat plain against an arboreal background, through which the Tummel winds, two redcoated surveying assistants lay out a measuring chain; in the foreground one holds a staff, with a counterpart in the background; and three soldiers tend to the horses. Hunched over a circumferentor on a tripod is a man attired in a blue coat, and this is almost certainly a unique image of the young William Roy. But like the surveys he conducted for Watson, Sandby's *View* was not an impartial representation. As a brilliant piece of visual propaganda, the young man's sketch advertised the indomitable force of the King's army. A craggy outcrop in the sketch's middle-ground is locally known as Craig Varr and in Sandby's painting it is being built over with soldiers' lodgings, as part of the occupation of the Highlands. Sandby's *View* also shows locals in plaid kilts and trews assisting the military engineers, possibly translators or guides or maybe just interested bystanders. Either way, the map-makers appear to have successfully converted the Highlanders to Enlightenment science and the Army's presence in Scotland.

What Sandby decided *not* to paint in his *View* was almost as telling. To the map-makers' backs, entirely ignored in the young man's watercolour, lies the best view in the entire region of a vast, symmetrical mountain called Schiehallion, which resembles an indelicate Mount Fuji. It may seem strange that a cartographic draughtsman chose to ignore such an aesthetically fascinating scene. But a background featuring Schiehallion would have dwarfed the surveyors, making their task to map the whole Scottish mainland appear comically absurd. Sandby's *View Near Loch Rannoch* was an image consciously constructed to demonstrate the might of the Hanoverian military and its relatively tame backdrop shows his colleagues clearly in control. Furthermore, the particular region in which Sandby set his *View* was scarred by a traumatic recent history of anti-Jacobite violence. Four miles north lay a Hanoverian barracks from which redcoats had descended on the surrounding area in the aftermath of the 1745 Rebellion, seizing Jacobite estates, executing Highland criminals and razing buildings to the ground. Sandby's painting was a barefaced celebration of Hanoverian power, its composition and location consciously chosen to reinforce that message. Displaying such an acute awareness of the political power of art and landscape, it is small wonder that in the decades following the Military Survey of Scotland Paul Sandby enjoyed a short spell producing satirical political cartoons. He also helped found the Royal Academy of Art (an institution designed to promote a 'British school' of painting), and pursued a career as a nationalist landscape painter – his obituary called him the 'father of English watercolour'.

IN THE THREE years that followed the formation of this small and inexperienced map-making team, its members surpassed all expectations. They worked on a standard pattern for surveyors that saw the men 'in the field' during the spring and summer months, when the days were long, the temperatures moderate, the winds and storms not too frequent or fierce, and the light bright. David Watson paid regular visits to his charges, reporting on their progress to the Board of Ordnance's senior officers in London. In autumn and winter the map-makers retreated to Scotland's capital where, in the Governor's House of Edinburgh Castle, they collated their measurements of the roads and waterways and their sketches and informed guesses about the intervening landscape to construct the intricate maps that comprise the Highland sheets of the Military Survey of Scotland.

The climate and topography of what Watson termed the most 'remote Corners of the Highlands' were a consistent challenge throughout this epic enterprise. The young men's notebooks, journals and correspondence during the project have not survived, but Sandby made an etching that captured the daily tribulations of Roy and his map-makers. He showed a small party huddled at the base of a treacherous mountain pass, raddled by gales, pitting themselves futilely against the forces of nature in their most extreme manifestation. In the repeated soakings that the men endured, their woollen uniforms must have become a heavy burden. Seventy years after the Military Surveyors' extraordinary tramp across Scotland, the Ordnance Survey followed in their footsteps. From one map-maker's descriptions of his ordeal, we can get some idea of the arduousness of his predecessors' experiences. This man described 'the really laborious part of the business', that of 'conveying the camp equipage, instruments and stores' across the violently undulating landscape. Scotland's notorious midge population contributed to the ordeal and 'it was our practice in walking to put our coats and waistcoats into our knapsacks,' the Ordnance Surveyor recounted.

> Thus, with our shirt necks thrown open, and our sleeves tucked up, we were exposed in a peculiar manner to the baneful attacks of those venomous insects. We suffered very severely; our arms, necks, and faces, were covered with scarlet pimples, and we lost several hours' rest at night from the intense itching and pain which they caused. Even at the inns we had frequently to smoke in our bedrooms and over our meals to drive those insects away.

9. Party of Six Surveyors, Highlands in Distance, *1750, by Paul Sandby.*

The weather added further trials. At 11 a.m. on 28 June, at the height of summer, the map-maker reported that his thermometer suddenly plummeted, closely followed by a hailstorm. With a surveyor's eye for detail, he noted the hail's 'large and conical' stones 'with smooth convex bases and striated sides'. When the storm had passed, this surveyor recalled how 'the men set-to snow-balling each other as a means of warming themselves – a rather unusual amusement at the latter end of June'.

Finally, by 1752, in the space of two years of solitary endeavour and three years of shared toil, despite indomitable hills, claggy bogs and hostile Jacobites, William Roy and his men had produced finished maps of the entire Scottish Highlands, a region comprising 15,000 square miles of mountains, lochs, beaches, cliffs, forests and glens. Back in Edinburgh, the surveyors reduced their sketches and calculations into maps on a scale of one inch to 1000 yards (so that a straight distance of 1000 yards on the ground was represented by one inch on paper).[2] The map-making project might easily have stopped there, at the outermost border of the Scottish Highlands. After all, the principal motivations for making the Military Survey were specific to this inaccessible home of Jacobitism. The Lowlands were less inclined to rebellion, easier to travel through and already possessed many good maps.

But the Military Survey did not wind down its operations in 1752. Instead, for reasons we shall see, it ramped them up. Large and regular donations of cash started to be pumped into the map-making project, facilitating the appointment of so many more surveyors, assistants, interpreters and draughtsman that, at the Military Survey's high point, it was staffed by around sixty personnel. Some of this money came from the map's progenitor, David Watson, and numerous instalments were also contributed by his brother-in-law Robert Dundas and Dundas's now 39-year-old son Robin.

---

2 Map scales can also be written as fractions to describe the ratio between the distance on paper and the distance on the ground. The Military Survey's scale of one inch to 1000 yards can be written roughly as the fraction of 1:36,000, as one inch on paper represents around 36,000 inches (there are thirty-six inches per yard) on the ground. Maps are also spoken of as being small- or large-scale. Large-scale maps are surveys of confined spaces subjected to a large magnification; small-scale maps 'zoom out' and represent a greater area in less detail. The Military Survey was a small-scale *survey* in the true sense of the word: an overview of a nation.

Their funding may have engineered the employment of yet another member of their family on the mapping team. David Watson had a fifteen-year-old nephew called David Dundas, Robin's cousin, who had been eagerly antici-pating a career in medicine until his persuasive and domineering uncle had manoeuvred him towards the Military Survey. (But he hardly did badly from this career change: within fifty years, David Dundas would become Commander-in-Chief of the British Army.) Principally, however, the injec-tions of funding into the Military Survey from 1752 onwards were designed to allow William Roy to lead his expanded team across the Highland–Lowland border into the calmer realms of southern and eastern Scotland.

The map-makers seem to have found surveying the Lowlands a delight compared with the Highlands. The scenery was flatter, lusher, densely inter-laced with roads and punctuated with luxurious manicured estates. And the men were received politely, even enthusiastically, by the residents. Robert and Robin Dundas opened their family seat at Arniston to the Military Surveyors, and David Dundas repaid his extended family's hospitality by mapping their fertile, elegant estate. The Dundases' neighbours proved equally welcoming. The proprietor of the nearby estate of Penicuik was a man called Sir John Clerk, whose portraits show him to have had an alert gaze and a stern jaw offset by a quick readiness to smile. Clerk was an extraordinary polymath: a politician, lawyer, judge, patron of the arts, occa-sional natural philosopher, composer and a committed antiquarian. At his mansion, Clerk gathered around him the leading lights of the Scottish Enlightenment: the poet and painter Allan Ramsay, the archaeologist Alexander Gordon, the philosophers Adam Smith and David Hume, and figures such as William Robertson, John Home, Alexander Wedderburn, Alexander Carlyle, John MacGowan, William Wilkie, John Blair, James Hutton, William Falconer, Adam Ferguson, 'and many others . . . whose superior taste and genius have since been displayed in elegant, useful works which have rendered their names immortal'.

The Hanoverian military had long been warmly received by the gentry of Lowland Scotland, who often welcomed the soldiers as a source of willing manual workers. In the 1740s one military engineer boasted to Robert Dundas that 'it has often been in my power, during the time that the

Reg[imen]t has been in Scotland', to oblige landowners 'with Ditchers & other Artificers', and the two duly struck a deal whereby a team of military engineers agreed to mend Arniston's internal roads. But the social status of the Board of Ordnance's employees extended from the working class right up to the aristocracy, and many were welcomed into Lowland society on more equal terms. For example, the extraordinary Adam family of architects worked in Scotland as Master Masons to the Board of Ordnance, repairing and building forts and barracks. They supplemented these military duties with lucrative commissions to design and construct Palladian mansions for the richest Scottish proprietors, such as Floors Castle in Roxburghshire, Hopetoun House in Linlithgowshire and the Dundases' own mansion at Arniston. The Adams were great friends of Sir John Clerk and his son, who referred to William Adam in particular as 'a man of distinguished genius, inventive enterprise and persevering application, attended with a graceful, independent and engaging address which was remarked to command reverence from his inferiors, respect from his equals and uncommon friendship and attachment from those of the highest rank'.

Clerk sought out the Military Surveyors during their mission to map the Lowlands. He employed their young draughtsman, Paul Sandby, as a private tutor in painting for his son, in whom Sandby succeeded in igniting a passion for art that stayed with him until the end of his long life. And Clerk particularly took to the reserved director of the survey, William Roy. Perhaps sparked by an early childhood visit to the site of a Roman encampment near Cleghorn in Lanarkshire, Roy harboured a fascination for antiquities. He was specifically interested in remnants of permanent or temporary army encampments dating from the Roman presence in Scotland between 55 BC and AD 410. A keen amateur antiquarian himself, Clerk promptly led the map-maker on a tour of his estate where he had recently found a Roman station, which was marked, Clerk explained, by 'a tumulus, where several urns, filled with burnt bones, have been dug up'. Delighted by the discovery, Roy duly 'pointed it out in his maps'.

In the years that followed, the influx of new personnel onto the Military Survey meant that Roy found more time to indulge his antiquarian foible. He felt that the 'ordinary employments' of a military map-maker, which involved minute attention to the landscape, were ideally suited to serious examination of antiquities. Roy also believed that the relatively unchanged Scottish landscape provided a line of communication and sympathy between the military men of his own time and the Roman generals who had stalked that same territory 1500 years previously. 'While the ranges of mountains, the long extended valleys, and remarkable rivers, continue the same,' he wrote, 'the reasons of war cannot essentially change.' In Roy's view, the military mapmaker was the ideal antiquarian, able 'to compare present things with past' and 'converse with the people of those remote times'. So he delegated the mapping of the Lowlands' interior to his colleagues and reserved for himself the border lands between England and Scotland. While surveying this territory between 1752 and 1755, this gentle young man in his mid-twenties investigated a host of Roman encampments, and made minute charts of those at Chew Green, Liddle Moat and Castle-o'er. Roy became particularly fascinated by the way in which Scotland's boundaries had shifted over time and how these changes were imprinted onto the land itself by the Antonine Wall (which 'begin[s] at the Clyde, and end[s] at the Forth') and Hadrian's Wall (running between Newcastle upon Tyne and the Solway Firth).

William Roy's obsession with Roman military antiquities in Scotland would last until the end of his life. Twenty years after it first gripped him, Roy became a prominent member of London's Society of Antiquaries, and the recipient of heartfelt pleas from colleagues to write up his researches for publication. What began as a smattering of essays gradually expanded into a lavish book entitled the *Military Antiquities of the Romans in North Britain*. Roy adorned his text with his own meticulous surveys of Roman sites, including a particularly beautiful map of the encampment at Cleghorn. He prefaced the entire volume with a national survey that caused quite a stir on its first publication. The *Mappa Britanniae Septentrionalis* was a minutely detailed map of north Britain, based on information that Roy had collected during the Military Survey of Scotland. When the reading public got wind of the existence of such an accurate map of the whole Scottish nation, demand was so great that, despite having no plans yet for publication of the text, Roy was persuaded to engrave and publish the *Mappa* separately. Three years after his death, the Society of Antiquaries finally published Roy's *Military Antiquities* as

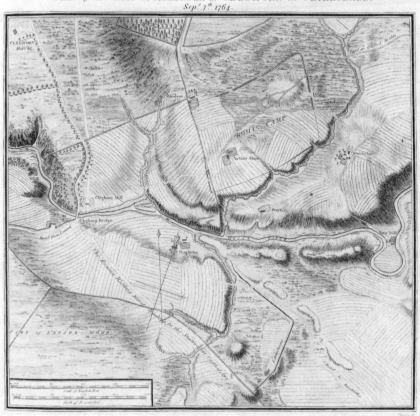

PLAN of the CAMP of AGRICOLA at CLEGHORN in CLYDESDALE.
Sep.r 7th 1764.

10. A sketch from William Roy's Military Antiquities of the Romans in North Britain, *1793, showing a 'Plan of the Camp of Agricola at Cleghorn in Clydesdale', not far from Roy's birthplace near Carluke.*

a spectacular elephant folio, a fitting memorial to his achievements in the fields of military history and antiquarian research.

THE EXPANDED TEAM of Military Surveyors swept through the Scottish Lowlands in three years. By 1755 they had made working drafts of maps of

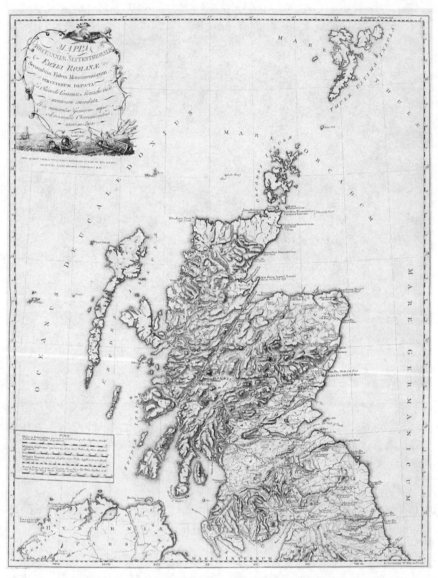

11. *A map of the Scottish mainland from William Roy's* Military Antiquities of the Romans in North Britain. *This incorporates measurements taken during the Military Survey of Scotland.*

the whole region, although they had not managed to produce finished ones. But by the late summer of that year, conflict on the Continent had become inevitable and the Board of Ordnance called the Military Survey's personnel away from Scotland to employments required by the impending Seven Years War. The mapping project came to an end.

In the period between the Military Survey's commencement in the summer of 1747 and its termination in 1755, Roy and his men had produced maps of varying degrees of completion of the entire Scottish mainland. The Military Surveyors proudly mounted their maps of the Highlands onto eighty-four brown linen rolls, of which twelve constituted the 'fair plan', and they prepared their Lowland maps on ten. These were later cut and reconstituted to make thirty-eight folding sheets which, when pieced together, form a complete map thirty feet high and twenty feet wide. As one witness aptly commented, it was 'immense'. David Watson initially kept hold of the individual sheets, but at some point in the 1760s they were transferred to the King's Library at Windsor. Finally the Military Survey became part of the 'King's Topographical Collection', the title attributed by the British Library to the Topographical and Maritime Collections of King George III, a breathtaking array of some 50,000 atlases, maps, plans, prospects and views assembled through that monarch's reign. Any reader of that public library can order sheets of this beautiful, nation-changing map to view in person. Anyone can spread out, across the Map Reading Room's enormous tables, this window onto a lost landscape.

The Military Survey of Scotland, in its final state, is a vast, gorgeous bird's-eye view of mid-eighteenth-century Scotland. One of its surveyors, Hugh Debbieg, described the project as 'the greatest work of this sort ever performed by British subjects' and one of 'the fine[st] Representations of the Country . . . in the World'. From the cliffs of Cape Wrath in the most north-westerly point of Scotland to the windswept coastlines and castles of Aberdeenshire in the east; across the glens of Angus, Fife's bays, the forests of Perthshire, and the rugged mountains and lochs that punctuate the Trossachs and the Western Highlands; to the burgeoning cities of Edinburgh and Glasgow, the Lothians' damp and fertile parks, the ancient stone circles in Ayrshire, and the deserted beaches of Galloway; right down to the gently undulating hills and plains of the Borders, the map-makers had drawn it all, in pen-and-ink and watercolour washes. An early witness of the survey noted that 'the Mountains and Ground appear shaded in a capital style by the pencil of Mr Paul Sandby, subsequently so much celebrated as a Landscape Draughtsman. The outlines were drawn and other particulars were inserted under the care of General Watson by sundry assistants.' And although the Military Survey of Scotland bore no legend to explain its 'vocabulary of symbols', the map was constructed in a language derived from contemporary Continental military surveying. Hedgerows were shown as lines of miniature trees. Enclosed fields were patches of green, sometimes containing scatterings of bushes. The roads over which Roy had trudged for eight years and six months were delineated by brown lines. Tilled land was represented by parallel hatching and rivers by turquoise streaks. Sand was denoted by stipple and moorland by infrequent patches of grass. Urban settlements consisted of loose grids and inhospitable mountains were signified by fierce strokes of brown-black paint, whose tone and direction indicated the shape and slope of the peaks. The Military Survey is a rare, delicate specimen of Enlightenment cartography that has been said to offer 'a picture of Scotland on the eve of great changes'.

Wonderful as it is, the Military Survey is not a uniformly accurate image of the landscape. It was the product of a small team of young and inexperienced map-makers, and the eight years and six months it took to complete was ridiculously swift for such an enormous, varied area of country. There are wild discrepancies in the levels of detail and completion between different sheets of the map, probably due to the changing practices of individual surveyors. Scotland's islands were completely omitted, the finished maps bore no information about longitude or latitude, and they failed to indicate the direction of magnetic or true north. More importantly, the map-makers' method of 'traverse surveying' resulted in the proliferation of errors over large distances, and a comparison of the Military Survey with a modern Ordnance Survey map of Scotland reveals serious divergences. The eastern end of Loch Leven, for example, is positioned on the Military Survey a good twenty miles south of its actual location. William Roy later justified his map's inaccuracies by explaining that it was 'carried on with instruments of the common, or even inferior kind, and the sum annually allowed for it [was] inadequate to the execution of so great a design in the best manner'. It was, Roy admitted, 'rather to be considered as a magnificent military sketch, than a very accurate map of a country'.

A question that troubles map historians to this day is what exactly the Military Survey was for. It began as one element of the post-Culloden project of Highland reform, after the soldiers charged with rooting out rebels from the deepest recesses of the Highlands had keenly felt the lack of a good map of that region. A small-scale chart of Highland Scotland was also useful to the engineers involved in the extension of roads and to officers tasked with the transportation of heavy artillery and troops from the Lowlands up to forts and barracks at places like Fort Augustus or Inversnaid. But there is no evidence that the Military Survey was ever actually employed. Locked away in libraries for half a century after its completion, the maps were only temporarily excavated from a dusty basement by a road surveyor when an Act of Parliament in 1803 decreed the building of more highways and bridges in Scotland. The Military Survey was never published, and it has only been exposed to a mass audience in exhibition spaces in the last century.

Moreover, none of these practical motives for making the map explain why in 1752 William Roy and his surveyors turned their attention to Lowland Scotland. That the project benefited from private funding implies that it was attractive to the nobility of Enlightenment Scotland, and this is supported by the fact that William Roy and David Watson produced a unique hand-painted presentation copy of the Military Survey's map of the area around Arniston especially for the Dundases. The sculpted geometry of Arniston's avenues and paths imprints a neat geometric cluster at the centre of this map and a scribbled note on the reverse explains that the survey was 'drawn by Roy, and given by him to Lord President Dundas, sometime about the year 1755. It is an Excerpt from the Survey of Scotland executed at that Time by the General & other Officers under the Direction of General David Watson.' Scotland's noblemen may have seen in the Military Survey a response to heated criticisms of the poor and fragmentary state of that nation's maps, such as those voiced by the surveyor George Mark back in the 1720s. Furthermore, the purchase, sponsorship or making of maps was likely to have been a badge of Enlightenment. In mapping Scotland, the King's engineers may have considered themselves vitally differentiated from 'barbarian' Highlanders who had no choice *but to be mapped* (pointedly ignoring the fact that the Jacobites were extremely cartographically literate – the exiled Stuart court in France had made a point of collecting maps of Scotland). There is also something enduringly powerful in an image of a complete map of a nation. Men like Dundas considered that Scotland's political disunity weakened the resistance it posed to England's 'unsufferable tyranny' in the Anglo-Scottish Union. A picture of a unified nation was a picture of a strong Scotland and this idea was so potent that the adventure-writer Robert Louis Stevenson later proclaimed that 'Scotland has no unity *except on the map*'.

The Enlightenment raised the bar for cartography higher than it had ever been before. But as the Military Survey's example showed, map-makers' attempts to attain the ideal of perfect measurement, a one-to-one replication of the physical world, could not take place overnight. The history of cartography's progress is a story of trial and error in which every inaccuracy, every shortcoming, every failure of expectations forms a stepping stone on the path to success. Despite its shortcomings, the Military Survey was one of the first national maps to be constructed through actual measurement of the ground, rather than by amalgamating existing charts. It became what we might call a 'public-private partnership', and we shall see how this marriage of military concerns with civil and idiosyncratic preoccupations was a feature of the mapping projects by which it was superseded. Perhaps most importantly of all, the Military Survey sparked in William Roy the dream of making a complete and accurate national map of the entire British Isles. But this was an ambition that would not begin to be realised for almost half a century.

CHAPTER TWO

'The Propriety of Making a General Military Map of England'

WILLIAM ROY PROBABLY greeted the news of the Military Survey's enforced end in 1755 with some trepidation. For the last eight years, this unassuming map-maker had been grateful to find himself accepted into a world of blue woollen uniforms, shiny buttons and strange habits of salutation. When Britain's politicians led the nation into war, and his friends were called away from their Scottish escapade to more pressing matters, the 29-year-old Roy may have feared he would be left in Scotland, unemployed and alone. If so, he need not have worried. His erstwhile mentor David Watson came to the rescue. By December, he had wangled a position for his young charge in the Corps of Engineers on the lowest rung of that establishment's hierarchy, which commanded a respectable annual salary of £54 15s. Many officers in the Corps of Engineers held positions in the regular army too, and on 24 January 1756 the *London Gazette* reported that the name of 'Engineer William Roy' had been included on a list of lieutenants appointed to a new regiment raised at Exeter. For the first time in his life, Roy enjoyed permanent employment with a regular income, and thrilling prospects. Over the next forty years he would work his way up the ranks to become Britain's most famous military map-maker.

In the months that followed the Military Survey, Watson was asked to conduct a reconnaissance of sections of England's south coast between Dover in Kent and Milford Haven in the far south-west corner of Wales. This was intended to help the Army prepare for a feared French invasion amid the turmoil of the Seven Years War, and Watson chose his two favourite assistants, his nephew David Dundas and Roy, to help him. Existing maps were particularly poor at showing what was known as 'relief', the landscape's three-dimensional characteristics, its undulations and declivities. Roy and Dundas were commanded to trace the rise and fall of southern England's scenery in words and images, and their resulting descriptions and maps were designed to help tacticians assess the strength and weakness of various locations, depending on their vulnerability to attack. Roy learnt these military techniques quickly and was soon reporting how 'the Position in front of Dorking' was 'very Strong indeed', as it 'is cover'd by a Ridge of Heights that takes the Shape of a Bow', which meant that 'the Circumstances of the Adjoining Country would Render a Combin'd Attack upon it, a most Hazardous . . . Enterprise'. Roy developed a habit of conjuring in his mind's eye a picture of southern England's abundant orchards, fertile pastures and tranquil hedgerows overrun by marauding French invaders. And he was later asked to do the same for Ireland too.

William Roy had grown up amid quietly handsome mansions separated by fields that gently sloped towards the wooded banks of the Clyde, and he had mapped the rugged Highlands too. He owned a more intimate knowledge of Scotland's extremes of landscape than anyone else alive and it is intriguing to reflect on how this reserved, inquisitive man might have approached the very different surroundings of London, Kent, Sussex and Hampshire. When Roy looked at the 'deep valleys and intervening ridges' of the South Downs and found a shape that resembled 'the fingers of a hand' (as he wrote in his 'Military Description of the South-East of England'), perhaps this account served as his way of offering a friendly handshake to his new home. Over the decades that followed, Roy would spend the greater part of his time here. On southern England's estuaries, marshes, plains, clearings and hills, and amid the press and bustle of its capital, Roy would make friends, achieve promotions and foster interests that would lead him to formulate a vision of founding Britain's first national mapping agency.

His mentor, David Watson, did not enjoy the same happiness, however. Immediately after the Military Survey, before heading south, he helped to found a Commission to confiscate the estates of Jacobite rebels and establish schools and manufacturing works on their sites. By the late 1750s, this military engineer was in his late forties and was living in Westminster. Watson had suffered with gout for a long while but it grew worse and the suffering eroded his good looks and charming bonhomie. By July 1761, Watson was forced to admit that his health was 'very indifferent' and that he was 'oblige[d] to keep [to] my Room from severe Attacks of the Gravel [gall-stones] that immediately follow'd my Gout'. Watson's housekeeper, a woman called Sophia Wilson, tended to him with great compassion as his malady worsened. He also had an illegitimate son around this time, on whom he bestowed his own name.

Shortly before 1 November 1761, David Watson died. His death was reported in the *Gentleman's Magazine*, a monthly digest of news for an educated reading public. Prior to his demise, a friend had celebrated what he considered to be Watson's greatest achievement: the pacification of the Highlands. Watson 'has made [Scotland] more his study than any man alive,' he reflected. 'He knows every Corner of the Kingdom, he is acquainted personally w[i]t[h] all Ranks & Degrees, knows their principles, views, Connections & importance, in what they can be trusted & where they cannot. In case of any disturbance in this Country, there is no person of what rank soever can be of such service to his majesties interest as Collonel Watson.' Shortly before his death, Watson's portrait had been painted by the Italian artist Andrea Soldi: it shows a world-weary soldier in the grip of disease, with tired eyes and an equivocal mouth. But although it is a picture of a dying man, it is also that of a proud lover of maps. Watson is shown pointing to his crowning glory: a military survey of a complex, beautiful and sometimes unruly landscape.

David Watson's last will and testament left his housekeeper a generous annuity of £30 which he specified was 'for her own separate use . . . and noways subject to the disposal of her Husband'. He left his 'natural son' an annuity of £12 'to assist in paying for his Maintenance & Education'. To his nephew David Dundas, Watson bequeathed half of his furniture and 'to

Captain William Roy' he left 'all my Mathematical Instruments'. He thus passed on his baton to a deserving heir.

THE EARLY STAGES of the Seven Years War went badly for Britain, but they proved a boon for William Roy. When military strategists decided a decisive blow was needed against France on its home soil, Roy was sent as a junior member of a unit to attack Rochefort, a port in south-west France. The mission fell to pieces when its adviser lost his nerve and its leader called the whole thing off. At the resulting court martial, Roy gave eloquent and frank evidence and earned a reputation as an expert in military engineering.

By March 1759 Roy had been promoted to Sub-Engineer, the second rung on the Corps of Engineers' professional ladder. By now the Seven Years War was well under way and Roy was sent to fight under the leader of the allied troops – Prince Ferdinand, Duke of Brunswick – to defend the Hanoverian homelands of Britain's king against the French. Roy industriously surveyed potential camps and battlefields, and distinguished himself in the allied victory at the Battle of Minden in August 1759. In the following year he was commended for his 'zeal', 'assiduity' and 'talents' at the siege of Munster. Roy produced a map of Minden: a beautiful pen-and-ink and watercolour survey, which incorporated a technique whereby the positions of both armies' regiments were marked by overlapping flaps that were fixed to the map's surface and which could be moved about to follow the troops. The French appropriately called these *papillons*, butterflies. After the battle, Roy's map was pored over by those involved in a controversial court martial to determine whether a British commander had been guilty of insubordination. His contribution to the trial, and the part he had played in the Seven Years War generally, brought Roy further military accolades. He was promoted to Deputy Quartermaster General of the forces in Great Britain, responsible for maintaining the troops' supplies. Roy pulled off this tricky logistical role with aplomb, and was later promoted to 'Commissary General of all Stores, Provisions, &c.'

Over the years Roy grew into a patient, gentle and well-liked man, who managed people well. His shyness seems to have metamorphosed into a measured self-control and dislike of frivolity, and his soldiers were fond of him. An anecdote told almost twenty years after Minden mocked anyone who might suggest otherwise. In 1778, when the prospect of a French invasion reared its head again during the War of American Independence, Warley Common in Essex was turned into a camp for the county's militia. One late September morning, Roy was riding on the road to Warley when he noticed a young officer a little way ahead and 'observed the poor condition of his horse, and asked what it was owing to?' The young man, wanting to shift the blame and impress the stranger with a controversial opinion, began to lay into a well-respected figure and laid the fault at the door of the Commissary General, William Roy. He claimed intimate acquaintance with Roy, accused him of embezzling rations, and concluded 'his observations with many hearty curses against the Colonel'. Rather bemused, Roy listened patiently, only commenting quietly that he hoped 'Roy was not so much in fault as was apprehended'. The two rode on for a while, side by side. As they drew near to Warley Common, two senior officers approached the pair, and upon recognising Roy, they saluted him as such. His young companion was thrown into 'great astonishment and confusion' at this exposure of his lie and although Roy tried to persuade him to travel on, he was unsuccessful. The *Gazetteer and New Daily Advertiser* delightedly ridiculed the dupe who had questioned Roy's competence and affability.

In September 1759, Roy had been promoted in the Army to captain and the Board of Ordnance had particularly congratulated him on the fact that he had achieved this 'extraordinary promotion' on his 'own Personal Merit'. Six years later, in July 1765, he was appointed to the prestigious position of Surveyor-General of Coasts. He was hired to 'inspect, survey and make Reports from time to time of the State of the Coast and Districts of the Country adjacent to the Coast of this Kingdom'. The instruction to 'survey' the coasts did not solely entail map-making: in this sense, 'surveying' referred to the general communication of the shape and quality of the land in words or images. Within three months of his appointment, Roy received a letter from one of his seniors informing him that 'it is His Majesty's pleasure that you should immediately repair to Dunkirk'. In February 1763, the Seven

Years War had ended when France, Spain, Portugal and Britain had signed the Treaty of Paris. The Treaty's thirteenth article had stipulated that the military improvements made by the French to 'the town and port of Dunkirk' during the war should be destroyed to ensure the security of Britain's south coast. Roy was appointed to oversee the demolition. It was not an enjoyable mission. A glutinous fog hung over the port and the French were hostile and obstructive. They harped on the Treaty's caveat that the engineers should protect the safety, health and 'wholesomeness of the air' of Dunkirk's inhabitants, by ensuring that the demolitions did not lead to flooding or compromise sanitation. Stalemate was reached and the work ground to a halt, leaving Roy and his colleagues sitting around in idle frustration.

In November 1765 the new British Ambassador to France descended on Dunkirk to negotiate between the parties. Roy may have already met Charles Lennox, 3rd Duke of Richmond, at the Battle of Minden, where both were present and shared an interest in military cartography. Now in Dunkirk, Lennox quickly singled out Roy, who was nine years his senior, as a like-minded compadre. Confidentially, and without mincing his words, this rather mercurial, egocentric aristocrat, who was dismissive of fools and rabidly protective of allies, solicited the engineer's opinion, and Roy voiced his suspicion that the French were indulging in groundless prevarications. Both men duly resolved that 'the Demolition . . . should . . . go on' regardless. Roy took this opportunity to confide in Lennox, confessing that his health had 'not been bettered by my winters Occupation at Dunkirk'. He begged of his influential friend that 'you will be so good as [to] procure me His Majesty's leave to return to England', and in mid February 1766 the two men boarded a boat in Calais together and set sail for England.

Upon his return, the middle-aged Roy took stock. He knew that his career in military map-making had given him 'a [more] thorough knowledge of the country' than anyone else alive. The story of his life could be plotted on a map: he had spent time in Scotland, south and western England, south Wales, Ireland (which he had surveyed in 1765), and even various Continental hotspots. But the British Isles's real maps were upsettingly patchy. There was no single national survey that could 'join the whole together', as Roy put it, and this shortcoming in the nation's self-knowledge worried him immensely.

On 24 May 1766 Roy committed his concerns to paper and drafted a memorandum to the King, George III. In these 'Considerations on the Propriety of making a General Military Map of England' Roy fretted that, in time of war, when the British Navy was called away, the nation's coastlines became dangerously vulnerable to invasion. He felt it was of the utmost importance that 'knowledge should be acquired, in as far, at least, as regards the Nature of the Coast, and the principal Positions & Posts which an Army should occupy, when called upon to defend the Country against the Invasions of its Enemies. The only Method of attaining this Knowledge seems to be, by making a good Military Plan or Map of the whole Country.'

Roy estimated that the expenses of the map's start-up year would reach £2778 12s. And then he guessed that around £2500 annually would be sufficient, over a period of six to eight years. He effectively promised George III that the nation could be given its first complete, accurate map for a sum in the region of £15,000–£20,000 (roughly between £1 million and £1.5 million today), and in the space of less than a decade. But his proposal was knocked back, probably on account of cost, combined with political apathy in the midst of a period of peace. Nevertheless, Roy's 'Considerations' were the first viable proposal for a national mapping agency in Great Britain. It was a momentous document, on which its creator considered that 'the honour of the nation' depended. Roy was certain that, when a national survey came into being, it would be executed by military map-makers.

The rejection of Roy's proposal did not mean that the progress of cartography had ground to a halt in Britain. A number of citizens from a variety of backgrounds turned their attention to map-making in the second half of the eighteenth century, and when the Ordnance Survey was finally established in 1791 it would owe much to these innovators, as well as to its more obvious military progenitors.

ON 22 MARCH 1754 a small group of 'Noblemen, Clergy, Gentlemen, & Merchants' met at Rawthmell's coffee-house on Henrietta Street in

London, between Covent Garden and the Strand. Amid the chatter of Rawthmell's comfortable surroundings, these men brought into being the Society for the Encouragement of Arts, Manufactures and Commerce. (In 1847 the Society was given a royal charter and became the Royal Society of Arts, a name it retains to this day.) Designed to encourage the application of those disciplines to the public good, the Society of Arts soon attracted a host of eminent members, including the charismatic botanist Joseph Banks, the furniture maker Thomas Chippendale, the future statesman Benjamin Franklin, the actor David Garrick, the musicologist Charles Burney, the writers Oliver Goldsmith and Laurence Sterne, the artist William Hogarth and the great lexicographer Samuel Johnson. What a 'Virtuoso Tribe of Arts and Sciences!' the Marquess of Rockingham exclaimed.

The Society of Arts was initially focused on 'the Encouragement of Boys and Girls in the Art of Drawing, [which] is necessary in many Employments, Trade and Manufactures', on the basis that 'the Encouragement thereof may prove of great Utility to the Public'. Charles Lennox, 3rd Duke of Richmond, opened the sculpture gallery at his mansion at Goodwood in West Sussex to enthusiastic young painters who needed somewhere to practise their skills. The Society's fervour was partly patriotic: its members wanted to turn London into a rival to Paris, 'a Seat of Arts, as it is now of Commerce, inferior to none in the Universe'. The Society of Arts offered incentives, 'premiums', in the form of healthy sums of money or medals to those who successfully met its challenges. And in 1759 its members set their sights on another practical application of art: map-making.

In the early 1750s, a Cornish antiquarian and naturalist called William Borlase had become aggravated at the shortcomings of the maps he was using. Borlase wrote frustrated letters to a friend, challenging him to deny 'whether the state of British Geography be not very low, and at present wholly destitute of any public encouragement. Our Maps of England and its counties are extremely defective,' he grumbled. The recipient of Borlase's letters was Henry Baker, a fellow antiquarian and naturalist, and a founding member of the Society of Arts. Borlase persuaded Baker to make the lamentable state of British map-making a priority for the Society. ''Tis to be wished, that some people of weight would, when a proper opportunity offers, hint the necessity of such Survey,' he wheedled, suggesting that

> if among your premiums for Drawings some reward were offered for the best plan, measurement, and actual Survey of city or District, it might move the attention of the public towards Geography. In time, perhaps, [this might even] incline the Administration to take this matter into their hands (as I am informed it does in some foreign Countries) and employ proper persons every year from actual surveys to make accurate Maps of Districts, till the whole Island is regularly surveyed.

On Baker's recommendation, the Society of Arts duly agreed to establish a regular prize of up to £100 to 'give proper surveyors such Encouragement as may induce them to make accurate Surveys of two or three Counties towards completing the whole'. Its members hoped that the Society of Arts' competition would result in the production of a series of accurate maps of Britain's counties that could be pieced together like a jigsaw into a full national survey. These maps would be created using the most up-to-date methods and instruments, on a uniform scale of one inch to one mile. And the Society also tried to accelerate the painfully lethargic progress of British cartography by stipulating that candidates must complete their surveys within one, or at most two, years. Its members were enthused by the possibility that the resulting maps would be 'of great use in planning any scheme for the Improvement of the Highways, making Rivers Navigable and providing other means for the Ease and Advancement of the National Commerce'.

The Society of Arts advertised its first prize of £100 for the map-making competition in 1759. The response was somewhat underwhelming. In its first six years, only eleven candidates applied for the prize, out of whom a mere two were successful. Some applicants were full-time map-makers or estate surveyors, who could use the products of their day-to-day work in the task; but others were teachers, publishers or map-sellers, who were forced to work around other commitments. In the event, only one professional county surveyor applied and many candidates were put off by the meagreness of the prize. £100 was not adequate to fund such maps as the Society envisaged. The cost of instruments, assistants and above all the time required to make a good county map could exceed twenty times that amount, as one of the winning candidates discovered. The Society's prize acted as a gratifying award once the work had been completed, but it was not a proper salary.

The competition's time-limit also dissuaded many potential applicants. One year to map an entire county from scratch was terrifyingly rapid, and many contenders were forced to apply for extensions. Benjamin Donn, a teacher of mathematics and natural philosophy from Bideford, began his survey of Devon shortly after the Society first advertised its competition, but he did not publish it until 1 January 1765. Another candidate took seven years to complete a map of Somerset. The productions of other applicants were rejected by the Society for a variety of reasons: some petty, some sensible. Unsuccessful candidates were refused on the grounds of their maps' inaccuracy; failure to pay the Society of Arts' annual subscription; incompletion of the map or insolvency; ineligibility on the grounds that they had not declared entrance to the competition prior to beginning surveying; and the submission of a map that had been begun for another purpose many years before the contest was advertised in 1759.

Between the beginning of the map-making competition and the last official prize offered in 1802 the Society of Arts paid out a mere £460 in cash, plus seven medals and a silver palette, in reward for only thirteen county surveys. The accuracy and topographical detail of these maps showed a considerable improvement on their predecessors, but the Society of Arts' competition had failed to compile Britain's first complete, accurate national survey on a uniform scale. Nevertheless, the event was a notable stepping stone in the progress of eighteenth-century cartography. Its members had emphasised the importance of accurately surveying the ground from scratch using the most innovative methods and sophisticated instruments available, and the competition had cemented the one-inch scale as the standard for county maps. It has been claimed that the improvement of English county surveying in this period could 'without much doubt be traced to the offer by the Society of a prize of £100'.

When the Society of Arts decided to close its competition in 1802, it was because it had been superseded by the Ordnance Survey. Although it would be a military establishment that was ultimately responsible for creating Britain's first national mapping agency, the Ordnance Survey's ambitions and methods owed not a little to the traditions of civilian county surveying that had fuelled the Society of Arts' competition. That contest's progenitor, William Borlase, had been remarkably prophetic in his hope that such an event would prompt the government to instigate a national surveying programme. And in the meantime, between the opening of the Society of Arts' competition and the foundation of the Ordnance Survey, a number of other developments were also involved in the gestation of Britain's national mapping agency.

IN 1763 WILLIAM ROY had signed the lease on his first London property, a four-storey brick residence on Great Pulteney Street, a small thoroughfare in London's fashionable West End on which both General Wade and Paul Sandby had once lived. Sixteen years later, in 1779, he upgraded, and moved a few streets further north to a handsome modern townhouse on Argyll Street, a road running parallel to Regent Street, in a quarter set aside for military officers. Today, a blue plaque marks the house that belonged to 'the founder of the Ordnance Survey', but its interior has changed beyond almost all recognition. Roy had used the fourth floor as an observatory and this entire storey has been demolished and replaced with an elaborate fire-escape. Offices now orbit a central staircase that spirals up to a radiant stained-glass ceiling and the ground floor is currently leased out to the clothing chain French Connection: an amusing serendipity, for, as we shall see, one of Roy's greatest contributions to the earth sciences was a measurement conducted between the Royal Observatory at Greenwich and the Paris Observatory, a different sort of 'French Connection'. The Argyll Street house has not shrugged off all traces of its map-minded proprietor, however. A few years ago, the property's current owner prised open a bricked-up fireplace to discover, stashed away for around two hundred years, a presentation copy of Roy's *Military Antiquities of the Romans in North Britain*.

Roy's lease of these prestigious quarters indicated a shift in his emotional landscape. In 1767, his brother James, who had been a Presbyterian minister in East Lothian, died. His elderly mother was still alive, living in Lanark, and Roy paid her a visit every few years, but London was gradually replacing Scotland as the map-maker's home. Its clubs and societies were the focus of Roy's perambulations around the capital. He was a committed and active member of the Society of Antiquaries, which was based at the site of Robin's Coffee-House on Chancery Lane until 1780, when it moved to Somerset House on the Strand, near the Royal Academy of Arts. In 1812 the satirical cartoonist George Cruickshank drew a caricature of 'the Antiquarian Society' that conjured up the rambunctious atmosphere of its meetings: in this print, greying sages confer in cabals with the utmost sincerity about items of dubious authenticity. A figure on the far right of Cruickshank's picture, attired in a natty red military jacket, with a pocket stuffed full of papers marked 'Ordnanance [*sic*] Affairs', looks covetously at a Roman vase in the middle of the table. It has been suggested that this balding gentleman is a rare representation of the late-middle-aged Roy. But fond as he was of the Society of Antiquaries, it was Roy's immersion in the activities of another prominent London organisation that was to introduce him to those in the vanguard of research into the earth sciences.

William Roy was admitted as a Fellow of the Royal Society, England's chief crusader in the 'Improv[ement of] Natural Knowledge', on 9 April 1767. He was recommended by the telescope-maker James Short, the astronomer John Bevis and the horologist William Harrison as 'highly worthy of that Honour' of fellowship and 'likely to become a very useful Member'. Roy threw himself into the life of the Royal Society, regularly attending the dinners that were organised by its official dining society, the 'Royal Philosophers' Club', which offered a forum for relaxed, intellectual chatter. It is interesting to reflect on whether Roy's Scottish accent ever proved a burden at this time: it was a difficult period to be a Scot in London. In the wake of the 1745 Jacobite Rebellion, the long-standing antagonism between the English and the Scots had grown more pronounced. And shortly after the coronation of the young King George III in 1760, his childhood tutor John Stuart, 3rd Earl of Bute, had been appointed First Lord of the Treasury (to all intents and purposes, Prime Minister). Bute was Scottish, and in his election many saw a similar abuse of kingly power to that associated with the Stuart monarchs. Scotophobic cartoons flooded London's print-shops, casting immigrants from North Britain as vagrants that were bleeding England's prosperity dry. Booksellers touted manuals that promised the dilution of a Scottish accent and the complete eradication of 'the expressions of a beggarly Scot'. In William Roy's case, however, his nationality did not hinder the progress of his career.

Four years after Roy's election to the Royal Society, in July 1771, Captain James Cook returned to England after a three-year voyage to the South Pacific in the ship *Endeavour*. Cook's return sent London's intelligensia wild with enthusiasm for tales of exotic exploration and the advanced empirical methods he had used to record his findings. Cook was a talented naval surveyor, and one contemporary commentator has defined his voyage as the moment that '*truth* became our central criterion' in the field of geography. But the public were even more taken with Cook's co-traveller, a 28-year-old aristocratic botanist called Joseph Banks.

A striking young man with dominant features and sparkling, dark eyes, Banks bubbled over with tales of his exploits in 'paradise'. One of the Royal Society's most illustrious members, he revelled in the attention he garnered at the dinners of the Royal Philosophers' Club. It was on one of these occasions that Roy first made Banks's acquaintance. We can imagine the pair sitting beside each other at the large oak table, Roy initially reticent, a little gruff even, but warming up, and occasionally smiling as the young man entertained him with stories of his attempts at surfing in Tahiti. We can imagine Banks, too, leaning in to catch Roy's hushed accounts of his mapping exploits in the Scottish Highlands, an area considered less known 'than either of the Indies'. But Banks soon tired of the Royal Philosophers' rules that restricted the number of non-member guests that could attend its dinners. He decided to start his own breakaway dining association with a more relaxed admissions policy, which he called the Royal Society Club. Roy was not the only one to follow Banks to this new dining club. The young Henry Phipps, Lord Mulgrave, who would later become Foreign Secretary, transferred his affections too, as did Charles Blagden, an eminent physician who soon became the botanist's right-hand man. On 23 November 1775 the Royal Society Club held its first meeting, at the Mitre Tavern on Fleet Street. It was a resounding success, and Roy became a fixture at the hearty, convivial dinners that followed.

These were not just free lunches: Roy well and truly earned his fellowship of the Royal Society over the course of the next decade, attending its social gatherings and presenting the results of his work. At that association he found a storehouse of talented scientists whose research prompted him to think in new and exciting ways about the process of mapping. Although Roy continued to work for the Board of Ordnance, in the period that followed the rejection of his proposal for a national military survey in 1766 it was his contact with the ripening field of geodesy, 'earth-measurement' (from *geo*, earth, and *daiein*, to divide), that edged him closer to the foundation of the mapping agency of which he dreamed.

IN THE EARLY 1770s, William Roy became fascinated by 'that curious and useful branch of philosophy, whereby vertical heights are determined to a great degree of exactness by the pressure of the atmosphere alone': that is, the use of barometers to measure altitude. Roy carried his barometer around London, systematically measuring and recording heights. He lodged himself at the top of the stairs in St Paul's Cathedral, reporting that 'I have sometimes found, particularly in frosty weather, that a thermometer placed on the pavement of the North-side of St Paul's Church-yard . . . would stand two degrees lower than that which was exposed on the North-side of the iron gallery over the dome.' And he enjoyed the experience of measuring the air pressure on the first floor of the dining room of the Spaniard's Inn, a sixteenth-century oak-panelled public house that still crowns the north-western corner of Hampstead Heath. Roy also enlisted the help of friends scattered across Britain to glean as great a variety of barometric observations as possible. He asked the extremely thin Edinburgh-based physician and amateur astronomer James Lind – 'a mere lath' who was married to a 'fat handsome wife', Ann Elizabeth Mealy, 'who is as tall as himself and almost six times as big' – to carry a barometer to 'the summit of Arthur's Seat' and 'the observatory of Hawk-hill westward'. Lind's 'spirit of the kindest tolerance and the purest wisdom' was praised by his adoring surrogate son, the poet Percy Bysshe Shelley (who, it was rumoured, was saved from the asylum by the physician's intervention), and Lind gladly agreed to Roy's request. In 1774 Roy had a friendship with a fellow member of the Royal Society whose ambitious project gave the map-maker an opportunity to take his barometer to Scotland in person.

The laws of motion that Isaac Newton had laid out in his *Principia Mathematica* had sparked a widespread interest in geodesy. Without proof or evidence, Newton had assumed that the earth had originally been a homogeneous fluid mass and he claimed that one of the effects of the law of universal gravitation was that the centrifugal force to which the earth was subject meant that it was not a perfect sphere. Instead, its rotation produced a bulging at the equator and a flattening at the poles, making the earth resemble a less extreme version of a giant Smartie or 'oblate spheroid', whose horizontal axis was longer than its polar one. But geodesists refused to take Newton's description of the shape of the earth on trust. If his claim was correct, they surmised, then the length of one degree of latitude would necessarily be shorter at the Arctic and Antarctic poles than at the equator. The truth of this supposition could be ascertained through the comparison of two meridian arcs – long lines travelling north–south along the earth's surface – at two different locations.

In 1735 and 1736 France's Académie Royale des Sciences dispatched two expeditions to each measure the length of one meridian arc, extending over one degree of latitude. The first, which was led by Pierre Bouguer, was sent to Peru, near the equator. The second went to Lapland under the direction of Pierre-Louis Moreau de Maupertius. Their measurements duly confirmed Newton's theory and indicated that the earth was indeed an oblate spheroid. However, during the course of the Peru expedition a persistent anomaly niggled at the geodesists. In the Andes the surveying instruments did not behave as expected. The plumb lines that were supposed to keep the instruments perfectly poised did not hang vertically: they were deflected off-course. In fact, Bouguer was expecting something of the kind. Newton had predicted that mountainous masses on top of the earth's surface possessed their own gravitational attraction, and that instruments used beside large mountains would find their plumb lines pulled away from the vertical towards the mountain itself. The internal constitution of the earth was then unknown and geodesists debated whether it was a homogeneous mass, a hollow sphere, or of varying density. Mountains were particularly tricky, as it was not clear whether former volcanoes, such as the Andes, were hollow inside. The results of the Peruvian expedition indicated that the mean density of the earth was considerably more than the density of the earth near its surface. But what perplexed Bouguer was that the deflection exhibited by his instruments was very different from his predictions.

In the early 1770s, some thirty-five years after the confusing Peru expedition, Britain's Astronomer Royal, the Reverend Nevil Maskelyne, announced to the Royal Society his intention to interrogate this conundrum of the 'attraction of mountains'. Maskelyne was a key figure in Enlightenment science, whose work at Greenwich Observatory made significant advances in astronomical, navigational and geodetic knowledge. Among other objectives, he hoped that his proposed expedition would allow the attraction that was exerted by mountainous masses to be quantified, thus establishing greater knowledge about the relative density of the earth. Maskelyne read before the Royal Society his 'Proposal for measuring the Attraction of some Hill in this Kingdom by Astronomical Measurements'. The Royal Society duly set up a Committee for Measuring the Attraction of Hills and began to search for a suitable astronomer, mathematician, or surveyor to oversee Maskelyne's experiment. They chose a man called Charles Mason, who had achieved fame in the mid 1760s by measuring, with his colleague Jeremiah Dixon, a line demarcating the borders of Pennsylvania, Maryland and Delaware as part of a dispute between these British colonies.[1] A few years after his return from this American adventure, the Attraction Committee employed Mason to hunt among the peaks of the Scottish Highlands, Yorkshire and Lancashire for a suitable mountain for Maskelyne's project. The astronomer was exacting: he specified that this

---

1 Now known as the Mason–Dixon Line, the geopolitical boundary has since entered the popular imagination as a stark delineation of the cultural differences between the northern and southern United States. The novelist Thomas Pynchon used Mason and Dixon's mission as the backbone for a book that reflects upon the fragile and elusive nature of history and Enlightenment ideals of 'truth': 'Who claims Truth, Truth abandons,' Pynchon's *Mason & Dixon* claims.

should be 'a broad-back'd massy hill', which is 'either an oblong hill (that has not been a Vulcano), or long valley, the long way of either running from East to West or not deviating 30 or at most 40° from that direction, & ½ mile high or deep or nearly so, the hill with as few scoops & hollows as possible & the valley as excavated as possible'. One consequence of the lack of detailed, accurate maps of Britain was that Mason was required to physically scour the landscape for such a perfect peak.

In early 1774 Mason hit the bullseye. He found 'a remarkable hill, nearly in the centre of Scotland, of sufficient height, tolerably detached from other hills, and considerably larger from East to West than from North to South, called by the people of the low country *Maiden-pap*, but by the neighbouring inhabitants, *Schehallion*; which, I have since been informed, signifies in the [Scots Gaelic] language, *Constant Storm*'. This was, Maskelyne commented wryly, 'a name well adapted to the appearance which it so frequently exhibits to those who live near it, by the clouds and mists which usually crown its summit'. Schiehallion is positioned in Highland Perthshire at almost the latitudinal and longitudinal centre of Scotland. Its stunningly symmetrical bulk had presided over brutalities in the wake of the 1745 rebellion and later over the Military Surveyors' efforts to tame Scotland through cartography.

12. Schiehallion at night.

Now the mountain was about to play host to another landmark in the history of human attempts to dominate, civilise and know the earth. But, unexpectedly, Mason pulled out of the experiment, ostensibly on financial grounds, although perhaps he had simply lost interest in fieldwork after a career of surveying in South Africa, America and Ireland. Maskelyne was reluctantly persuaded by the Royal Society to conduct the measurements himself and before departing for Scotland he sent an assistant to prepare the ground. Reuben Burrow had previously acted as Maskelyne's assistant at Greenwich and he would later work as a mathematics teacher at the Tower of London. Now Maskelyne instructed him to establish sites on Schiehallion from which astronomical observations and land measurements could be made. The project's numerous instruments included a ten-foot zenith sector, which was a long, fragile telescope within a wooden frame, that could be positioned with a plumb line. Burrow transported these delicate artefacts up the conical and stumbly sides of the mountain, and he erected tents and a bothy near the mountain's summit to accommodate those instruments and Maskelyne himself.

On 30 June 1774, Maskelyne arrived, with rather bad grace, to a scene of impeccable order. He spent the next four months living at the exposed and rocky summit of Schiehallion, where he tried to carry out a nightly series of celestial observations and measurements at various points on the mountain's side to quantify the deflection of his instruments' plumb lines and to test the accuracy of his newly improved zenith sector. Living a mostly nocturnal existence, the astronomer became infuriated by 'the badness of the weather, which was almost continually cloudy or misty', and which retarded the observations. In mid July Maskelyne wrote in frustration that 'it is not but a week ago since I was first able to see a star in the sector (so bad has the weather been) and I have yet had only one good day'. Meanwhile Burrow and a local land-surveyor called William Menzies together conducted detailed surveys 'to ascertain [Schiehallion's] dimensions and figure' and 'to determine the position and distance of the two stations of the observatory'. But the expedition was not unmitigated torture. Maskelyne recollected how he 'was honoured . . . by visits from many learned gentlemen who came from a great distance' to see him perched on his mountain top. In

August Maskelyne's trial was ameliorated by the arrival of a friend he knew from the Royal Society: William Roy.

Roy had been to Schiehallion before, of course. His first engagement with national map-making had taken place in its shadow during the Military Survey of Scotland. Now, almost thirty years later, Roy was returning as an experienced and respected member of the military, scientific and anti-quarian establishment. At the summit of Schiehallion he used his barometer to measure the air pressure and, from that measurement, to calculate the mountain's altitude. Roy then compared his results with those garnered by Maskelyne. Satisfyingly for both parties, the difference between Roy's and Maskelyne's calculations of the height of this 3554-foot mountain was a mere three feet. Roy later wrote up the results of his experiments and his conclusion that barometers could provide reliable measurements of alti-tude, and read his article before the Royal Society's fellowship in four stages, in June and November 1777. Shortly afterwards, he published his report in that institution's prestigious journal, the *Philosophical Transactions*.

Maskelyne's time on Schiehallion was not as lonely as one might expect. He found the inhabitants of the settlements that orbited the mountain gen-erous, polite and interested. He remembered how 'they often paid me visits on the hill, and gave me the fullest conviction that their country is with jus-tice celebrated for its hospitality and attention to strangers'. Today Schiehallion is classified as a Munro:[2] that is, a mountain in Scotland over 3000 feet, a member of one of the loftiest classes of peaks. When I climbed Schiehallion's tricky upper slopes a few years ago, it was an adventure. I packed a space blanket and Kendal Mint Cake in case of emergency, relied

---

2 This category of mountains is named after a Victorian mountaineer called Hugh Munro, who produced the first catalogue of Scotland's highest summits. Other classes of hills include Marilyns (hills in the British Isles with a relative height of at least 150 metres), Corbetts (peaks in Scotland between 2500 and 3000 feet), Donalds (hills in the Scottish Lowlands over 2000 feet), Grahams (mountains in Scotland between 2000 and 2499 feet, owning a clear drop of at least 150 metres), Murdos (Scottish peaks over 3000 feet, which have a relative height of at least 30 metres), and Nuttalls (hills in England and Wales over 2000 feet, with a relative height of at least 15 metres). I was childishly pleased to discover that there is also a class of mountains known as HEWITTs: a rough acronym for Hills in England, Wales, and Ireland over Two Thousand feet.

upon my spring-mounted walking stick and compass, and derived enormous satisfaction from reaching the eroded trig point that marks its summit. But the couple I stayed with in Kinloch Rannoch thought nothing of trotting up Schiehallion to watch the sun rise or set, or simply to walk the dog, carrying nothing with them, not even water. In the mid 1770s the locals seem to have exhibited an equally relaxed attitude. They took regular meals up to Roy and Maskelyne, and a local boy called Duncan Robertson, known as red-haired Duncan, *Donnaeha Ruadh*, whiled many of the nights away playing the fiddle and singing Scottish folk songs to the strange men at the top of the peak.

Roy did not stay long on Schiehallion and Maskelyne himself came down off the mountain in late October. Before he departed, he held a farewell party for the local people who had been so welcoming, in the bothy in which he had lived for nearly four months. The story goes that after copious quantities of local whisky, somehow the revellers managed to set the hut alight. Maskelyne's instruments and observations remained safe, but Duncan Robertson's precious fiddle, which had helped to pass so many of the astronomer's lonely nights on the mountain, was reduced to ashes. Maskelyne promised Duncan that, on his return to London, he would 'seek you out a new fiddle and send it to you'. The instrument reached Kinloch Rannoch a few months later: Maskelyne had bought Duncan a Stradivarius, whose price, even at the end of the eighteenth century, was well beyond the reach of the average musician. The boy was ecstatic and composed a song to the new violin and to Maskelyne's generosity. 'On the trip I took to Schiehallion,/ I lost my wealth and my darling', it began. But 'Mr Maskelyne, the hero,/ . . . did not leave me long a widower,/ He sent to me my choice treasure/ That will leave me thankful while I live.'

THE SCHIEHALLION EXPEDITION was a success. Maskelyne proved that 'the mountain Schehallien [*sic*] exerts a sensible attraction' and his experiment contributed to a body of research that proved Newton's laws to be true. Moreover, Maskelyne had deduced the reason why plumb-line deflections were not so great as predicted: their attraction towards mountains was counteracted by another attraction towards the earth's crust and core, and this meant that 'the mean density of the earth' must be 'at least double of that at the surface'; he happily recorded how this was 'totally contrary to the hypothesis of some naturalists, who suppose the earth to be only a great hollow shell'. Maskleyne's Schiehallion adventure was an important landmark in the progress of geodesy in the eighteenth century. It was also a feat of human endurance. When Maskelyne won the Royal Society's highest accolade for his research in 1775, its president, Sir John Pringle, described in awe to his audience how the astronomer had spent 'a residence of four months in a mean hut, on the side of a bleak mountain'.

Maskelyne's achievement had significant repercussions for map-making. While he had been busy with his astronomical observations, his assistant Reuben Burrow made a meticulous survey of the mountain's bulk. He and Maskelyne passed on the results of this survey to the Professor of Mathematics at the Royal Military Academy in Woolwich, Charles Hutton, who agreed to calculate the mountain's volume. To do so, Hutton adopted a map-making technique that was a novelty in Britain. Contours trace the crinkles and furrows of landscape by joining up points of equal altitude on a hill or valley with a thin line. Otherwise known as isopleths, contours are now a defining feature of Ordnance Survey maps. It is thanks to these thin orange lines and their accompanying numbers that map-readers can conjure up a three-dimensional landscape in their minds; that hikers can brace themselves for the steep ascent around the next corner; and that the promise of a downward slope can be dangled before the tired members of a walking party. Contours had been used by Continental surveyors since the early eighteenth century, but British map-makers had traditionally tended to prefer a more pictorial mode of representation called hachuring: tiny parallel lines in black ink, whose density and darkness increased according to steepness (something similar had been used during the Military Survey of Scotland). Faced with the complex task of deducing Schiehallion's volume,

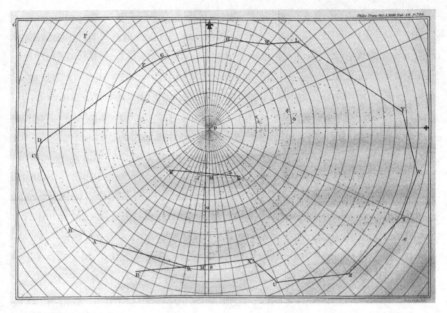

13. A simple contour map of Schiehallion by Charles Hutton, 1778.

Hutton produced a map on which points of equal altitude were linked by swirling lines. This was an important moment in cartographical history. Although contours would not become a standard component of map-making in Britain for another sixty years, Hutton's pioneering use of the technique helped pave the way.

The Schiehallion expedition was a vital landmark in Enlightenment attempts to find the earth's concealed 'pattern' and to reveal the obscure 'plan' of its 'mighty maze' through empirical measurement. It also occurred at a point in William Roy's life when it had become evident that he was as committed to scientific institutions and preoccupations as he was to military ones. In the decades that followed the Military Survey of Scotland, county surveyors had contributed fragmentary pieces towards a jigsaw map of the nation, military engineers had closely assessed the south coast of England for its vulnerability to invasion, and geodesists had measured portions of the earth to discover whether the planet's shape bore out Newton's laws. These events had taken place against a backdrop of frequent global wars and an apparent upsurge in nationalistic feeling in Britain. Institutions as diverse as the Society of Arts, the Board of Ordnance and the Royal Society had been the chief sponsors of cartographical progress in this period. Although William Roy had only possessed tangential links to the county map-makers of the Society of Arts' competition, he had cut his teeth among the type of civilian estate-surveyors who proved its chief applicants, and he was deeply preoccupied with both military and geodetic traditions of map-making. His involvement in all these projects significantly strengthened his desire to put into place a national mapping agency.

CHAPTER THREE

The French Connection

'LE MINISTRE DES AFFAIRES philosophiques' was how a friend half-jokingly referred to Joseph Banks. He was not far wrong. By the early 1780s Banks's magnetism and talents had elevated him to the most influential position in British science: President of the Royal Society. On the morning of 8 October 1783, we can imagine that he drew back his curtains with an eager anticipation of the day ahead. Only a month previously Britain had been released at last from conflict in the American War of Independence. Even the skies seemed to reflect a national feeling of release. All summer an ashy fog, the result of a volcanic eruption in southern Iceland, had filled the skies above Britain, dimming the stars and sun. By October the fog had lifted, the war was ended and the sun shone out as brightly as Banks's mood.

It was a good day to receive the letter that was awaiting him on the breakfast table. The politician Charles James Fox had forwarded to Banks a 'Mémoire' that he had received from the director of the Paris Observatory, César François Cassini de Thury. This sought to persuade its reader that 'il est interessant pour le progrés de l'Astronomie que l'on connaisse exactement la différence de Longitude et de Latitude entre les deux plus fameux observatoires de l'Europe'.[1] Cassini de Thury explained that, although the

---

1 'It is interesting for the progress of astronomy that we should know the exact difference between the longitude and latitude of the two most famous observatories of Europe.'

exact longitude and latitude of the royal observatories at Paris and Greenwich were considered well known, in practice the English and French astronomers' calculations of their locations diverged to a worrying degree. He particularly called into question the determination of Greenwich's situation. Banks did not have to be told that it was of the utmost importance that the positions of these sites were known with precision and confidence. Both observatories were essential to national and global navigation: for sailors, the problem that Cassini de Thury had raised could be a matter of life and death. Banks had no choice but to take his memorandum seriously.

César François Cassini de Thury was the third generation of a dynasty of astronomers who had presided over French cartography since 1669. In that year, a brilliant Italian star-gazer called Giovanni Domenico Cassini had moved to Paris and began to call himself 'Jean Dominique Cassini'. He was granted membership of the newly founded institution for natural philosophy, the Académie Royale des Sciences, and in 1671 King Louis XIV gave Cassini a grant to establish the Paris Observatory, of which he became essentially director. By the end of the century Cassini had become involved in a project that had been initiated by the French astronomer Jean Picard to measure a meridian arc (a line travelling north–south across the earth's surface) across the vertical length of France. This arc extended from Dunkirk in the north, through Paris, all the way down to the border with Spain, a little east of Perpignan, in south-west France. Picard and Cassini's revelation of the country's true extent was said to have provoked the King to comment that he had lost more land to his astronomers than to his enemies.

Cassini and Picard had measured their meridian arc using the most up-to-date technique of the day. Triangulation had first emerged as a map-making method in the mid sixteenth century when the Flemish mathematician Gemma Frisius set out the idea in his *Libellus de locorum describendorum ratione* (*Booklet concerning a way of describing places*), and by the turn of the eighteenth century it had become the most respected surveying technique in use. It was a similar method to plane-table surveying, but its instruments conducted measurements over much longer distances. The French meridian arc's triangulation began when Cassini and Picard measured a baseline, a straight, level and unbroken line on a plain stretching from Villejuif, south of Paris, seven miles further south to Juvisy (traversing the site of what is now the Orly airport). This baseline served as the cornerstone for the entire triangulation, and required rigorous and painstaking measurement, by laying measuring chains or sticks along the ground with great care. Cassini then positioned his theodolite exactly at one end of the baseline and trained its sights on a prominent nearby landmark, a church steeple or the top of a hill, for example, on which a colleague stood clutching a surveying staff, usually a flagstaff. Such targets in a triangulation are now known as 'trig points' or 'trig stations', which was an abbreviation of 'trigonometrical' and referred to the mathematical basis of the process. Once the theodolite was positioned so that the flagstaff filled the centre of its sights, Cassini stood back to read the angle between this sight line and the base line from the theodolite's compass. He then repeated this process at the other end of the baseline, training the theodolite's sights onto the same trig point and noting down the angle.

An imaginary triangle had thus been created, whose sides extended between the two ends of the baseline to the trig point. Cassini knew the length of the base from actual measurement, and with his theodolite he had discovered the two angles between the baseline and the two sight lines that stretched from its ends to the trig point. This one length and two angles were enough to allow Cassini to calculate the length of the other two sides of the triangle, using relatively simple trigonometric formulae. In doing so, he could discover the trig point's distance from the baseline's two ends. Cassini then repeated this process, using each side of this first triangle as the base for a new triangle that extended to a new trig point. Cassini replicated these measurements over and over again, building triangles on triangles, until a complex network of sight lines and measurements extended over the landscape.

Triangulation was extraordinarily useful because it discovered the relative distances and bearings of a series of landmarks scattered over a terrain, using sight alone. By ascertaining those lengths and angles, triangulation allowed the surveyor to calculate the longitude and latitude of the trig points, once the longitudinal and latitudinal positions of the baseline's ends had been discovered through astronomical observations. Triangulation was a far quicker way of establishing large measurements than doing so with a chain on the ground, but it could be expensive, as it depended on instruments sophisticated enough to precisely measure the angles of sight lines that extended over tens of miles and map-makers proficient in such observations and the complex mathematical equations that came afterwards.

The first national triangulation, or 'trigonometrical survey', in Britain had been proposed in the years directly following the instigation of the French one. By the end of 1681 a cartographer called John Adams had measured a twelve-mile baseline on King's Sedgemoor in Somerset, with the backing of the newly founded Royal Society. But by 1688 Adams had complained to the Archbishop of Canterbury of the lack of finances that meant the project had folded. Similarly lacking the necessary instruments, personnel and finances, William Roy had not based his Military Survey of Scotland on a trigono-metrical survey either, and consequently there had been an accumulation of errors in that map's measurements. As this suggests, the principal purpose of triangulation was to form an accurate skeleton for a map of a large piece of land, a structure around which the finer details of the interior landscape could be arranged. It could also be used to create a simple chain of triangles, running north to south, with which to measure a meridian arc. Triangulation was the surveying technique *du jour* during the Enlightenment. Measurements taken from the ground itself could be distorted by imperfect measuring chains, bogs and thorny undergrowth, roads and fences, or interfering locals. But triangulation promised to realise the Enlightenment ideal of perfect measurement. Because each corner of each triangle was also at the corner of further triangles, any errors could be quickly identified. And apart from the initial ground-level measurement of the baseline, the surveyor's sight lines seemed to shoot, unimpeded, above the landscape, ricocheting smoothly from steeple to steeple, summit to summit. As we shall see, these aspirations to perfection were not quite so easily realised, but the technique was never-theless more sophisticated than any other.

Jean Dominique Cassini began by using triangulation to measure his meridian arc through France. Before long he had expanded this project to attempt a complete national map. Thanks to its unprecedented accuracy and scope, this became the first survey of its kind in the world, and a model for all future national maps based on triangulation, including the Ordnance Survey. The French mapping project would occupy four generations of the

Cassini family: Jean Dominique, his son Jacques (known as Cassini II), his son César François Cassini de Thury (Cassini III), and *his* son, again called Jean Dominique Cassini (Cassini IV). In 1756 the first of the eventual 181 sheets of the Cassinis' national map of France was published. Known variously as the *Carte de Cassini*, *Carte de France*, *Carte de France de Cassini*, or *Carte de l'Académie*, the endeavour sealed the family's fame. The Cassinis' achievements in the fields of astronomy and map-making were such that their name is commemorated now by a host of memorials: the 24101 Cassini asteroid, lunar and Martian craters named Cassini, a mathematical curve called the Cassini Oval, the Cassini Web Server and a British publisher of historic maps named Cassini.

BY OCTOBER 1783 the third generation of the Cassini dynasty, in the shape of César François Cassini de Thury, was in charge of the Paris Observatory and overseeing the progress of the *Carte de Cassini*. It was then that he approached the British with the idea of verifying the precise longitude and latitude of the two royal observatories at Paris and Greenwich. But whereas English astronomers typically preferred to discover these positions by treating the stars as celestial compasses, Cassini de Thury recommended eschewing such 'positional astronomy' in favour of using triangulation to discover longitude and latitude by 'joining up' the meridian arc that bisected Paris with the Greenwich Observatory. His motives for suggesting the project were manifold. The check on Paris's longitude and latitude would confirm the accuracy of the *Carte de Cassini*. And such a 'junction' between the Greenwich and Paris observatories would put global navigation on a firmer footing. Since 1767, Britain's Astronomer Royal Nevil Maskelyne had been compiling an annual collection of tables of astronomical observations made at Greenwich, known as the *Nautical Almanac*. These tables allowed sailors to determine their longitude at sea, thus avoiding shipwreck and disorientation. A miscalculation of Greenwich's position therefore jeopardised the safety of mariners across the globe. And there were other merits to the project too. The extension of the French meridian arc a few degrees north would contribute further knowledge about the precise shape of the earth, a question that continued to trouble geodesists up until the twentieth century. Finally, in the wake of almost a century of intermittent hostilities between France and Britain, a 'connection' between 'the two most famous observatories in Europe' could forge a reconciliatory bond.

France and England had long been one another's arch-rivals, scapegoats and punchbags. By 1783, a prolonged series of wars had repeatedly set Britain against its neighbour over the course of a century: the War of the Spanish Succession, the War of the Austrian Succession, the Seven Years War and the American War of Independence, not to mention the Jacobite rebellions that had rocked Britain and Ireland between 1688 and 1746. There was an entrenched cultural hostility between the Protestant island nation and Catholic France. The English considered the French to be effeminate, debauched, weak and seduced by fashion and luxury. They were thought to be the diametric opposite of the archetypal Englishman: 'John Bull', the hard-working, honest, down-to-earth, beef-eating, beer-swilling patriot. Much of this xenophobic hostility was expressed through culinary metaphors. The French called the English 'les rosbifs', and considered them crass, uncultured and bigoted. The English referred to the French as 'frogs', after their notorious penchant for frogs' legs, and they ridiculed the gastronomic affectations behind their 'fricassees' and 'ragouts'. In the face of this deep-rooted nationalist animosity, collaboration and communication between British and French scientists continued throughout the eighteenth century, but they rarely remained entirely free from cultural antagonism.

When Joseph Banks shared Cassini de Thury's proposal with a number of Royal Society fellows, many interpreted it as an enormous national affront. How dare the French question whether the English were entirely certain of the position of their own observatory. It was the Astronomer Royal Nevil Maskelyne's *job* to use the stars to ascertain precisely where he was on earth. Charles Blagden, secretary to the Royal Society and Banks's right-hand man, was sceptical about the French astronomer's claim for the superiority of triangulation over astronomy. 'I believe that it is not strictly the case,' Blagden wrote cautiously, 'and that occultations of fixed stars by the moon, carefully observed at both places, would give their differences of longitude nearer than any actual measurement.' A journalist in the *Annual Register* huffed that Cassini de Thury's claim for the uncertainty of Greenwich's location was made 'without sufficient reason'. Maskelyne was also shown the memorandum, so 'that you might consider it fully and at your leisure'. Although he was asked to word a formal response to the French geodesist, Maskelyne was too piqued to reply. A small, marbled notebook in Greenwich Observatory shows that he actually instigated his own secret experiment to measure the longitudinal difference between the two observatories, using chronometers, hoping to prove Cassini de Thury's surveyors wrong.

But Joseph Banks was not so offended by the French proposal. He considered Cassini de Thury's approach 'as doing honor to our scientific character'. Banks passed it on to his great friend William Roy, whom he knew was considerably better qualified to judge its merits. Roy himself was an enthusiastic advocate of triangulation. He had spent much of that summer measuring beneath ashen skies a baseline of just under one and a half miles in central London, extending from the middle of what is now Regent's Park, to Black Lane, a street north of St Pancras Station (now buried beneath railway yards). In the wake of the rejection of his proposal for a 'General Military Map of England', Roy had decided personally to take Britain's poor state of cartography in hand. Ultimately, he wanted to use this baseline as the backbone for a triangulation around London, which he hoped would ascertain the longitude and latitude of certain important locations, including Greenwich Observatory.

When Roy received Cassini de Thury's memorandum shortly after his own baseline measurement, he was surprised and delighted 'that an operation of the same nature, but much more important in its object, was really in agitation'. Roy foresaw that the proposed triangulation between Paris and Greenwich would be 'the first of the kind, on any extensive scale, ever undertaken in this country'. But although Cassini de Thury played up the scheme's value in cementing amicable international relations, Roy was principally excited by what it had to offer Britain. He saw the triangulation as the basis for the 'accurate Survey of the British Dominions' of which he had dreamed for so long. Roy promised that the proposed 'French Connection' would 'redound to the credit of the Nation'.

SIX MONTHS AFTER Cassini de Thury's note had landed on Joseph Banks's breakfast table, William Roy found himself on Hounslow Heath, to the south-west of London. Once George III had approved a sum of £2000 for its expenses, the 58-year-old Scot had been chosen to direct the British side of the project under the Royal Society's auspices. The King was glad to extend an amicable hand to the French *savants* in this brief moment of peace, and to further the cause of British science. Now Roy's first task was to measure a baseline, and he knew the exact spot. He had been fantasising about potential baselines since 1766, when he had originally thought that a base in 'one of the open level Counties, such as Cambridgeshire or Wiltshire', would be an ideal foundation for a national survey. But these locations were clearly unsuitable for a measurement between Paris and Greenwich. So Roy wrote to Banks that 'in all the Tours I have made, no place, at least near the Capital, seemed to me so proper for the measurement of a Base as Hounslow Heath'.

In Roy's time, the Heath encompassed a vast stretch of land. Now it is reduced to a small golf course, just over a mile wide: a meagre piece of grass north of Feltham that gasps for air amid the frenzied arteries of the A315, A314, and A312 roads. The old Hounslow Heath is now hidden under Heathrow Airport, buried beneath the unlovely roads that run between Hampton Hill and Hatton Cross, and swallowed up by the suburban housing and industrial estates that populate North Feltham, Hanworth and Hampton Hill. Roy, however, saw in the old Heath ideal surroundings for the beginnings of the Paris–Greenwich triangulation. The Heath was perfect for the triangulation's baseline for the same reasons that it makes an ideal home for an airport today. 'Because,' Roy explained, 'of its vicinity to the Capital . . . its great extent, and the extraordinary levelness of its surface, without any local obstructions whatever to render the measurement difficult.' Heathrow's barbed-wire perimeters, pavement-free roads and cramped tunnels put the site of the old Heath almost completely out of bounds for the pedestrian trying to follow in Roy's footsteps. This was one of the least beautiful hikes that I undertook in the course of researching this book.

But even in Roy's day, Hounslow Heath was not particularly savoury. Ridden with thorny scrub, ant hills and gravel pits, it was notorious for the highwaymen who plagued its internal paths. The Heath was also where convicted criminals came to rest: bodies suspended from gibbets were silhouetted against the horizon. Although attempts were made to gentrify the area in the 1780s, many despaired of success. The *World and Fashionable Advertiser* reported despondently that improvers faced 'natural obstacles at every turn – a clay soil – a confined boundary – bad country beyond it – and no collateral idea from contiguous objects, but what is gloomy and disgusting! – The Heath of Hounslow, with all its gibbets and powder mills!'

On 16 April 1784 preparations for the baseline measurement got under way. Assisted by Banks and Blagden, Roy first made a reconnaissance, identifying a dead-straight tract extending across almost the whole Heath. Its south-western extremity was on the site of Hampton Poor House, now long gone. (In its place, amid the leafier, prettier end of the old Hounslow Heath, only a couple of miles from Hampton Court Palace, there is instead a memorial to Roy, a small cul-de-sac named Roy Grove and a plaque celebrating his achievements.) The base then extended just 'upwards of five miles' to the north-west, to a small settlement once known as King's Arbour, which has been swallowed up by the world's second busiest airport. Roy marked the baseline's ends with wooden pipes that he drove deeply into the ground and which he hoped would form permanent markers.

In appointing William Roy to direct the British side of the Paris–Greenwich triangulation, Joseph Banks and the Royal Society benefited from his military affiliations. Instead of hiring 'country labourers', Roy found trained and 'attentive' soldiers to assist the measurement. Their job was to clear 'a narrow tract along the heath' to 'permit the base to be accurately traced out thereon'. But the soldiers found themselves stumbling through brushwood and thick brambles, and coming dangerously close to 'certain ponds, or gravel-pits full of water'. Through this unpredictable landscape, they had to pick out a length of five miles that was entirely free of such obstacles. The summer of 1783 had been unusually hot and dry, but that of 1784 was distinguished by 'extreme wetness', in Roy's words. The soldiers sat up all night through

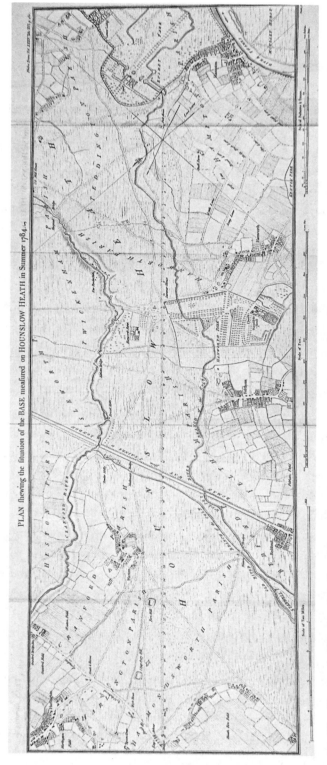

14. 'Plan Shewing the Situation of the Base Measured on Hounslow Heath in Summer 1784'.

75

torrential downpours, 'guarding such parts of the apparatus' against thieves and grazing cattle.

On 16 June the base measurement began. Roy first conducted a 'rough' survey of the distance with a 100-foot steel chain that weighed eighteen pounds, laying it along the space that had been cleared by his soldiers. But the accurate determination of the base's length was to take place with wooden measuring rods. These were dowels of, in this case, deal or Scots pine, which extended to twenty feet and rested on stands, 2 foot 7 inches high. From a side view, the instruments resembled suspension bridges, with an elaborate network of supports above and on either side of the central rods to prevent them from buckling. The day, 15 July 1784, that these deal rods arrived from the instrument-maker Jesse Ramsden's workshop was momentous. Joseph Banks hurried to the southern end of the baseline, accompanied by Blagden and three other eminent Royal Society fellows. Ever the charming host, Banks ordered tents to be pitched nearby, where Roy recorded in wonderment that visitors to the site 'met with the most hospitable supply of every necessary and even elegant refreshment'.

The baseline measurement began when Roy placed the first rod at the base's end at Hampton Poor House, and arranged it to coincide with the direction of the roughly traced base. He had found rods measuring between twenty-five and thirty feet to be 'rather too unwieldy', and it is likely that the twenty-foot rods were also initially tricky to position in a straight line, and perhaps liable to tip over too. Roy gradually shifted a second rod into alignment, end to end, with the first. Once this second rod was firmly stabilised on the ground, the first could then be detached and brought round to the front of the second, where it was again placed nose to nose. Carefully noting the number of times he alternated the rods, Roy inched along the length of the baseline, hoping to produce the most accurate assessment possible of that stretch of land. It took five hours to measure a 'short distance of 300 feet', the length of only fifteen rods.

Despite the initial excitement surrounding the commencement of the base measurement, Roy quickly became troubled. The incessant rain and humidity of that summer wreaked havoc with the wooden measuring rods; they expanded and warped so much as to become almost unusable. Depressed and disappointed, Roy was at a loss, fed up with the relentless rain and perhaps beginning to question whether a man almost sixty years old should be exposing himself to the elements in this way. But he rallied when a young friend from his days in Scotland, William Calderwood, who was assisting the baseline measurement, hit upon a solution: rods made of glass. After a brief delay to allow Ramsden to construct these new instruments, the base measurement began again at 8.30 a.m. on 2 August. Calderwood's rods acquitted themselves well. Roy delightedly found them to be 'so straight that, when laid on a table, the eye, placed at one end looking through them, could see any small object in the axis of the bore at the other end'.

On 30 August the baseline measurement was finished, after seventy-five consecutive days on the Heath. The sight of a group of uniformed soldiers manoeuvring strange, twenty-foot glass rods around the brambles and pools of the Heath must have bemused passing tradesmen and the homeless people who congregated around Hampton Poor House. And the night-time glow of the soldiers' encampment may have temporarily deterred the Heath's highwaymen from their illegal pursuits. The baseline measurement was also cause for the arrival of more salubrious guests from further afield. Due to the sophistication of Roy's techniques and the political significance of this Anglo-French collaboration, many eager voyeurs flocked to the event on Hounslow Heath, especially on its first and last days. Journalists for national newspapers reported Roy's valiant 'undertaking'. The frenzy came to a head on 21 August when King George III and his entourage paid a visit. (They had delayed it from 19 July, a day that experienced exceptionally heavy rain.) The King stayed for 'the space of two hours, entering very minutely into the mode of conducting [the measurement], which met with his gracious approbation'. The *St James's Chronicle* reported George III's encouragement of the Paris–Greenwich triangulation in admiring terms, praising 'his Majesty (who is ever ready to patronise useful Schemes)'. In the wake of the American War of Independence that had seen the King publicly vilified for his intransigent refusal to be reconciled with the rebellious colonists, this approval must have been a pleasant novelty for him.

The final moments on the Heath lived up to the hype of the previous days. A host of excited celebrities flocked to the baseline to celebrate Roy's achievement, including the art collector Sir William Hamilton, the mineralogist Charles Francis Greville (who was then the lover of Emma Hamilton, later Nelson's mistress), the Scottish physician and military engineer Charles Bisset, and Trinity College Dublin's astronomy professor, the Reverend Dr Henry Usher. The urbane figure of Joseph Banks, who had given 'his attendance from morning to night in the field, during the whole progress of the work', presided over the closing festivities. When the party broke up the following day, Banks took the wooden and glass rods back to his house. Roy disappeared, to calculate the baseline's length. He found it to be 27,404.7 feet long (5.190 miles). Compared with its modern GPS-defined distance of 27,376.8 feet (5.185 miles), and taking into account the limitations of his time, Roy's baseline measurement was astonishingly accurate. He himself stated proudly that 'there never has been so great a proportion of the surface of the Earth measured with so much care & accuracy'.

NOW THAT THE baseline had been measured, the triangulation could begin. From his first perusal of Cassini de Thury's memorandum back in 1783, William Roy had realised that the proposed connection between Paris and Greenwich required surveying apparatus far more accurate than any currently in existence. Over the Channel, French surveyors extolled the merits of a new instrument called a 'repeating circle'. Constructed from two telescopes fixed to a circular scale, the repeating circle worked by taking multiple observations of angles on different parts of the scale and averaging these out to produce a single accurate measurement. But the British were proud of the theodolite, their national instrument, and refused to adopt this foreign innovation. So in 1784 Roy approached Jesse Ramsden and placed an order for the most sophisticated theodolite in the world.

Ramsden was an innkeeper's son, born near Halifax in Yorkshire, who had risen from modest beginnings to become the most sought-after mathematical-instrument-maker in London. It was an exciting time to be involved in the design and creation of such apparatus. Before the mid eighteenth century, it

was unusual for surveying instruments such as theodolites to incorporate telescopic lenses, which were dogged by 'chromatic aberration', giving rise to
distracting coronas of colour that clung to major outlines. But in 1757 John
Dollond, the son of a Huguenot silk-weaver and one of the founders of the
still extant opticians Dolland [*sic*] & Aitchison, reinvented telescope technology by patenting an 'achromatic lens' which limited this irritating
phenomenon. By fixing a convex to a concave lens, Dollond's device brought
wavelengths of colour to a single focal point, thus eliminating the coronas
that had been caused by the divergence of these waves. Although he was not
the first person to invent the achromatic lens, Dollond's patent made him a
very wealthy and famous craftsman. The lens revolutionised the worlds of
mathematical instrument-making and surveying, and allowed map-makers to
see further and more distinctly than ever before.

In 1766 Ramsden married Dollond's youngest daughter, and worked
alongside his father-in-law for the next seven years. In 1773 Ramsden set
up his own premises on Piccadilly, 'at the Sign of the Golden Spectacles'.
The following year, in this busy workshop at the heart of London's fashionable West End, he perfected a machine called a 'dividing engine' that
engraved scales, such as degrees of an angle or inches in length, onto instruments more minutely than almost any in existence. A circle can be divided
into 360 degrees, each degree is subdivided into sixty minutes, and each
minute into sixty seconds. Back in the seventeenth century, the natural
philosopher Robert Hooke had surmised that the average human eye
could not properly distinguish any angular divisions smaller than around
one 'minute of arc', a sixtieth of a degree. If the scale was any more minute,
Hooke suggested that the eye became dizzy and unable to read the precise
value of the angle. But Ramsden's 'dividing engine' engraved instruments
with scales that measured angles to one *second* of arc, a 3600th of a degree.
These tiny divisions could only be read with the help of a 'micrometer': a
device that had been invented in the 1640s by the instrument-maker William
Gascoigne and which amplified the most minute divisions of the scale to a
level comprehendable by the eye by means of a screw. Thanks to the unprecedented precision of Ramsden's instruments, astronomers, mathematicians,
geodesists and map-makers flocked from all over Europe to buy and com-

mission apparatus from his workshop. Ramsden's innovating imagination and consummate skill set him leagues apart from his colleagues.

Ramsden was a good-natured man, said to be 'full of intelligence and sweetness'. When Roy approached him in 1784, the instrument-maker was about to turn fifty. He was said by the diplomat and writer Louis Dutens to possess 'an expression of cheerfulness and a smile the playful benevolence of which will not easily be forgotten by his friends. His whole manner had a character of frankness and good humour which he well knew to be irresistible.' Eight years his elder, Roy was of a very different temperament: serious, reticent, but every bit as much of a perfectionist. One fatal flaw in Ramsden's character would set the two at loggerheads. Ramsden was late for almost everything. His insistence on perfection meant that instruments were kept at his warehouse for years beyond their deadline, while he tinkered and refined. One client commented crossly that 'there are few lively persons with such an open countenance as Ramsden, but one can also say that not withstanding this same countenance he has no scruples about failing to meet his promises'. There is an apocryphal story that when summoned to attend George III, Ramsden was chided by that monarch that, while he was 'punctual as to the day and the hour' he was 'late by a whole year'. When Roy commissioned Ramsden to make 'the best Instrument [he] could within a limited price, for measuring horizontal angles', the instrument-maker saw this as a thrilling opportunity to produce a theodolite that 'might be rendered superior in point of accuracy to any thing of whatever radius yet made if the expence necessarily incurred could be allowed'. Enthused by Ramsden's eagerness, Roy and Banks agreed to meet the costs. But they were not prepared for the toll in time.

Early theodolites consisted of sights, without lenses, that were attached to a circular scale to measure the angles of observations. The instrument that Ramsden began constructing for the Paris–Greenwich triangulation was of a new generation. It was huge. An enormous 'brass circle, three feet in diameter' lay at its heart. On this circle a scale was engraved from which the angle of a sight line's horizontal position could be read. This scale was more minute than any in existence: the circle's vastness and Ramsden's skill meant that it boasted divisions down to 'about 1/24,000 part of an inch', measuring angles

15. The Great Theodolite.

to one second of arc. Micrometers were fixed to the scale to render these infinitesimally tiny angular measurements clearly visible. And the theodolite was not just restricted to the measurement of horizontal angles. It boasted *two* telescopes, each containing Dollond's famous achromatic lenses. The lower telescope was positioned just beneath the brass circular scale and could only move horizontally. But the upper telescope was positioned above the scale, and it swivelled both horizontally and vertically beside an upright semi-circular scale that measured the angle of its vertical elevation. This meant that the theodolite could measure heights as well as distances, and even look to the stars.

Numerous further refinements ensured that Ramsden's theodolite was the most accurate in the world: spirit levels, internal lanterns, and joints so well constructed that the instrument was found to 'turn round very smoothly, and is perfectly, or at least as to sense, free from any central shake'. And the theodolite came with an entourage of accoutrements: a stand, a scaffold and crane to lift the instrument's enormous bulk as high as thirty-two feet, pulleys, ropes, a portable observatory made from oak and canvas, a travelling case, and a tent and canopy to guard against the elements. Later it would even be given its own customised carriage with spring-suspension to protect it from jerks and jolts.

Such unprecedented sophistication took Ramsden a long time to produce, much longer than either he or Roy had predicted. Years passed and Roy became infuriated at the delay, while Ramsden complained that his time and effort was quite disproportionate to the remuneration. An accident in the final year of the theodolite's construction postponed its completion even further. One of Ramsden's workmen dropped a 'large brass ruler' on the central circular scale 'when the divisions were nearly all cut', damaging it to such a degree that the instrument-maker was 'obliged . . . to take out the whole of the dividing and to begin entirely a new'. On 1 June 1787, Charles Blagden excitedly informed Joseph Banks that 'Ramsden is hard at work upon General Roy's instrument which it seems may be finished by the end of next week.' But Ramsden's health suddenly took a downturn and the theodolite was not finished until August. This delay and the bitterness it engendered would come back to haunt Ramsden and Roy both.

DESPITE THESE FRUSTRATIONS for William Roy, the years after the completion of the base measurement were not devoid of activity. He wrote up a detailed report of his experiences on Hounslow Heath: the trauma of the wooden rods, the innovation of glass replacements and the resulting baseline measurement whose accuracy surpassed any previous such operation in Britain. He read his account before the Royal Society between 21 April and 16 June 1785, and was delighted to find himself the recipient of the Copley Medal, the Society's highest accolade, for his efforts. And the fame that Roy had enjoyed during the baseline measurement continued. Newspapers reported the promotion of this 'incomparable Engineer' from lieutenant-colonel to colonel; his attendances at St James's Palace, where he 'kissed the King's hand'; his election as a member of the Royal Society's council; and the sole exemption of 'our trusty and well-beloved Lieutenant-Colonel William Roy' from a pay cut that affected every other military engineer in the kingdom.

Since Cassini de Thury had first mooted the idea of a triangulation between Greenwich and Paris, the French and English surveying parties had both found themselves treading on nationalist eggshells, repeatedly coming up against xenophobia and jingoism, despite Joseph Banks's plea that scientists 'ought not to hate one another [just] because the armies of our respective nations may shed each other's blood'. Cassini de Thury's initial memorandum had been careful to praise previous British achievements and flatter the monarch. But, replying in French, Banks had managed to offend him with an ill-phrased remark, 'owing I suppose to the difficulty I have in Expressing my Ideas correctly according to the Idiom of your Language'. After this faux pas, Banks begged Blagden to take over the task of communicating with the French savants in their own language. Even Roy himself was not very friendly at first to his 'fellow Labourers', as he termed the French astronomers. Like many British scientists, he had a patriotic attachment to the theodolite and went to 'some pains to investigate the degree of accuracy of the French trigonometrical operations' as he felt 'they certainly were not executed with the best Instruments'. By his own admission he was not fond of small talk, and Roy's communications with the French astronomers were perfunctory and devoid of niceties.

These nationalist irritations were considerably diminished when in September 1784 the progenitor of the whole project, Cassini de Thury, contracted smallpox and died. From the beginning Blagden had been sceptical about the French astronomer's merits. 'The *younger* Cassini,' Blagden mused, referring to Cassini de Thury's son, Jean Dominique (Cassini IV), 'seems to possess more acuteness' than his father 'and has, I think, good ideas.' Cassini de Thury's death cleared the way for his son to take over direction of the French side of the project, and much better relations ensued between the British and French parties. Patient in the wait for Ramsden to finish the theodolite, Jean Dominique Cassini was charming and cooperative when the British were finally ready to begin the triangulation in 1787. The French had little to do themselves, as their side of the triangulation between Paris and the north coast of France had already been measured during the making of the *Carte de Cassini*. Some triangles needed confirming, but Cassini's chief concern was coordinating the cross-Channel observations that would join together the French and English triangulations. The surveyor Pierre François André Méchain was appointed as Cassini's assistant and both declared themselves charmed at the prospect of visiting London, to meet with its 'illustres savans' and discuss their instruments and methods. Cassini made the most of his trip, taking a shopping list to Jesse Ramsden's Piccadilly workshop, where he commented to a bystander that 'this man is an electrical machine which has only to be touched to emit sparks'. Blagden, for his part, proposed that Cassini and Méchain should be made foreign members of the Royal Society.

In July 1787 the British finally began their triangulation. After preliminary arrangements, the theodolite was in use by August. The surveyors planned to observe triangles that stretched from the wooden pipes that marked the ends of the Hounslow Heath baseline right down to the south coast of Kent. But the triangulation got off to a bumpy start. A telescopic surveying instrument like Ramsden's three-foot theodolite greatly enhanced the distance and precision achievable by naked eyesight. But such sophistication often meant that surveyors found that they needed to relearn to see. We can imagine

William Roy positioning the instrument over the eastern extremity of the baseline, training the telescope in the general direction of the spire of Banstead Church ten miles away, and placing his right eye against the eyepiece to find . . . only black shadow. Finding the eye's optimum position in front of the telescope took time. And when the sights had been carefully focused on their target, and Banstead Church spire had come into view in the centre of the grid of wires that bisected the lens, it was all too easy to lose sight of one's prey with the slightest nudge of the instrument. It was also an acquired skill to manipulate the micrometer that allowed the sight line's angle to be deduced to the minutest degree. Surveying in the era of such superior instruments and methods was a highly developed craft.

But soon Roy had tamed the theodolite, and at the eastern end of the baseline its sights were being successfully trained on trig points at Windsor Castle in the west, Hanger Hill Tower to the north-east and the summit of St Ann's Hill in Chertsey. Roy and his military assistants then carted the three-foot, 200-pound, nerve-janglingly intricate instrument to each of these

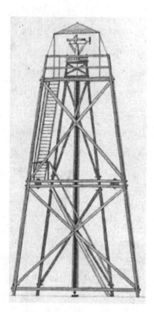

16. *The scaffolding used to hoist the theodolite to the required height, during the Paris–Greenwich triangulation.*

trig stations. Where necessary, they winched it to the top of towers or spires or up its own thirty-two-foot scaffolding. Roy then clambered up to the theodolite and trained its telescopes back onto Hounslow Heath, to verify the earlier observations. In a notebook he carefully recorded each angle and the temperature of the air, to allow him to deal in his equations with any expansion or contraction in the instrument's metals. He directed the theodolite's sights over towards Hundred Acres near Banstead, then to Norwood, Greenwich Observatory and Severndroog Castle. The same rigmarole was repeated at trig point after trig point, until triangles of sight lines were plotted across Sussex and Kent to Botley Hill and Wrotham Hill.

The process was desperately slow. Some of Roy's observations required him to identify trig points more than thirty miles away, requiring exceptionally clear weather conditions and consummate skill at manipulating the theodolite's telescopes and scales. After a month or so, Blagden wrote apologetically to Cassini: 'General Roy est à quelque distance de Londres . . . mais, à ce qu'on me dit, il n'est pas encore fort avancé.'[2] Uncomfortable at having kept the French surveyors waiting so long, Blagden suggested to Roy that he halt the mainland triangulation at Wrotham Hill and head straight for the coast to begin the cross-Channel observations with the French. Roy and Cassini both proved amenable to this idea (Roy even tried to pass it off as his own), and in the last week of September they met face to face in Dover. Roy proved a magnanimous host. He gave his guests a guided tour of the 'antique and half-ruined' Dover Castle and proudly showed off the workings of the 'Great Theodolite'. It was an efficient and productive visit too. Leaning over maps spread out on a large oak table, with Blagden translating, Roy and Cassini finalised dates, sites and techniques for the cross-Channel observations.

Cassini's party left Dover at the end of September to position themselves along the northern French coastline. Roy and Cassini had agreed that 'reverbaratory [sic] lamps' or 'white lights [with] extraordinary brilliancy' should be attached to the surveying flagstaffs to illuminate them through the thick

2 'General Roy is at some distance from London . . . but from what I hear, he has not reached very far yet.'

fog that often hung over the Channel. To distinguish their own lights from any others on the horizon, they 'placed [one lamp] over the other'. On 29 September, Cassini lit his signals at the top of the tower at Dunkirk. Roy grumbled that he did so 'somewhat earlier than the times appointed', taking the British surveying team at Dover by surprise. When the flares were lit again, Roy was ready but the air was 'not sufficiently clear to permit us to intersect it with the Wires of the Telescope'. Over the next three weeks, the French and British alternately lit signals and conducted observations that criss-crossed over the Channel, between Dover Castle and the windmill at Fairlight Head on England's south coast, and Montlambert (a hilltop near Boulogne), Cap Blanc Nez, Dunkirk tower and Calais's Notre Dame Church, all on the north coast of France. The lamps were a great success. Cassini reported incredulously that 'he hardly expects to be believed, when he tells that he observed one from Calais, which was fired on the opposite shore, about 40 miles off, and in bad weather'. Stormy weather on 2 October was 'very unfavourable for Observations', but it did not prevent Roy from successfully observing the flares at Calais and Blanc Nez. The sight of such bright lights seen irregularly through the vapour above the Channel must have been intriguing to locals, somewhat confusing to mariners and extremely worrying to smugglers.

By mid October the cross-Channel observations were complete and Britain and France had been trigonometrically bonded. The exercise acted as a double-check on the accuracy of the triangulation that underpinned the *Carte de Cassini* and it also allowed the relative merits of the French repeating circle and the British theodolite to be compared. Both were found to measure angles with similar precision, to within a couple of seconds of the truth. Roy's team now needed to complete the half-finished triangulation extending from Hounslow Heath down to the coast. But first he decided to measure a second baseline, known as a 'base of verification'. The actual measurement of this second base on the ground could be compared with the length provided by trigonometrical equations to test the accuracy of the triangulation so far and expose any errors. In August 1787 Roy and his assistants headed to Romney Marsh, a flat stretch of wetland just inland from the south coast of Kent and East Sussex, sheltered by the shingle beach of Dungeness (now

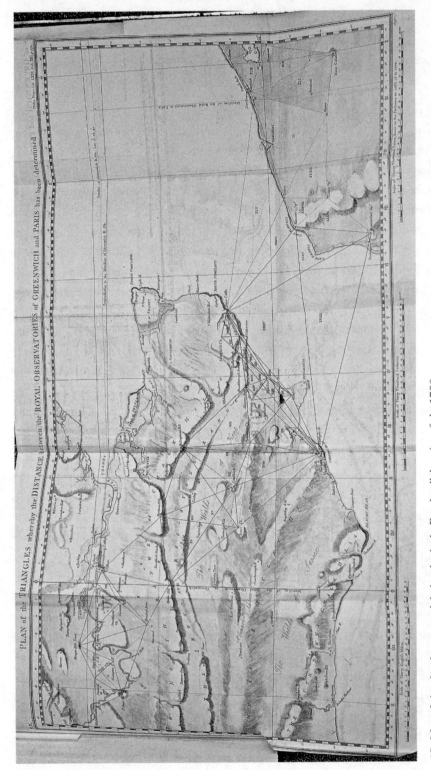

17. *Map of the triangles measured during the Anglo-French collaboration of the 1780s.*

home to a nuclear power station). Roy explained that 'from its levelness, as well as other advantageous circumstances attending its situation', Romney Marsh 'seem[ed] to me to afford the best base of verification for the last triangle'. But the soggy ground was 'so much intersected with ditches full of water' that 'the laying of bridges for the tripod stands' (for the glass rods) was 'a very troublesome and tedious operation'. Despite these hazards, the Romney Marsh measurement was a success. It gave a distance of 28,532.92 feet and showed that Roy's triangulation was extraordinarily accurate. As an impressed journalist reported, 'the *line of verification*, measured for that purpose on *Romney marsh*' showed that 'there was only an errour of *four inches and a half*' between the actual measurement of the base and that calculated by trigonometric equations from the theodolite's angular observations.

The task now for Roy and his assistants was to finish the triangulation from London to the south coast. As they hopped through Kent, local onlookers were intrigued by the Scottish surveyor hoisting his strange squat apparatus up churches, castles and hills, and the flashing lights on the horizon that accompanied its appearance. One Kentish man described in wonderment: 'as I passed through Tenterden, I observed a kind of tent on the top of the steeple, and on enquiry was informed that General Roy . . . and a party of the Artillery from Woolwich, with a large instrument made by Ramsden, were making observations of distant objects'. 'I know there was a measurement on Hounslow-heath two or three years ago,' he continued, 'but what connection it can have with signals; or what is the purport of those signals, I cannot imagine.'

Soon the nights closed in, the light faded, the temperature plummeted and the triangulation had to be suspended for another year. The short days and muted light of winter made long-distance observations impossible and Roy had to wait to fill in the few remaining pieces of the jigsaw puzzle until August 1788 – which would turn out to be this map-maker's last surveying fling. Once the observations were complete, Roy spent the remainder of the year and the beginning of the next making sense of his calculations. The French wrote up and published their side of the story in 1790. Roy crafted his experiences and calculations into two lengthy articles for the Royal Society's journal, the *Philosophical Transactions*. He calculated that the longitudinal separation between the two observatories at Paris and Greenwich amounted to a time difference of nine minutes and nineteen seconds. (The sun moves around the globe, passing through the imaginary lines of longitude at a rate of fifteen degrees per hour.) Although Maskelyne's secret experiment to measure this distance had actually come up with something very similar (nine minutes and twenty seconds), in a rather devious twist Roy misrepresented the Astronomer Royal's calculations of Greenwich's longitude in one of his articles as nine minutes and 30.5 seconds, presumably in order to prove the superior worth of the triangulation over astronomy and to demonstrate the value of the Paris–Greenwich connection in finding the true position of England's royal observatory. Importantly, Roy's results largely agreed with those of the French astronomers. The two parties' measurements of the twenty-six-mile distance between Dover Castle and Notre Dame's spire at Calais differed by only seven feet. This accord confirmed the accuracy of the teams' triangulations and the precision of their instruments. Compared with modern GPS data, the results of the Anglo-French collaboration are not spot-on, but in the 1780s the accuracy of the project's methods and instruments seemed miraculous.

Observers of the operation were given little choice but to regard it as a triumph of accuracy. 'Truth' was a particularly favourite word of Roy's, who hammered home 'the last exactness' and 'mathematical exactness' of his calculations and lauded the 'extremely perfect' instruments that had been used. He also showed a passionate allegiance to decimal points, although it was an inconsistent one: some calculations were shown to four decimal places, others to one or two. Roy's enthusiastic conviction of the accuracy of his work was infectious. Journalists repeated the claims that 'the accuracy of this operation is very considerable' and that the Paris–Greenwich triangulation was 'conducted in a manner which confers the highest honour on the abilities and attention of major-general Roy'. A writer for the *St James's Chronicle* wrote that 'among the improvements of the present age, may be reckoned, the extreme nicety in the construction of mathematical instruments, and the wonderful accuracy of trigonometrical mensuration', and he concluded that Roy's operation 'was prosecuted, at every step, with niceness of observation and accuracy of calculation'. Another journalist claimed that 'no measurement of a similar kind has hitherto, we believe, been carried on with so much accuracy'.

The Anglo-French venture seemed to have realised the highest ideal of the Enlightenment: perfect measurement of the ground beneath our feet. William Roy came to be popularly regarded as the embodiment of the 'Enlightenment method'. When the accuracy of Murdoch Mackenzie Sr's *Nautical Survey of the Orkney Islands and Hebrides*, which was based on maps made in the 1770s, came under fire, its 'uniform agree[ment]' with Roy's much earlier Military Survey of Scotland seemed an incontrovertible settlement of the debate in Mackenzie's favour, despite the obvious flaws of the earlier work. Credit for William Calderwood's glass measuring rods was sometimes given to Roy instead, who was thought to have 'invented an instrument, that from the materials of which it is composed is not subject to any variation whatever'. The supposedly supreme accuracy of Roy's triangulation also became a patriotic weapon. The *New Annual Register* called the triangulation 'a national work of great importance' and the astronomer Francis Wollaston felt it 'may be considered in some sort as a national concern'. They both based their patriotic pride on the truth and accuracy of Roy's instruments and methods. Rather than celebrate the enterprise for the amicability between the French and British surveying parties, most people in Britain, including Roy himself, seem to have been more interested in the project for its national merits.

And yet all this remarkable acclaim was not quite enough to satisfy William Roy. As he aged, his reticence turned to grouchiness. The delay that Jesse Ramsden had inflicted on the Paris–Greenwich triangulation gnawed away at him. 'No consideration upon earth would ever make me go through the same or such another operation again, merely from the drudgery of having to do with such a Man!', he fumed to Banks. Roy vented his rage in his last article for the *Philosophical Transactions*, where he accused Ramsden of intentional negligence and sloppy workmanship, and elsewhere of being 'too remiss and dilatory'. Blissfully ignorant of the contents of Roy's article, the equanimous Ramsden was in the audience when it was first read aloud before the Royal Society. 'Nothing could equal my surprize on hearing the charges brought against me, and misrepresentations contained in [the] paper,' Ramsden exclaimed. 'I was the more affected by it as coming from a Gentleman with whom I considered myself in Friendship, and, who had many obligations to me for my assistance.' In an open letter to the Royal Society, Ramsden tackled Roy's charges one by one. He accused him of 'unskilfulness' in the handling of the theodolite, and demanded that it be acknowledged 'that every part of the Instrument and Apparatus is of my invention'.

Roy's cantankerousness may have been due to his failing health, caused by chronic lung disease. In the spring of 1789 he was forced to retreat to the mountains of North Wales to convalesce. Although temporarily better, by winter he was much worse again and started to suffer frightening episodes of spitting blood. At the close of the Paris–Greenwich triangulation, Roy made an extended visit to Lisbon to recuperate from the toll the project had had on his well-being. Feeling much better, he returned to London in the spring of 1790 and applied himself to writing an 'Account of the Trigonometrical Operations by which the Distance between the Meridians of the Royal Observatories of Greenwich and Paris has been determined'.

On 30 June 1790 Roy spent the afternoon at the War Office before returning home to correct the proofs for his article. Whilst hunched over his desk, pen in hand, he suffered a sudden seizure at nine o'clock in the evening. In the early hours of 1 July William Roy died, at the age of sixty-four. The news spread quickly throughout Britain's military and scientific elite. Five days after his death *The Times* mourned that 'the Republic of Letters will experience a great loss in the death of General Roy. As a draughtsman, his pencil was universally admired [and] he was a great favourite with his Majesty.' This innovative, imaginative man, who had transformed the fields of military map-making and geodesy in Britain and appeared to have realised the Enlightenment's ideal of creating a perfectly truthful image of the natural world, had died without seeing his life's ambition realised: the creation of a complete, accurate, national survey. But the Paris–Greenwich triangulation had laid the groundwork, and many would later consider it as the true antecedent of that map.

CHAPTER FOUR

The Aristocrat and the Revolution

BEFORE LONG, ACCOUNTS of William Roy's death reached the drawing room of Goodwood House, a semi-octagon of brick and flint at the centre of a vast estate just inland from the Sussex coast. Goodwood belonged to Charles Lennox, 3rd Duke of Richmond, the great-grandson of Charles II and that king's mistress, the 'young wanton' Louise de Kéroualle, Duchess of Portsmouth. By July 1790 Charles Lennox was fifty-four years old and somewhat worn out. He retained the muscular figure of his youth, but his jowls were sagging and his hair was greying and thinning. News of Roy's demise is likely to have affected him deeply. The two men had experienced wildly different upbringings: Roy was the son of a land-steward, Lennox the offspring of a duke, the employer of numerous estate managers of his own and the owner of vast property in London, Sussex and the Loire Valley in France. But the two men's lives had often converged and Lennox had publicly celebrated Roy during his lifetime as an 'excellent and universally esteemed officer'.

In the months that followed Roy's death, Lennox took up the cause that Roy had owned ever since his experience on the Military Survey of Scotland: the conviction that 'the honour of the nation' depended on creating 'a map of the British islands' that was 'greatly superior in point

18. Portrait of Charles Lennox, 3rd Duke of Richmond, by William Evans, after George Romney, 1808 (1776).

of accuracy to any that is now extant'. William Roy's own recommendations and supplications had mostly fallen on deaf ears during his lifetime. But by 1790, Charles Lennox – aristocrat, politician and Master-General of the Board of Ordnance – possessed sufficient power to transform those recommendations into reality. It was through Roy's imagination and ambition and Lennox's clout that the Ordnance Survey finally came into being.

CHARLES LENNOX HAD been fascinated by maps since childhood. His father had encouraged all his children to read widely in geography, travel-writing, history and natural philosophy. Inspired by this education, Lennox's sisters went on to encourage their own offspring to enjoy surveying and topography, often using the wooden jigsaw maps that were a popular early aid to geographical education. (Jigsaws as we know them were invented in Britain in the 1760s by the London map-maker John Spilsbury. All these first jigsaws took the form of maps, and were designed as educational aids. In Jane Austen's novel *Mansfield Park* (1814), the heroine Fanny Price was mocked for her inability to 'put the map of Europe together' using one of these jigsaws.) The Lennox sisters' love of maps lasted a lifetime. By 1808 Charles's younger sister Sarah had lost her sight, but she was thrilled when her son constructed an early form of Braille map that allowed her to visualise the events of the Napoleonic Wars. He depicted Spain and Portugal by fixing raised patches of cardboard to the map and 'contrived by little pebbles to mark out the different places by feeling, very cleverly; the rivers by bits of twist. The result is that she feels out any place she wants to find in a minute and diverts herself for hours with it.'

Charles Lennox was even more actively engaged with cartography than his sisters. Practical map-making skills were an important component of eighteenth-century aristocratic male education and in Lennox's case this was augmented by a thorough education in trigonometry at the University of Leiden. In the 1750s, when Lennox was on the Grand Tour, his chaperone and tutor Abraham Trembley helped him to practise surveying on Europe's major battlefields. On his return to England, Lennox straight away commissioned a map of Richmond in Yorkshire.

On coming of age, Lennox began to buy up the properties that surrounded his estate. Originally 1100 acres, Goodwood gobbled up its neighbours:

Halnaker House and Park, Westhampnett House and its adjoining land, West Lavant, Raughmere, Stoke, Singleton, Charlton, East Dean and Selhurst Park. Soon the estate was almost seventeen times its initial size. Lennox then ruthlessly 'improved' his new monster territory. He envisaged pleasure-gardens, a 'high wood' in which exotic birds sang, a working mill to produce mortar from the estate's sandpits, a secluded 'pheasantry', heated lodges for tenants and employees, and acres of forests, rare trees and imported plants. (Lennox's improvements were precursors of the current multifunctional estate, still owned by his heirs, which incorporates a hotel, spa, farm, forest, aerodrome, golf course and sculpture park, and hosts vintage car rallies and horse races.) With the help of an architect, Lennox also set about converting the old Goodwood House. Later he turned his attention from his own crea-ture comforts to those of his beloved hunting dogs and erected ornamental kennels within sight of the house, at a cost of £6000.

Lennox's early interest in cartography and the ambitious plans he enter-tained for his estate's improvements prompted him on 1 November 1758, shortly after his marriage to Lady Mary Bruce, to offer a young Dutch surveyor called Thomas Yeakell full-time, permament work as his estate surveyor. Landowners frequently commissioned map-makers to make one-off charts of their estate but it was extremely unusual to employ a surveyor on a long-term basis. Lennox arranged for Yeakell to be tutored by the Royal Military Academy's mathematics professor and then, at the beginning of 1759, the aristocrat set off on a four-month tour of Holland and Germany, taking his map-maker with him. On his return to Britain in July, Yeakell was prolific. Over the next four years, this young surveyor made hundreds of maps and sketches of Goodwood, perfecting his methods and draughtsmanship and paying careful attention to every fold and indentation of the landscape.

IN 1763 LENNOX decided that his vast estate was too big for Yeakell to survey alone, so he employed James Sampson, who 'gave proofs of an extraordinary genius for drawing'. Lennox found Sampson stimulating company. He was a charismatic young man who spent his leisure hours in the British Museum copying its artefacts and chatting to visitors. Lennox was so taken with his new estate surveyor that he introduced Sampson to his father-in-law, Henry Seymour Conway, a senior military officer and politician. Conway also responded to the map-maker favourably and before long Sampson had married one of his senior servants and was enjoying the free run of his enormous house.

But James Sampson had a dark side. As newspapers would later report, he 'maintained an illicit intercourse with some women of debauched principles, whose extravagances involved him in many embarrassments'. Short of funds, and learning that Conway hid money and valuables in a desk drawer in his library, Sampson hatched a plan. He plotted to steal Conway's cash and set fire to the library to disguise his crime. At six o'clock one morning in August 1766 Conway awoke to the cry of 'Fire!' and the smell of singed vellum. When a team of fire-fighters arrived, he frantically directed them to recover the bureau. Most of the desk had been reduced to ash but Conway's secret drawer was miraculously intact and he gratefully retrieved its contents. But when the confusion had died down and Conway examined these papers properly, he found that a banknote for the enormous sum of £500 and four £100 notes were missing. Conway realised at once that the fire was designed to conceal the robbery.

Conway contacted his bank, who informed him that his large note had already been exchanged. But the two clerks who had managed the transaction clearly remembered its bearer and, from the physical description they offered, Conway suspected Sampson. He passed this information on to Lennox, who summoned the map-maker to his home 'on business'. Chattering away to Lennox 'on different subjects', Sampson was unaware of the clerks' concealed presence in the room. Recognising Sampson instantly, they made a signal to Lennox, who then accused his young surveyor of theft and arson. Sampson initially prevaricated but soon 'confessed all the particulars of his guilt'. He was committed to Newgate and eighteen months later, on 11 March 1768, this rash young map-maker was executed at Tyburn.

Charles Lennox's bad experience with James Sampson did not, however, stop him from employing more surveyors at Goodwood. He replaced Sampson with a 24-year-old map-maker called William Gardner, and this appointment was an entire success. Lennox's long-serving estate surveyor Thomas Yeakell struck up a professional and personal relationship with Gardner which endured until Yeakell's premature death in 1787. The alliance would be recognised as one of the most important in British cartographical history.

Throughout the late 1760s and 1770s, Yeakell and Gardner's joint responsibility was to make accurate maps of Lennox's property. They covered the full seventy-two square miles of his enormous estate with loving intimacy, on the expansive scale of six inches to a mile. But it soon became apparent to Yeakell and Gardner, and to Lennox too, that mapping Goodwood was no longer enough to occupy two surveyors on a full-time basis. Yeakell and Gardner instead began to dream of making a really accurate map of a substantial portion of the British landscape, based on a triangulation and draughted on the scale of two inches to a mile. This would be twice as detailed as the maps submitted to the Society of Arts' competition. Yeakell and Gardner anticipated that their map of Sussex would

> not only contain an accurate plan of every town and village, but every farmhouse, barn, and garden, will have its place. Every inclosure . . . with the nature of its fence, whether bank, ditch, pale, or wall, will be described; every road public or private, every bridle way will be traced. The hills and vallies will be clearly distinguished from the low lands, and their shape and even height made sensible to the eye.

This was an almost unprecedented level of information for a county map to display.

Yeakell and Gardner's resulting 'Actual Topographical Survey of Sussex' is a stunningly intricate image of mid-eighteenth-century Sussex. Field boundaries lattice the county's plains and promontories stretch into the sea. The settlement of Chichester is a red weal on a tea-coloured landscape. As the northern part of the county rears up into the South Downs, black hachures trace the shadow of its slopes and declivities. And nestling in the crook of these hills, depicted in resplendent green, is the delicate, elaborate outline of Lennox's Goodwood estate. Only the four southern sheets of the projected eight of Yeakell and Gardner's 'Great Map' were eventually published, between 1778 and 1783. Sales were rather poor (probably due to its costliness), but Yeakell and Gardner's 'Great Map' undoubtedly raised the bar for subsequent map-makers.

MAPS DOMINATED Charles Lennox's profession as much as his personal life. In his early twenties, he had become passionately interested in the relationship between geography and military defence. This athletic, handsome, rather arrogant young man was particularly obsessed by the topography of the region he knew best – south-east England, especially the coast – and he became preoccupied by its potential as a theatre of war. Lennox had many personal investments in the area's security. He was the third generation of his family to be based at Goodwood, only fifteen miles from the crucially important naval base at Portsmouth. He also commanded local army and militia regiments, which he led back and forth across south-east England in reconnaissance and training marches. His fervour was driven by visions of this southern landscape overrun by French invaders. 'The peaceful plains of England and the habitations of its industrious people [might] become the scene of bloody war and desolation,' he warned. As the century progressed and Lennox aged, these dreadful possibilities seemed, if anything, more likely to materialise. The military conflicts that dominated the second half of the eighteenth century provided the background against which Lennox rose to political prominence and took the security of the nation into his own hands.

But this rise to respectability took a long time. Lennox was not a popular man. In the past, he had alienated himself from George III by resigning from the post of Lord of the Bedchamber when his brother was overlooked for a promotion. This was an enormous slight and the King openly referred to 'the Duke of Richmond's blackness' thereafter. The initial rift between Charles Lennox and the monarch was widened by a further controversy, a serious flirtation between Lennox's younger sister Sarah and George III, shortly after his accession to the throne in 1760. The King strongly hinted at marriage to Sarah, whom the politician Henry Fox described as 'different from and prettyer than any other girl I saw'. But on the advice of his former tutor John Stuart, 3rd Earl of Bute, George III opted instead for the 'plain' Princess Charlotte of Mecklenburg-Strelitz. This humiliating public rejection of Lennox's younger sister intensified his dislike for the King.

George III's animosity towards Lennox would only increase in the decades that followed. In the 1780s, Lennox was among a number of Whig politicians who began to seriously entertain the idea that a Reform Bill might be passed through Parliament. The term 'Reform' referred to the restructuring of the nation's electoral system by extending voting rights and making government more accountable through frequent elections. In fighting for these changes, eighteenth-century Reformers often felt that they were upholding the spirit of the Glorious Revolution and the Enlightenment. In the 1780s Lennox became involved in the foundation of the popular Society for Constitutional Information, designed to 'diffuse throughout the Kingdom . . . a knowledge of the great principles of Constitutional Freedom'. In collaboration with politicians such as William Pitt the Younger and the 2nd Earl of Shelburne, he drafted a Reform Bill to present before Parliament. His ambitions were far more radical than those of many of his contemporaries. In a period in which suffrage was restricted to owners of significant property, Lennox asserted that 'it is the Right of every Commoner of this Realm (Infants, Persons of Insane Mind, and Criminals incapacitated by Law, only excepted) to have a Vote'. He advertised his bill as a measure to combat the 'manifold Abuses which in Process of Time have been suffered to take Root in the Manner of electing the Representatives of the Commons'. George III read this as a shameless gibe at the election of Bute as First Lord of the Treasury back in 1763. The King interpreted Lennox's bill as evidence of 'his Unremitted personal ill conduct to Me' and added, 'it cannot be expected that I should express any wish of seeing him in my Service'.

Lennox's breach with George III did him few favours. Sycophants knew that he could not help them gain royal patronage or a sympathetic ear. Lennox was also a difficult man. Brash and egotistical, he had a notoriously blunt manner and a hot temper which put paid to many of his political proposals and ambitions. Promotion was slow to arrive. It was not until 1765, at thirty years of age, that Lennox was appointed British Ambassador to France, whereupon he was sent to Dunkirk to resolve a contretemps between British military engineers and the French, and came to form a long-lasting friendship with William Roy. Over the next decade, he edged his way from the outskirts of respectability to the centre of a prominent faction of Whig politicians. But his tenacity paid off. In March 1782 Lennox was given the prestigious post of Master-General of the Board of Ordnance, head of the branch responsible for the distribution of armaments, munitions and fortifications, and for the creation of military surveys. Until 1828 this post came with a Cabinet seat, so Lennox suddenly found himself one of the government's principal military advisers. The Master-Generalship also brought him into frequent contact with William Roy.

Defences had been a major concern for Lennox for a long time. Upon his appointment as Master-General, he set about getting approval for a bill that granted the Board of Ordnance £400,000 in eight annual instalments to improve fortifications and dockyards along England's south coast. In early 1783 when he first presented it before Parliament, Lennox's scheme met with little resistance. Britain and France's horns were still locked in the American War of Independence, there was widespread anxiety about a French invasion and the utility of such fortifications was self-evident. But by the end of that year the war was over, and when Lennox tried to claim the second massive instalment for his project, Parliament objected and demanded that his 'Fortifications Bill' be rigorously investigated. Lennox then set about forming an investigating committee, but he composed it of close friends and professional associates, including Roy, and dismayed onlookers cried 'foul'. Furious that Lennox had crammed the committee with his cronies, appalled at the fortifications' expense, and distressed by the wider implications of Lennox's bill (namely, that Britain must 'change our system and become a military nation'), his opponents defeated the scheme in Parliament. But within two decades, with the advent of the French Revolutionary and Napoleonic Wars, Lennox's ambition would not seem so reprehensible. In 1804 the first 'Martello tower' would be built, after Captain W.H. Ford of the Engineers, Brigadier-General Twiss and David Dundas – then General

Sir David Dundas – had recommended the idea to the Secretary of State for War. These coastal towers were small defensive forts designed to house around twenty-five soldiers plus ammunition, and they were based on the design of a similar fort at Mortella Point in Corsica ('Martello' is a mis-spelling of *Mortella*). The first decades of the nineteenth century saw a chain of Martello towers constructed along the south coast of England and then along the coastlines of East Anglia, Ireland and the Channel Islands, until they numbered over a hundred. So although his own scheme was stopped, Lennox's vision of a chain of forts was realised within his lifetime.

In the wake of the failure of his Fortifications Bill in 1784, Lennox resigned himself to erecting minor defences in southern England. He also engineered places at the Board of Ordnance for his two estate surveyors, Thomas Yeakell and William Gardner, working with the cohort of civilian map-makers who were lodged in the Drawing Room at the Tower of London. Yeakell was granted the prestigious post of Chief Draughtsman, and when he died in 1787 Gardner replaced him. Between the late 1780s and early 1790s, Gardner's mapping projects included surveys of the Channel Islands and Plymouth. In early July 1790, Lennox learned of the death of his old compadre, William Roy. The reminder of Roy's lifelong efforts to create a national military survey appears to have spurred Lennox on in his role as Master-General. He moved quickly to put Roy's plan into action. In June 1791 he recorded that he had secured the consent of the King, who had been fond of the Scottish map-maker, to proceed 'with the Trigonometrical Operation begun by the late Major General Roy' during the Paris–Greenwich triangulation, and the Board of Ordnance's Expenses Ledger duly shows that £373 14s was paid for a 'Great Theodolite' to begin the measurements.

At this initial stage of the project, Lennox established only a national tri-angulation and not the means to flesh it out into a full map. The triangulation was the most scientifically advanced aspect of map-making, but on its own it simply produced data. Its results functioned like a precur-sor of the modern National Grid Reference System, a unique division of the kingdom into progressively smaller squares, designated by letter and number, through which the exact location of every spot of the nation's landscape can be identified. This sort of information had a wealth of utility to the scientific community but it was less immediately helpful to the military commanders who needed an accurate map of the south coast to organise their troops against a French invasion. To fill out a national triangulation into a complete map, a team of so-called interior surveyors were required to map the intervening countryside around the trigonometric skeleton, which 'fixed' these smaller measurements. It is unthinkable that, in his proposal to the King on the 'Propriety of Making a General Military Map of England', William Roy had meant anything other than the completion of a country-wide map, underpinned by a triangulation. We can imagine him in subsequent decades passionately discussing such an idea over the oak tables of the Mitre Tavern with other members of the Royal Society Club. It also seems inconceivable that Lennox did not intend that the minutiae of the British landscape be mapped in tandem with the triangulation he instigated in June 1791, but initially it was unclear whether he envisaged the mapping would be done by the Board of Ordnance's own team of civilian draughtsmen from the Tower, or by external map-makers. In the Ordnance Survey's early days, its surveyors were clearly conscious that the triangulation would later be supplemented by an interior survey and they marked their trig points on the ground with 'small stakes' and piles of stones to allow 'some individual' to produce 'more correct maps of the counties over which the triangles have been carried'.

Because of the Ordnance Survey's initial focus on triangulation, the project was popularly referred to in its first two decades by the name 'the Trigonometrical Survey'. It was also called the 'General Survey', the 'British Survey' or sometimes 'The Duke of Richmond's Survey'. 'Ordnance Survey' was not used at all until 1801, when its director wrote the term on one of the draught interior surveys. The term first appeared in print in 1809, eighteen years after the project's foundation, in a memoir written by a northern map-maker called Aaron Arrowsmith who was also 'engaged in constructing a large Map of England'. The institution's own map-makers first used the name on a publication a year later, in 1810, on their 'Ordnance Survey of the Isle of Wight and Part of Hampshire'.

On 22 June 1791 Lennox made the first appointment to the national triangulation. Isaac Dalby, a Gloucestershire clothworker, had taught himself mathematics in his spare time, later becoming a teacher in a country school and then in London. Whilst in the capital, Dalby had met the 'wit, rake, and dope-fiend' Topham Beauclerk, who had hired him to act as librarian, astronomer and chemist in his private Highgate laboratory. After Beauclerk's death in March 1780, Dalby was recommended to Roy as a reliable number-cruncher by Jesse Ramsden, and he was hired to assist during the Paris–Greenwich triangulation. When Roy retreated to Lisbon in the winter of 1789 it was Dalby who had tidied up his calculations and prepared his article for publication in the *Philosophical Transactions*. Boasting such familiarity with Roy's methods, Dalby was the ideal candidate for the post of assistant to the Ordnance Survey. He was offered a starting salary of a hundred guineas per year. On 12 July 1791, Lennox chose the project's two directors. He recorded that he had selected 'Major Williams and Lieut. Mudge, of the Royal Regiment of Artillery, to carry on the Trigonometrical Survey with the assistance of Mr Dalby, and desired that they might receive an Extra Allowance equal to their pay and half-pay whilst actually in the field'. With these words, the Ordnance Survey was born.

SHOFTLY BEFORE THE Ordnance Survey's foundation in 1791, Joseph Banks had presented the Royal Society's prestigious Copley Medal to Major James Rennell, who had made an accurate map of Bengal, based on a trigonometrical survey. Banks's presentation speech was bittersweet: 'I should rejoice could I say that Britons, fond as they are of being considered by surrounding nations as taking the lead in scientific improvements, could boast a general map of their island as well executed as Major Rennell's delineation of Bengal and Bahar: a tract of country considerably larger in extent than the whole of Great Britain and Ireland.' Banks concluded with a final dig at the achievements of British map-makers: 'The accuracy of [Rennell's] particular surveys stands yet unrivalled by the most laborious performance of the best county maps that this nation has hitherto been able to produce.' As Banks indicated, the Ordnance Survey's arrival in 1791 was long overdue.

But it is reasonable to wonder why it came into being at that precise moment, when William Roy had spent over forty years failing to achieve the same.

The answer partly lies in events that began over the English Channel in the summer of 1789. In France, long-simmering resentment at the absolutist power of the monarchy combined with anger at the Catholic Church's disproportionate privilege. Humiliating military defeat during the Seven Years War was followed by serious national debts accrued during the American War of Independence. A sweltering summer, a failed harvest and financial and ministerial collapse united to spark a revolution. On 14 July 1789 a crowd stormed the Bastille prison. This mob was joined by soldiers and the uprising acquired an armed militia. The old administration retreated and in its place a National Assembly emerged, rewriting France's constitution and removing power from the monarchy.

At first, the French Revolution drew a mixed response in Britain. British Whigs were generally rather bullish. Celebrations had been held the previous year to mark the centenary of the 1688 Glorious Revolution, and 101 years later there was widespread support in England for the French Revolution as a long-due equivalent. Both events were thought to embody enlightened principles of government, based on mankind's inalienable rights rather than the power of an arbitrary monarch. Addressing the Society for the Commemoration of the Revolution in Great Britain in November 1789, the Nonconformist minister Richard Price celebrated the event in impassioned tones:

> I have lived to see a diffusion of knowledge, which has undermined superstition and error – I have lived to see the rights of men better understood than ever; and nations panting for liberty, which seem to have lost the idea of it. I have lived to see 30 MILLIONS of people, indignant and resolute, spurning at slavery, and demanding liberty with an irresistible voice; their king led in triumph, and an arbitrary monarch surrendering himself to his subjects.

He offered an eloquent warning to proponents of despotic government: 'Tremble all ye oppressors of the world! . . . You cannot hold the world in darkness. Struggle no longer against increasing light and liberality. Restore to mankind their rights; and consent to the correction of abuses, before they and you are destroyed together.'

The French Revolution had important implications for cartography on both sides of the Channel. The French had begun a national triangulation over a century before the Ordnance Survey, and their scientists and military officers boasted a far more sophisticated acquaintance with maps than their contemporaries in Britain. Over the course of the eighteenth century the French army had become large, mobile and extremely skilled in battle manoeuvres and movements, which were often based on its officers' innovative use of maps. It has been said that it was 'the French who had made the greatest strides in the application of cartography to the logistics of an army on the march and who, in any case, had now assumed the mantle of European leadership in the realm of scientific map-making'.

These military and geodetic manifestations of cartography were underpinned by a firm philosophical basis. French *philosophes* had embraced maps as embodiments of the 'esprit géométrique' of the Enlightenment and, as a language of reason, cartography seemed a perfect way to communicate the Revolution's emphasis on *liberté, égalité,* and *fraternité*. And so from 1789 onwards the revolutionary Assembly set about reorganising France's administrative boundaries. The provinces of *ancien régime* France were dissolved and in their place a more 'rational' structure was proposed, with the intention of weakening local loyalties and homogenising regional idiosyncrasies into a single national identity. The Assembly segmented France into ever smaller subdivisions: *régions, départements,* cantons and communes. In numerous cases, the *départements* were named after communal geographical attributes, such as mountains and waterways, to which every citizen enjoyed the same access. The plan of this new 'rational' France looked like a cartographer's dream: a reorganisation of the French nation according to equal geometric units. And this redrawing of France's internal boundaries was entirely reliant upon the work of map-makers, especially on charts produced by Robert de Hesseln and Philippe Hennequin, and the *Carte de Cassini*. Back in Britain, the conservative Whig politician and polemicist Edmund Burke condemned this rational restructuring of France as the malady of 'a geometrical and arithmetical constitution'. He dismissed the Revolution and its consequences

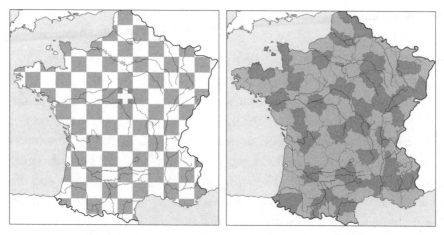

19. Outline for administrative reorganisation of France following the French Revolution.

as the products of a nonsensically map-minded set of politicians. 'Nothing more than an accurate land surveyor, with his chain, sight, and theodolite, is requisite for such a plan as this,' he wrote.

In the following years, the Revolution continued to concern itself with measurements. In 1790 the National Assembly and King Louis XVI, who was a great enthusiast for cartography, both authorised the Académie des Sciences to design a new system of weights and measures. It has been estimated that prior to this project, '*ancien régime* France contained a staggering 250,000 different units of weights and measures'. The revolutionaries wanted to instigate national homogeneity by coming up with a universal standard measurement. This would not be based on arbitrary authority like the 'foot' unit, which was said to be based on the length of King Henry I's twelve-inch foot, but on reason and the earth itself, on which all citizens could stake a claim. This unit of measurement would be 'for all men, for all time'. The resulting idea of the 'metre' unit was intended to measure one ten-millionth of the distance from the North Pole to the equator. To extrapolate this length, the geodesists Jean Baptiste Joseph Delambre and Pierre François André Méchain (the latter of whom had worked alongside Jean Dominique Cassini and William Roy during the Paris–Greenwich triangulation) extended the Cassinis' Paris Meridian down to Barcelona in Spain. The arc now spanned the distance between Dunkirk and Barcelona and represented one-tenth of the distance between the equator and North Pole. By dividing the resulting measurement by one million, the value of the metre unit was found.

Méchain and Delambre had been members of the Académie Royale des Sciences. After the Revolution, this organisation was reconstituted as the Académie des Sciences, part of the Institut de France. Previously the concern of a royal institution, surveying was now working on behalf of the Revolution and its love of reason. As we can see, politics and space had become deeply intertwined in this period. The redrawing of France's boundaries and the measurement of the metre show how space became politicised, and at the same time politics were spatialised: that is to say, political concepts were described in a language of space. Our well-worn terminology of right- and left-wing political principles derives from the positioning of conservatives and radicals on the right and left sides of the president in Assembly debates in France from 1789. And this use of space to describe a political world was also evident in more metaphorical ways, in Britain as well as in France.

Since the English Civil War, British satirists had used images of upside-down worlds as 'emblems of these distracted times'. A 1642 print called *Mad Fashions, Odd Fashions, All Out of Fashion* showed how 'A horse erect upon his hind-legs drives the cart, a church is inverted, fish sail in the atmosphere, a candle burns with the flame downwards, a labourer is wheeled by his own barrow, and several timid animals chase the more ferocious.' The artist had designed this 'topsy-turvy' world as a metaphor for a nation whose king had been ousted by a commoner during the Civil War. After the French Revolution, similar images proliferated as depictions of the republican climate across the Channel. The politician and man of letters Horace Walpole wrote of 'that topsy-turvy-hood which characterizes the present age', and the novelist Fanny Burney described a radical character as 'an adept in turning the world upside down'. A children's book called *Signor Topsy-Turvy's Wonderful Magic Lantern; or, the World Turned Upside Down* was a thinly veiled attack on a world in which 'the servants [had] turned masters' and 'the fish [had] turned fishers'; it ended with the conservative 'simple moral' that 'whate'er thy station, be content'. In this atmosphere, when the politics of space was so topical, the act of mapping possessed particular resonance.

THE NATURE OF the French Revolution changed dramatically in the years after its eruption in 1789. On 13 August 1792 King Louis XVI and his wife Marie Antoinette were arrested. Five weeks later a new constitutional and legislative assembly declared France a republic and abolished the monarchy. In mid December charges of high treason and crimes against the state were levelled at Louis XVI. Although not all the Assembly was in favour of the death penalty, the motion was carried nonetheless. On 21 January 1793 in the Place de la Concorde, the King was executed by guillotine.

As the spring progressed, Paris was engulfed by increasing violence as rival revolutionaries clashed. Initially a minority, the Jacobins soon gathered strength. Fuelled by the Paris mob and led by Maximilien Robespierre, this faction took over the Assembly and established a revolutionary dictatorship. The newly founded Committee of Public Safety fell under their control. Through this Committee, which was designed to suppress counter-revolutionary activity, and along with the Revolutionary Tribunal, the Jacobins instigated a policy of execution of anyone accused of 'crimes against liberty'. Suspects included royalists, aristocrats, the Catholic clergy, members of the middle class and thousands of workers accused of draft-dodging and desertion. Between 16,000 and 40,000 people lost their heads to 'la veuve', the widow (or, more aptly, widow-maker): the guillotine.

In the midst of this bloody 'Reign of Terror', European war broke out. At the very beginning of the Revolution, only a few voices had predicted a martial outcome. But by 1791 various Continental monarchies were debating whether to intervene in France's political upheavals. In the summer of 1792, the French Revolutionary Army declared war on the rulers of Austria and Prussia. Superficially this conflict was driven by ideology, but there were practical considerations too, involving territorial disputes. And it was primarily for such pragmatic reasons as the need to protect the nation's overseas trading networks that Britain also drifted into the conflict in February 1793.

Even in the early days of the Revolution, there were those in Britain who had predicted its transformation into something more threatening. When in

109

January 1790 Edmund Burke had read Richard Price's sermon extolling the Revolution's virtues as an exemplar of enlightened government, based on 'the right to choose our own governors', he had replied with an ardent, eloquent defence of tradition, gradual change and social stratification. Price had viewed Britain's Glorious Revolution as a move towards republicanism and a direct antecedent of the French Revolution. But Burke's pamphlet interpreted the events of 1688 very differently, as instrumental in '*settling* the *succession* of the [Hanoverian] crown'. He described how it was 'natural' to 'look up with awe to kings' and worried that the French Revolution might prove dangerously contagious, 'drawing us into an imitation of the conduct of the National Assembly'. Burke also gloomily predicted the 'assassination' of Louis XVI and Marie-Antoinette, two years before that event actually materialised. His *Reflections on the Revolution in France* sold astonishingly well: it went through eleven editions and sold 32,000 copies within its first year of publication. The book functioned as 'the manifesto of a Counter-Revolution', shoring up Britain's defences against the eruption of republican urges on this side of the Channel.

The combined circumstances of war, the Terror and Burke's eloquent rebuttals of radical politics all helped to turn a great deal of British public opinion against the French Revolution. Ideological objections combined with weighty military anxieties. A rumoured uprising in Britain in late 1792 sparked fears about a French invasion of England's south coast with the aim of fomenting a revolution here. When war broke out between Britain and France in 1793, the possibility of invasion became a grave concern. The most serious threat was posed in October 1797, shortly before the French Revolutionary Wars became the Napoleonic Wars. The Corsican artillery leader Napoleon Bonaparte was rising to prominence as a charismatic and efficient military leader, and he was put in charge of what was known as the French 'Army of England'. A force of 120,000 soldiers was stationed on the northern coast of France, waiting to cross the Channel. The danger was averted when Napoleon decided to go into Egypt instead, but this scenario had nearly made England's worst nightmare a reality.

Strategies to defend England's south coast against invasion were many and varied. Since 1757 the Militia Act had ordered each county in England and

Wales to maintain a quota of able-bodied men, who were willing to fight on a part-time basis. In the 1790s local militia regiments were raised and trained along the south coast. From 1794 corps of volunteers were also raised. In the event of an invasion, some statesmen suggested that a 'scorched earth' policy should be pursued whereby all livestock in the vicinity of the coast would be driven inland with the aim of leaving French invaders no incentive to remain in the country. Some suggested that the southern peasantry should be educated in local defence with the assistance of the '*carte du Paÿs*'. Lennox's own obsession with fortifications was reinforced by the outbreak of the French Revolutionary Wars on the Continent, and he embarked on a lengthy tour of the nation's coastal defences.

On the face of it, it looks as if the Ordnance Survey was established as one of a series of military reactions to the threat of a feared French invasion of the south coast. But the truth is not quite so simple. Britain was at war with France by the mid 1790s, but when the Ordnance Survey was founded in 1791 the Anglo-French relationship was more benign. So although the mapping project later gained some legitimacy as a defensive military scheme, the Ordnance Survey's foundation cannot entirely be explained in these terms. It took a decade to produce the first maps and by the time that a number of different surveys were rolling off the press, the threat of invasion had largely passed. Furthermore, the military utility of such a map only really applied to those spots most vulnerable to invasion – the coasts and the country leading to London – and did not justify the creation of a complete national survey. The Army already possessed such military maps, which were inscribed with potential landing points and which categorised the roads according to the quantities of men and artillery that they could accommodate.

In the early 1790s, a rather shadowy figure called Robert Edward Clifford returned to Britain from France where he had been attending an Academy for English Catholic Youths, followed by a position in a regiment of Irishmen. As the Revolution became hostile for Catholics, Clifford returned to Britain where, because of anti-Catholic penal laws, he could not obtain an official commission in the Army. But his acquaintance with French tactics and surveying methods was invaluable, and General John Graves Simcoe asked him to help produce a military map of the whole of southern England as far north as the line running between Anglesey and the Wash, where Norfolk meets Lincolnshire. At this time the civilian map-maker John Cary was producing maps of England and Wales on a scale of one inch to five miles, which showed inhabitations, roads and rivers, and which were even recommended to Napoleon as the most suitable for planning an invasion. Clifford produced skeletons of Cary's maps that showed only the coastline and rivers and were said to resemble 'the anatomy of the Veins'. He intended that these would 'be very easy to fill up [with] just what one pleases' and he hoped that his maps

> would be of great use for officers going to the outposts, as in 5 minutes they may take the position of all the roads & passes within three miles of their post from the general map & keep it in their orderly book. This would form officers to understand the value of positions, & give them a desire of taking & drawing plans, hence they would acquire that coup d'oeil which blind commanders seldom acquire.

'Coup d'oeil' literally meant 'stroke of the eye' and denoted the talent of discerning the military strength or weakness of the land at a glance. Clifford hoped that officers in the field could nurture these skills and, in doing so, create their own maps of the territory.

This is not to say that the Ordnance Survey had no military utility: generals certainly saw the value of more accurate maps of the coastal areas than currently existed, and skeleton maps of the early sheets of the Ordnance Survey would be used for defensive planning similar to General Simcoe's. Its progress would also be structured according to sites of greatest military utility: the Ordnance Survey would map the coastal areas and the land surrounding London first. But that the Ordnance Survey was the product of a number of quite different other pressures is reflected by the fact that it derived funding from George III as well as from the Board of Ordnance. The King was an enthusiastic sponsor of Enlightenment and nationalist endeavours, such as the Royal Academy of Art, and the mapping project pricked his interest. The Ordnance Survey was also indebted to the Society of Arts' attempt to 'incline the Administration . . . to make accurate Maps of Districts, till the whole Island is regularly surveyed', and to the dramatic rise in the quality of county maps and estate surveys in the same period.

Instrument-makers' improvement of the precision of surveying equipment and map-makers' widespread acceptance of triangulation as the most accurate technique for large areas provided the Ordnance Survey with its methods and apparatus. Surveys conducted by geodesists to ascertain knowledge about the shape of the earth and institutions like the Royal Society contributed to its ambitions and patronage. Most of all, the Ordnance Survey was a product of the Enlightenment ambitions of William Roy to create an accurate image of the natural world. When it was finally founded in June 1791, the Ordnance Survey was a culmination of the efforts of literally hundreds of map-makers and organisations over the previous century, deriving from a wide variety of surveying traditions. The French Revolution served to bring these to a head. The hopes and expectations of all these ghosts now rested on the shoulders of four men: the Ordnance Survey's progenitor Charles Lennox, its assistant Isaac Dalby and its directors, Edward Williams and William Mudge.

CHAPTER FIVE

Theodolites and Triangles

CHARLES LENNOX FELT unequal to the task of choosing the Ordnance Survey's first directors on his own. He made a special trip south of the river to the Royal Military Academy at Woolwich, where he sought out the Professor of Mathematics, Charles Hutton. Hutton did not hesitate. William Mudge and Edward Williams were, he said, the best mathematicians among the Ordnance corps and 'the fittest officers' to oversee the new map of the nation. Ranking major to Mudge's lieutenancy, Williams was duly appointed the Survey's director in July 1791 and Mudge became his deputy.

Little is known about Edward Williams. A member of the Royal Regiment of Artillery, he appears to have relished pomp and circumstance. Three years before the Ordnance Survey's foundation, Williams staged a mock battle before Lennox, King George III and the Prince of Wales. Its showiness and sophistication provoked a news reporter to rhapsodise that, 'of all the sham battles we ever witnessed, we think that this, in point of shew and interest, was the best'. During the social season, the press enthusiastically covered Williams's attendance at glittering gatherings. Shortly after the Ordnance Survey's foundation, in the winter of 1792, he acted as Steward for a 'Ladies Night' at the town hall in Bath. Williams also enjoyed regular afternoon assemblies at St James's Palace, where he was presented to George III or, in the event of the King's indisposition, due to recurrent bouts of porphyria-induced insanity, to his flamboyant son George, the Prince Regent.

Williams readily adopted a role as the Ordnance Survey's figurehead and he largely eschewed its practical day-to-day activities. This substitution of celebrity for surveying did not endear him to his closest colleagues. The Ordnance Survey's first assistant, Isaac Dalby, described how Williams 'never made an observation or calculation' and 'proved a dead weight in the undertaking by frequently retarding its progress; and the only time he benefitted the service, was when he took his departure to the next world'.

Fortunately William Mudge was the opposite of Williams in terms of both attentiveness and gravity. In 1804 the artist James Northcote painted a portrait (the original is now missing) of Mudge, in which thoughtful, kindly eyes hinted at their bearer's solicitude and a smile played around his lips. Mudge protested modestly at his depiction and playfully accused Northcote of having 'put too much brains in it', but even he acknowledged 'the likeness to be extremely strong'. These qualities of kindness, conscientiousness and quiet humour rendered Mudge a well-liked and respected member of the Ordnance Survey from the start. From the moment of his appointment Mudge declared himself 'impressed with just ideas, as to the importance of the task and responsibility of my situation'.

Mudge's disposition was perhaps a reflection of his family: an extraordinary conglomeration of high-achievers with a wide variety of interests, temperaments and friends. He had grown up in Plymouth in the 1760s and 1770s as part of an old Devonshire clan. His grandfather Zachariah had been born in Exeter in 1694, and although he was raised a Presbyterian dissenter, this serious, thoughtful man turned to the Church of England in his late twenties. Zachariah Mudge took Holy Orders and was rapidly made vicar at St Andrew's Church in Plymouth, a post which carried one of the richest ecclesiastical salaries in England. Zachariah's politics were conservative and his sermons eloquently asserted 'the necessity of Government' and urged 'Obedience to Authority'. He was certain that 'there has been ever acknowledged something sacred in the Persons of Princes, a kind of Divine Cloud hovering round their Heads, to which we are naturally prompted to pay a Veneration'. It is likely that William Mudge inherited from his grandfather a distaste for republicanism that perhaps sharpened his efforts to fend off the French Revolutionary Armies.

From childhood, Zachariah's second son Thomas (William's uncle) exhibited 'strong indications of mechanical genius'. As a teenager, he was apprenticed to a famous London watchmaker and showed sufficient talent to set up his own business. Soon Thomas Mudge was making clocks and watches for King Ferdinand VI of Spain. Where Zachariah was said to be 'very fond of that method of philosophizing' that prioritised the general idea over individual detail, his son was instead transfixed by the mechanical minutiae of timekeepers. In the last decades of the eighteenth century, Thomas threw himself into a dispute that split the scientific community in two.

Back in 1714, a governmental body had been established 'for the Discovery of Longitude at Sea'. Longitude and latitude, those conceptual lines that criss-cross the globe, were the primary means by which mariners oriented their vessels. Latitude was easy to discover according to the height of the sun or various stars above the horizon. But in the eighteenth century longitude was much trickier to establish on board ship. There were two principal ways: because the earth spins around the sun, the latter's light shifts longitudinally as time passes and therefore a calculation of the time difference between the boat and the home port allows the determination of the distance between them in longitude. Proponents of a technique known as the 'chronometer method' of discovering longitude suggested that if a mariner could carry a clock set to the time of the home port, the time difference between the port and the ship could be simply discovered. Advocates of the alternative 'lunar distance method' of determining longitude used the moon as a clock from which sailors could derive the time on board ship. By consulting a volume of 'lunar tables', they could then discover the time over in Greenwich and derive the longitudinal difference between the two locations. But there were serious problems with both methods. There were neither lunar-distance tables nor chronometers accurate enough to cope with this task: the rolling of a ship, changes in temperature or variations in air pressure played havoc with the rate of timekeepers. So in 1714 the Board of Longitude set up a competition that offered a prize of £20,000 to anyone who could determine longitude on board ship to within thirty nautical miles of accuracy.

Famously a man called John Harrison, who had built his first clock in 1713 at the age of twenty, dedicated himself to the chronometer method of overcoming the longitude problem for over forty years. Between 1730 and the 1770s, this talented craftsmen made five beautiful timekeepers for the competition, for which he eventually received a total of £23,065. Harrison's payment was not won without a battle. Britain's Astronomer Royal Nevil Maskelyne was passionately supportive of the lunar-distance method. He worried, with some justification, that even if a supremely precise and reliable chronometer was constructed, it would be impossible to mass-produce such a minutely accurate instrument cheaply enough to allow every mariner in the world to benefit. Maskelyne was a prominent member of the Board of Longitude, and his sometimes combative insistence on the thorough testing of Harrison's timekeepers and his general scepticism about the chrono-meter method has been well documented (often in ways that malign the astronomer's reasonable doubts as snobbery).

When Harrison died in 1776 it was William Mudge's uncle Thomas who picked up the baton. Harrison had produced unprecedentedly accurate chronometers but there still remained much room for improvement: his clocks were found to become unreliable over a period of years. In the mid 1770s Thomas Mudge produced three timekeepers for the longitude competition and over the next decade these were subjected to numerous trials under Maskelyne's watchful eye. Although Mudge's chronometers proved much more accurate in the long term than Harrison's, he too became infuriated by the astronomer's apparent attempts to thwart his success and with-hold remuneration. His eldest son Thomas Mudge Junior was a lawyer and in 1793 he fought for his father's case to be brought before Parliament, with support from the prominent Whig William Windham, the Hungarian astronomer Francis Xavier Baron de Zach and Admiral John Campbell. But Joseph Banks, President of the Royal Society and an influential member of the Board of Longitude, was strongly opposed to Mudge's efforts to claim the prize. He flatly rejected the superiority of Mudge's timekeepers and protested that a financial reward would 'discourage the advancement of knowledge, or check the spirits of emulation among artists, by neglecting those who have a superior, and rewarding others who have an inferior, claim to patronage and liberality'. A parliamentary committee was established to look into the case and eventually awarded Thomas Mudge the relatively modest sum of £2500. He died two years later, but his son continued to fight on behalf of him and his timekeepers, publishing pamphlets and establishing a factory to show how the chronometers were suitable for mass-production (which proved unsuccessful).

In the first years of William Mudge's employment on the Ordnance Survey, the name of 'Mudge' was firmly associated in the public consciousness with the longitude conundrum, and also with the antagonism with the Astronomer Royal and the nation's foremost scientific institution, the Royal Society, both prominent supporters of British surveying. Probably for these reasons William Mudge remained notably reticent about the spat between his uncle and cousin, and Banks and Maskelyne. When the furore had died down, he inserted a brief aside into a published article that commended Thomas Mudge's chrono-meters: a small mark of family loyalty. If William Mudge had inherited his grandfather's ability to see the big picture, then his uncle Thomas gave him the means to temper it with an eye for detail. William enjoyed clock-making as a hobby and was delighted when the King of Denmark presented him in 1819 with a gold chronometer as an award for his achievements. But it was his father and his father's friends who arguably played the greatest role in William Mudge's development and helped him to foster a kind tolerance and quiet humour, and an eloquent love of literature and art.

WILLIAM MUDGE'S FATHER John had grown up in Plympton St Maurice, a pretty village north-east of Plymouth that is overlooked by a ruined motte-and-bailey castle. (The Mudges' presence in Plympton is now commemorated by a busy thoroughfare called 'Mudge Way'.) John Mudge had found his headteacher at Plympton Grammar an inspirational presence. Samuel Reynolds had a number of hobby-horses, among them medicine and astrology. In one class he terrified and enthused his students by present-ing them with a human skull for dissection; and he was frequently spotted star-gazing among the ruined turrets of Plympton Castle. An apocryphal and distressing anecdote recounts how, using astrological observations to calculate his children's horoscopes, Samuel was appalled to discover that his young daughter, Theophila 'Offy' Reynolds, was faced with 'very great danger' around her fifth birthday. When that time came, Samuel forbade his daughter to leave the house and took 'every precaution' to ensure her safety. But 'at the very time predicted', Offy's nurse was walking by an open window on the top floor of the house and accidentally let the child slip from her arms. Offy plummeted through the window to her death.

Samuel Reynolds also had a son called Joshua, a short, stocky young boy with a ruddy complexion, rounded cheeks and a slight harelip. Only two years John Mudge's junior, the two boys formed a close friendship during their time at Plympton Grammar. But while John hung, entranced, on his teacher's descriptions of gruesome medical ailments, Joshua's schoolbooks were crammed with sketches and architectural plans. By the age of eight, he had taught himself the rudiments of perspective and was often found outside rather than inside his classroom, drawing 'the schoolhouse according to rule'. He was an avid fan of his schoolfriend's father's sermons, and later it was said of Joshua that he owed 'his first disposition to generalize, and to view things in the abstract, to old Mr Mudge'. In his teens, Joshua found himself torn between a career as a general practitioner in medicine and chasing his artistic ambitions. Reasoning that if he were only destined to be 'an *ordinary* painter' then he would rather follow a medical career, Joshua made up his mind to become a *great* artist.

And so he did. After an apprenticeship to one of the principal portrait painters in Britain, Joshua Reynolds set up his own practice. By 1748 the press was hailing him as one of the youngest of fifty-seven best 'painters of our own nation now living'. Despite his drawling Devonshire accent, his 'coarse' features, 'slovenly' dress and a hunger for gambling, Joshua became one of the most sought-after portraitists in mid-eighteenth-century Britain. He was the hub of a vibrant intellectual circle that included the lexicographer Samuel Johnson, the politician Edmund Burke, the actor David Garrick and the poet and playwright Oliver Goldsmith. When King George III granted his 'gracious assistance, patronage, and protection' towards the founding of a Royal Academy of Arts in 1768, Joshua was elected its first president. Four months later he became Sir Joshua Reynolds.

Throughout this meteoric rise to fame, Joshua remained close to his Devonshire family and friends. Various portraits testify to his ongoing friendship with the Mudges. His old childhood companion John Mudge pursued the career in medicine that Joshua had himself rejected. John trained at the nearby Plymouth Hospital, becoming a distinguished doctor who specialised in the treatment of smallpox and respiratory illnesses: 'Mudge's Inhalers' were popular chemists' items. When in 1752 Joshua found himself suffering a period of poor health, he travelled to Devon to consult John and gratefully painted the doctor's portrait while he was there. A few years later, John's eldest son was employed in the Navy Office in London. Unwell on his sixteenth birthday, the young boy was so upset at having to forgo his intended celebratory trip home that Joshua sympathetically reassured him with the promise 'Never mind! I will send you to your father!' He painted a portrait of the boy peeping out from behind a curtain and sent this to John Mudge in place of his son.

20. *John Mudge, by Samuel William Reynolds, after Joshua Reynolds, 1838 (1752).*

By the late summer of 1762, John Mudge had remarried after the death of his first wife. When his second wife Jane was heavily pregnant with her second child (William), Joshua decided to pay a visit to his old schoolfriend and the now elderly Zachariah Mudge, and he brought along his good friend Samuel Johnson. On 31 August, the pair arrived at John's Plymouth home, where they stayed for almost four weeks. Joshua painted John's portrait again and the group made excursions to local friends and landmarks, including the newly reconstructed lighthouse on the treacherous Eddystone Rocks, fourteen or so miles offshore from Plymouth: one of the shining achievements of John Mudge's friend, the civil engineer John Smeaton. Samuel Johnson's distinctive physical appearance and idiosyncratic behaviour startled many of the Mudges' associates. A towering six feet tall, his body riddled by convulsions, with poor eyesight and hearing, sudden grunts, erratic head-rolling and a face disfigured by scrofula, Johnson's mannerisms – combined with his 'excesses in new honey, new cider, and clouted cream, at one of the Devonshire tables' – were said to 'alarm his entertainer[s] much'. John Mudge and Joshua Reynolds took Samuel Johnson to meet Zachariah Mudge and hear him preach. Rumour has it that, on returning to Zachariah's vicarage after the sermon, the Mudges offered Johnson afternoon tea. Blissfully innocent of the polymath's enormous appetite, when he presented his cup for an eighteenth helping, Zachariah's wife exclaimed 'What! Another, Dr Johnson?!' 'Madam, you are rude,' he retorted.

Nevertheless, a series of close friendships was formed. In the years that followed the visit of Reynolds and Johnson to Devon, Zachariah and John Mudge were welcomed into their urban circle of friends. William Mudge's grandfather subsequently visited London at least once a year until his death in 1769 and became a valued companion of Reynolds, Johnson, Garrick, Goldsmith and Burke. Zachariah was said to be 'esteemed an idol' by the men, a 'learned and venerable old man', and he was even considered by Reynolds to be 'the wisest man he had ever met with in his life'. In the 1790s, after the outbreak of the French Revolution, Burke would arrange for Zachariah's sermons to be republished in Britain to bolster the same royalist message as his own *Reflections on the Revolution in France*. William Mudge was still in the womb during Reynolds and Johnson's Devon sojourn, but after greeting the world for the first time on 1 December 1762 he grew up in regular contact with this illustrious circle of family friends. It is recorded that when Burke was introduced to one of Zachariah's grandsons, perhaps William, he remarked happily: 'I have lived in intimacy with two generations of Mudges, and have much pleasure in making the acquaintance of a third.' Samuel Johnson is said to have become William's godfather and when at the age of fifteen the young man was accepted into the Royal Military Academy in Woolwich, Johnson came to visit and bestowed upon him a guinea and a book. Although recognised as 'a sharp boy', Mudge was 'not very attentive' as a cadet. Nevertheless, his abilities were enough to gain a commission in the Royal Regiment of Artillery in 1779 and then to be sent to South Carolina to fight in the American War of Independence.

William Mudge returned to Britain from America in 1783 at the age of twenty-one. He had always been close to his older sister Jane, affectionately nicknamed 'Jenny', and in that same year she married Richard Rosdew, the heir to a prosperous estate called Beechwood, near Plympton, 'one of the most comfortable and pleasant residences in the neighbourhood'. The Rosdews were already closely connected to the Mudges by a series of marriages, but Richard was a particularly impressive catch as a freeman of Plymouth, the town coroner and a good-humoured, gentle man to boot. In his new brother-in-law, William found a firm friend, and in the same year as Richard and Jenny's wedding took place, his own life took an exciting turn. He was sent for training in the Drawing Room of the Tower of London, amongst the Board of Ordnance's surveyors. Here William Mudge found himself falling in love with the daughter of one of his superiors, a Major-General Williamson of the Royal Artillery. We know little about Margaret Jane Williamson, but in the late 1780s she and William were married and in 1789 their eldest son Richard Zachariah Mudge was born.

At the Tower, William made up for his early neglect of his studies, and mathematics particularly began to entrance him. It was a preoccupation partially driven by rivalry. When his contemporary Henry Shrapnel was rumoured to have 'made considerable progress in mathematics', William Mudge refused to be outdone and applied to his old mathematics professor at the Royal Military Academy, Charles Hutton, for extra tuition. Shrapnel

21. William Mudge, by James Northcote, 1804.

went on to invent a particularly gruesome artillery shell. Mudge became 'a first rate mathematician' and in 1791 Hutton recommended him to Charles Lennox as an ideal candidate to help direct the Ordnance Survey.

It was partly thanks to his remarkable family that William Mudge's temperament was well suited to this responsibility. From his father he had learned to take seriously the foibles and anxieties of others, and a later biographer described him as 'a man of the nicest feelings of honor, [and] of the strictest integrity'. Mudge was cultured and eloquent and sensitive to civilian concerns as well as to military preoccupations. As we shall see, he was acutely aware of the power that maps wielded in the popular imagination. He came from a close-knit family and was said to be equally devoted to his own. But the task of directing a national map was hardly compatible with a full and rewarding personal life, and William Roy and David Watson had, as far as we know, both sacrificed the prospect of fulfilling relationships to the demands of the military. William Mudge's passion for maps and his love for his wife and children would eventually pull him in conflicting directions.

THE ORDNANCE SURVEY began straight away after Charles Lennox had appointed Edward Williams and William Mudge as its directors on 12 July 1791. The two men selected the same baseline that Roy had used seven years earlier during the Paris–Greenwich triangulation, the line that traversed 5.19 miles of Hounslow Heath to the south-west of London. But in order that 'this operation might not rest on *data* afforded by any former one' and their calculations be as trustworthy as possible, Mudge and Williams insisted on remeasuring Roy's base.

This decision appears to qualify William Roy's earlier emphatic statements of the Paris–Greenwich triangulation's superior exactitude. It certainly reveals the uncertain nature of accuracy in this period. It is accepted by historians of science that 'accuracy' was a relatively novel concern for eighteenth-century astronomers, geodesists and map-makers, who became preoccupied with the idea of the 'quantifying spirit' in attempting to emulate the certainty of Newton's *Principia Mathematicia* and realise the Enlightenment ideal of perfect measurement. The historian of science Thomas Kuhn has termed this quest for accuracy a 'second scientific revolution', after the first scientific revolution that is usually taken to refer to the publication of Copernicus's *De revolutionibus orbium coelestium* (*On the Revolutions of the Heavenly Spheres*) in 1543. But numerous factors could threaten the precision of a measurement, such as the minuscule erosion of instruments' parts, often simply through use; the expansion and contraction of apparatus according to changes in temperature; imperfect eyesight; or mistakes in calculations. The eighteenth century witnessed strenuous attempts to counteract many of these distortions, including the development of unprecedentedly accurate instruments and the formulation of new mathematical methods designed to reconcile a number of conflicting measurements or observations into an accurate 'mean'. But perfect accuracy could not be realised in an instant. As the century progressed, scientists and mathematicians moved slowly closer to the goal of precision by dealing with minor flaws and hindrances that had dogged earlier projects. Every failure of accuracy could help the next undertaking draw closer to the ideal. So when in 1790 the astronomer Francis Wollaston noticed that Roy's measurements for the Paris–Greenwich triangulation were tarnished with a few errors, this set the scene for Mudge and Williams to redo that experiment and, in learning from Roy's mistakes, to establish a measurement more accurate than any before.

On 23 July Mudge, Williams and Isaac Dalby, accompanied by eleven members of the Royal Artillery, travelled up the river Thames to Hounslow Heath. After initial checks and reconnaissances, on 15 August the re-measurement officially began. It took almost two and a half months to complete. As in Roy's time, the events on the Heath were cause for excited festivities, especially on the first and last days of the base measurement. 'While the sun shone out' on 15 August, Joseph Banks, Nevil Maskelyne and several other members of the Royal Society clustered breathlessly around the measuring chains, which were used for the preliminary stages of the measurement. On 28 September, more celebrities gathered, including Charles Lennox, Jesse Ramsden and William Mudge's old mathematics teacher, Charles Hutton, to witness the base measurement's completion. Edward Williams was in his element among these luminaries. After Banks sent the surveyors a 'Keg of small Beer' for sustenance, Williams purred his 'Thanks, for the kind Attentions you shewed to us whilst on the Heath'. In some respects, these festivities marked 'an official "handing over" ceremony' from the Royal Society, which had overseen the Paris–Greenwich triangulation in the 1780s, to the Board of Ordnance, which was now responsible for the new national triangulation. But Banks would certainly not lose interest in the Trigonometrical Survey once the Royal Society was no longer responsible for its progress; nor would the Surveyors reject Banks's future offers of assistance.

The Hounslow Heath baseline is, cartographically speaking, the most important spot in the British Isles. Until theodolites were replaced by satellites in the second half of the twentieth century, it was the bedrock of the national triangulation that underlay every single Ordnance Survey map. One cartographic historian has remarked that 'seldom if ever had there been a line measured with so much care'. Williams, Mudge and Dalby were acutely aware of the significance of this piece of land. They noticed that the wooden pipes that had been 'laid down by the General [Roy] for the termination of his Base' at its two ends at Hampton Poor House and King's Arbour in 1784 were now, seven years later, 'in a very decayed state'. Keen that the baseline's extent should be preserved 'in a more permanent manner', Charles Lennox arranged for Roy's old wooden markers to be replaced by 'heavy iron cannon'. Two decommissioned guns were shipped from Woolwich to the baseline's southern and northern ends. Up-ending such heavy artillery into the precise spots was described as 'an operation of a delicate nature, and attended with some difficulty', but eventually the cannons were 'fixed at the extremities of the base'. There they survive, remnants of cartographic history half-buried in two patches of ground whose surroundings have changed around them beyond all recognition.

THE BASELINE RE-MEASUREMENT produced a result of 27,404.3155 feet, 0.3845 feet less than Roy's original attempt. This discrepancy was caused by 'a small oversight': his failure to compensate for temperature change. Once the remeasurement was completed, Williams and Mudge were faced with the daunting responsibility of planning the Ordnance Survey's progress. Amid fears of a French invasion, a map of the south coast was a priority, so the two men decided to begin the national triangulation on the Sussex coastline and then progress to the counties that separated it from the prime target of London – namely, Surrey.

The Ordnance Survey was given a new, improved version of the theodolite that had accompanied Roy during the Paris–Greenwich triangulation. The original remained in the Royal Society's possession. After that instrument's triumph, the East India Company had commissioned Jesse Ramsden to make an identical model for a survey of India. William Mudge described the ensuing creation as 'of similar construction to that which was used by General Roy, but with some improvements'. Ramsden's characteristic perfectionism had driven him to rebuild the entire instrument from scratch, tinkering and refining until every minor criticism, mostly concerning the theodolite's microscopes, had been resolved. The result was a surveying instrument of unsurpassed accuracy. It also meant a bill presented to the East India Company wildly exceeding the original quote. They refused to pay and cancelled the order, and the instrument languished in Ramsden's showroom until, in the excitement around the Ordnance Survey's foundation, Charles Lennox caught wind of its existence and bought it for the new Trigonometrical Survey. Almost every member of the early Ordnance Survey reverently referred to the instrument as the 'Great Theodolite'. Protective of his expensive new toy, Lennox urged Mudge 'to avoid towers and high buildings' when executing the triangulation: a near impossibility.

The national triangulation began in the spring of 1792. Mudge had decided to tack the new Trigonometrical Survey onto the triangles that Roy had created during the Paris–Greenwich triangulation, and to progress west from there. (For this reason, the commencement of that earlier project in 1783 is often suggested by map historians to be the real beginning of the Ordnance Survey.) At both ends of the baseline, portable scaffolding was erected to a height of thirty-two feet, to the top of which the 200-pound theodolite was precariously winched by a crane manoeuvred by five or so artillerymen. Mudge then gingerly inched his way up a ladder to the observation platform at the top of the scaffolding. The instrument was protected from the weather by a canvas canopy, but Mudge was not so lucky. Surveying from this great height in a high wind, which caused the scaffold to sway alarmingly, must have been nerve-racking. Even the sight of the landscape through the sights of such a sophisticated theodolite was dizzying. In an instant, the panorama before Mudge's eyes disappeared and was replaced by an enlarged image of a minute portion of land, magnified almost beyond imagination and bisected by a grid of wires that helped to pinpoint an exact spot. The experience was like swapping the eye back and forth between a small-scale map of the nation and a large-scale estate survey. Right at the beginning of the eighteenth century the philosopher George Berkeley had described the strange effect of switching one's gaze from normal, unassisted eyesight to that provided by a lens. 'The same object' is not 'perceived by the microscope, which was by the naked eye,' he claimed.

But soon Mudge had got the hang of his theodolite and his sight lines were

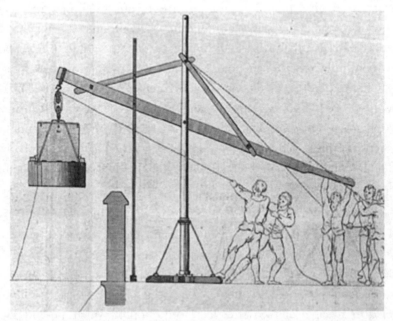

22. A team of men lifting the 200-pound theodolite using a crane.

ricocheting from the two ends of the Hounslow Heath baseline to St. Ann's Hill in Chertsey; to Banstead Church spire in the north-eastern corner of Surrey where it now merges into Greater London; to Shooter's Hill near Greenwich, one of the capital's highest points; to Bagshot Heath in north-west Surrey; to a tower on Leith Hill, in mid Surrey; to the chalky summit of Butser Hill, one of the highest points in Hampshire; and finally down to Rook's Hill, a few miles west of Charles Lennox's mansion at Goodwood in West Sussex. Up in his eyrie at the top of the scaffolding, Mudge jotted down in a small bound notebook the angles of his horizontal and vertical observations, the date, the exact time and even the temperature of the air, probably using either a goose quill pen or one of the new metal pens whose nibs did not require constant sharpening and were particularly useful outdoors.

Mudge also needed to measure the relative heights of the trig points, and he did so by a process called 'levelling'. A spirit level was used to provide the position of a sight line that travelled horizontally from a 'levelling telescope' at trig station A. Trig station B was at a different height from A, say fifty feet higher, and there a surveyor manipulated a 'levelling rod' (a staff marked with measurements of length) so that it dropped vertically towards the horizontal sight line from A. The observer at that first trig point noted where the vertical rod met the telescope's horizontal sight line, and used the rod's measurements to ascertain the vertical difference between the two trig points. In May 1793 Mudge used this levelling method to ascertain the height of a trig station at Dunnose, on the Isle of Wight, above the low-water mark. Using a spirit level, he positioned a telescope (on an instrument called a 'transit') on the horizontal from the trig point and trained it towards the coast at Shanklin. There his assistants used a measuring tape and rods, which they positioned vertically up a cliff face, to discover the 792-foot vertical distance that separated the height of the trig station from 'the water's edge'. Over longer distances, levelling could be done trigonometrically. The theodolite's telescope could be trained from one trig point to another at a different height, and the angle that separated the telescope from the horizontal could be used in a trigonometric equation to determine the vertical distance between them.

Mudge was not the only one to be transfixed by the eminently mappable qualities of the landscape of southern England. In 1793 the local poet Charlotte Smith described its 'boundless, yet distinct' hills that spread out before her 'even as a map'. After a spell in London the American author John Neal wrote in 1830 of the breathlessness that was triggered by the view from 'the top of a great hill (such as Leith Hill) with an empire lying under [its] feet like a map'. Eighty-five years later the mystic poet James Rhoades imagined this same view spread out 'like a map' before two observers who were elevated above it like map-makers:

> Where Leith Hill tower the landscape crowns,
> And points a stony finger,
> On Sussex, Surrey, Kent, the Downs,
> Our eyes have loved to linger.
>
> From Reigate round to Shoreham Gap
> We've marked the spires up-peeping,
> Fields, hamlets, hedgerows, like a map,
> In mellow sunlight sleeping.

Thanks to the Ordnance Survey, the landscape that these writers compared to a map became the subject of a real survey.

Once Mudge, Williams and Dalby had completed the observations from the two ends of the Hounslow Heath baseline, they began to move south through Sussex. With the assistance of men from the Royal Artillery, the Ordnance Survey's instruments and papers were transported from trig point to trig point by road, in carriages and on horseback. The Great Theodolite was given its own 'spring waggon' with improved suspension 'to preserve it from injury'. The Sussex coastline is a varied stretch of land, on which beaches are backgrounded by impressive ruined castles and high grasses are interspersed with wetlands and play home to a variety of birdlife. But the Ordnance surveyors' first experiences of the county were likely to have been unpleasant. The thoroughfares south of London, especially in Sussex, were notorious. Constant traffic churned their surfaces into mud and it was said that because of this mire 'respectable Sussex women went to church in ox-drawn coaches, and Sussex men and animals had grown long-legged through pulling their feet through the clay'. Horace Walpole warned a friend: 'if you love good roads . . . be so kind as never to go into Sussex' as 'the whole country has a Saxon air, and the inhabitants are savage'.

William Mudge and his colleagues carried out their observations in the morning and evening, 'when the air was free from vapour, and without that quivering motion, which, in summer, it generally has in the middle of the day'. They followed the standard annual surveying pattern that had been adopted by the Military Surveyors before them: late spring, summer and early autumn were spent 'in the field'. Late autumn to early spring were spent back at headquarters, processing the calculations. To protect the Great Theodolite from adverse weather, it was housed in a portable observatory: an octagonal tent with windows through which its telescopes surveyed the surrounding scenes. This observatory, according to a surveyor, 'consisted primarily of an internal skeleton of eight iron pillars, bound together by oak braces, and supporting a roof consisting of eight rafters united together at the top, and clamped at the bottom to the iron pillars. The sides and roof were composed of frames covered with painted canvas, and the whole covered with a strong tent.' Wherever the theodolite went, so did its shelter. When the Ordnance Survey turned its attention to hillier ground, the additional weight of the iron and oak frame and the canvas covering would prove an excruciating burden to carry up peaks and over the uneven ground of moors and heaths.

Many of the coastal landmarks on which Mudge trained the theodolite's sights as it travelled further south were old signal stations. Crowborough Beacon in East Sussex, Penn Beacon on Dartmoor and Ditchling Beacon on the South Downs were all part of a historic, now redundant network of warning stations. When one beacon flared in the distance, the next was lit by watchful residents, and that was seen by the next settlement along, who in turn lit their beacon, and so on. In event of an emergency this alarm system could transmit intelligence of a crisis or invasion from coast to capital. As each beacon was distant yet clearly visible from the next, they made ideal trig points for the Ordnance Survey's triangulation.

In the Survey's early days, Williams, Mudge and Dalby decided to illuminate their surveying staffs at the trig points with lamps, just as Roy
and Cassini had done during the cross-Channel observations of the
Paris–Greenwich triangulation. Lamps effectively illuminated the surveying staffs over long distances or through hazy weather. The map-makers
commissioned three such lanterns from a 'Mr Howard of Old-Street',
and they consisted of flares and reflectors inside 'strong tin cases, having
plates of ground glass in their fronts, which effectually prevented the bad
effects of an unequal and unsteady light'. These lamps were 'found to
equal everything which could be expected from them' and the largest,
which had a diameter of twenty-two inches, 'was lighted on Shooter's
Hill, and [was] clearly distinguished at the distance of 30 miles'. But as
these lanterns were lit beside the south coast's warning beacons, many
locals interpreted the flares as signals of French invasion. In April 1793
the *Sussex Weekly Advertiser* was forced to reassure its readers that 'a General
Survey of the Kingdom being now carr[ied] on by Government' would
necessarily involve the use of 'White Lights' at several stations along the
south coast. The journalist explained that the surveyors' lights could be
easily distinguished from emergency flares 'by their peculiar brilliancy,
and short duration, which will not exceed four or five minutes'. The
Gazetteer also had to assure the inhabitants of Ditchling that 'they need
be under no alarm for the visits lately paid them by some of the Artillery',
who were only 'making a trigonometrical survey of this county'.

LATE IN THE summer of 1792, Mudge, Dalby and Williams found themselves camped beside the trig point at Hindhead in the thickly wooded
south-western corner of Surrey. On the outskirts of the breathtaking geological scoop known as the Devil's Punch Bowl, the best view of the
surrounding landscape and the most apt location for a trig point was positioned '22 feet north-west' of a gibbet. In this spooky spot, the surveyors
received a letter from Charles Lennox, the Master-General of the Board of
Ordnance, with a command that altered the entire remit of the project:

Lennox wanted to begin fleshing out the Trigonometrical Survey into a map. He commanded Mudge to 'furnish Mr Gardner, chief Draftsman to the Board of Ordnance, with materials for [making] a Map' of Sussex, which Lennox emphasised was 'intended, at some future period, to be published'.

The Board of Ordnance boasted its own map-making department in the Drawing Room of the Tower of London, which was quite separate from the Ordnance Survey. These draughtsmen were civilians, not military men, and since 1787 they had been co-directed by Lennox's old estate surveyor William Gardner. Now Lennox had come up with the idea of reinvigorating the 'Great Map' of Sussex that Gardner and Lennox's other estate surveyor, the late Thomas Yeakell, had begun in the mid 1780s. Only four out of eight projected sheets had originally been published and Lennox decided that it was time to fill out the triangulation into a full map by completing the Sussex survey and conforming the initial measurements to those provided by Mudge's Trigonometrical Survey. Williams and Mudge had previously entertained the possibility, perhaps reluctantly, that map-makers from outside the Ordnance might use their triangulation to make maps. But now Lennox was recommending that the Board of Ordnance itself should be responsible for such an achievement.

Mudge and Dalby duly spent much of 1793 on the south coast of England, in East and West Sussex. They got drenched and frustrated in the Isle of Wight, where astronomical measurements to find the precise directions of north and south were thwarted by consistent fog. They also encountered a weird phenomenon known as 'terrestrial refraction', in which the atmosphere acted as a prism and bent rays of light to produce mirages. Strolling around the Isle of Wight, Dalby noticed that the top of a faraway hill 'seemed to dance up and down in a very extraordinary manner'. He described how, 'when the eye was brought to about 2 feet from the ground, the top of the hill appeared totally detached, or lifted up from the lower part, for the sky was seen under it'. And that same summer the surveyors also camped at Beachy Head, the perilously high chalk cliff in East Sussex. Fourteen years later, the poet Charlotte Smith described the awe-inspiring view from its summit:

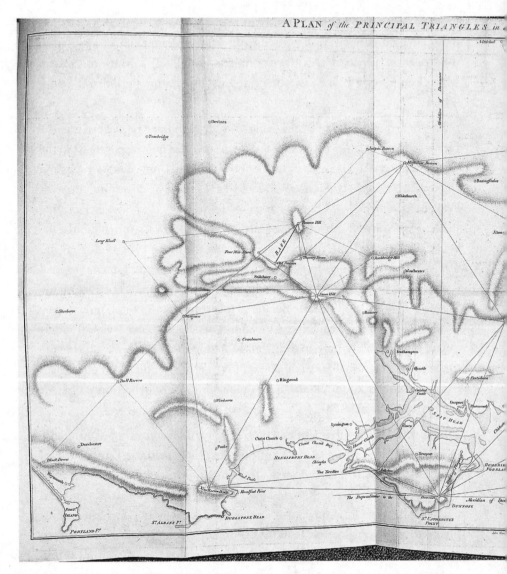

23. The 'Principal Triangles' measured during the first three years of the Ordnance Survey, between 1791 and 179[...]

. . . how wide the view!
Till in the distant north it melts away,
And mingles indiscriminate with clouds:
But if the eye could reach so far, the mart
Of England's capital, its domes and spires
Might be perceived – Yet hence the distant range

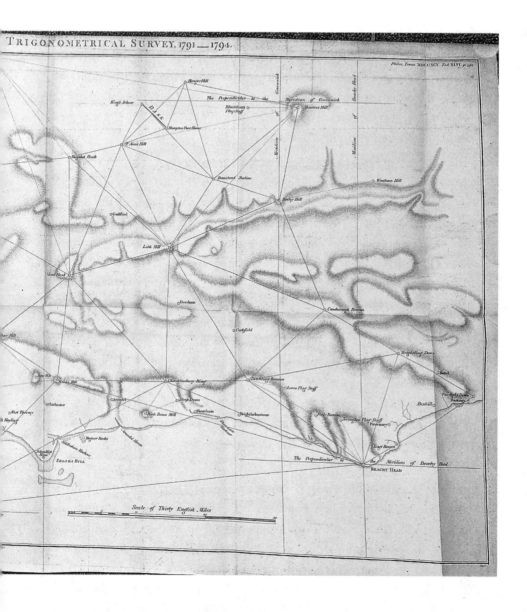

TRIGONOMETRICAL SURVEY, 1791—1794.

Of Kentish hills, appear in purple haze;
And nearer, undulate the wooded heights,
And airy summits, that above the mole
Rise in green beauty; and the beacon'd ridge
Of Black-down shagg'd with heath, and swelling rude
Like a dark island from the vale . . .

Mudge and Dalby's time at Beachy Head was not without incident. Whilst encamped one night, they 'had the mortification to hear plainly the cries of the poor English sailors' on a ship below the cliff as they were boarded and attacked by passing French pirates. 'For want of cannon', possessing merely telescopes and compasses, and poised 530 feet above the sea, the surveyors could only listen in impotent horror.

By the end of 1793's surveying season, Mudge and Dalby had completed the Trigonometrical Survey of Sussex. Responsibility for the creation of the map now lay with the Tower draughtsmen. The co-director of the original 'Great Map' of Sussex, Thomas Yeakell, was deceased, but its surviving author, William Gardner, selected Thomas Gream, who was also from the Tower Drawing Room, to assist him. Gardner and Gream 'filled in' Mudge and Dalby's primary triangulation by observing a network of smaller triangles between trig points such as church spires or the turrets of stately homes with a much smaller and less sophisticated theodolite. They even used the 'north-west chimney' of Lennox's mansion at Goodwood, which possessed its own observatory, as one of these secondary triangulation stations. Then the pair traced the course of Sussex's major roads and rivers, using roughly the same 'traverse' methodology that William Roy had adopted half a century previously in Scotland. The small theodolite was used to place features and landmarks, such as forests and buildings, that punctuated the intervening land between these networks.

Gardner and Gream's map of Sussex was published commercially by the Royal Geographer William Faden in 1795, with a dedication to Charles Lennox. In black and white and on a scale of an inch to a mile, the map showed the gentle undulations of the Sussex coastline in twisting lines of black hachures that in some places could look as if a series of small furry caterpillars had fallen asleep on the sheet. The depiction of the prickly groundcover of gorse was almost tangible, shown by clumps of tiny vertical lines interspersed with the occasional dark knot of scrub. The forests north of Rotherfield were represented by black segments of miniature trees, whose dense clustering evoked the claustrophobia of the woods themselves. A neat mosaic of fields sprawled over the intervening land: a vision of order and harmony.

As the first joint production of the Trigonometrical Survey with another team of Ordnance map-makers, this survey is a contender for the accolade of the 'first Ordnance Survey map', and Mudge and Williams lauded the 'Survey of Sussex' as one of 'the only maps which have passed under our notice, worthy [of] commendation'. However, at this stage there was no long-term scheme to create a series of county maps of the whole nation from a formal collaboration of the Trigonometrical Survey with the Ordnance's draughtsmen. The Sussex survey was therefore chiefly celebrated then and since as a unique event, albeit a particularly accomplished one, but not as the first of the Ordnance Survey's First Series of maps.

AFTER THE COMPLETION of the triangulation across Sussex, William Mudge wanted to take the Trigonometrical Survey west through Hampshire, Devon, Somerset and Cornwall, still concentrating on the coast and the land that led from it to London. He started to keep his eyes open for a new site that would act as a 'base of verification'. A second baseline in south-west Britain would check the accuracy of the triangles that had already been observed. It would also act as a reliable foundation for future series of triangles that were to cascade into the West Country and Wales.

But a suitable site proved elusive. Mudge visited Longham Common near Poole in Dorset, but the stretch of land was too populated with obstacles such as barns, stables and outbuildings to be freely measured. He also trekked to King's Sedgemoor, a peat basin in the centre of Somerset, where the baseline for the very first British triangulation had been measured in 1681. But the land had recently come under an Enclosure Act by which it was being drained and then perhaps partitioned and cultivated. Over a period of about seventy-five years, England's landscape was redrawn through such legislation. Although it had been a legal phenomenon since the early modern period, 'enclosure' accelerated between the second quarter of the eighteenth century and the first quarter of the nineteenth. Previously labourers had enjoyed traditional mowing and grazing rights over pieces of common land, even if they were

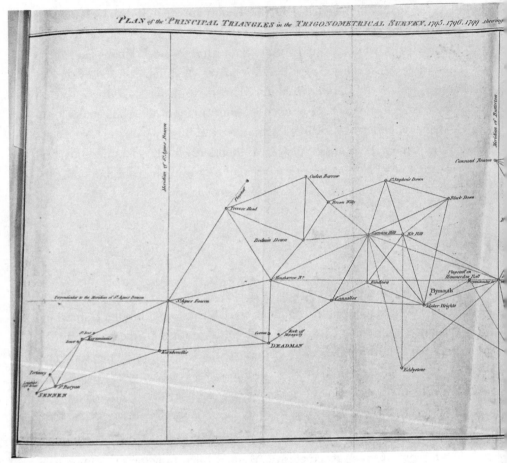

24. The progress of the Ordnance Survey's triangulation into the West Country and Cornwall.

owned by another person, and the 'open field system' gave over four very large unfenced fields per manor or village to labouring families to farm in strips. But the Enclosure Acts consolidated these farmed strips and pieces of common land into fields that were privately owned by landlords, thus creating England's present-day patchwork landscape of agriculture. Over six million acres of English soil were 'enclosed' in the eighteenth century, and one historian has described how the hedges and fences that demarcated the new distribution of land became 'the symbols of the new England'.

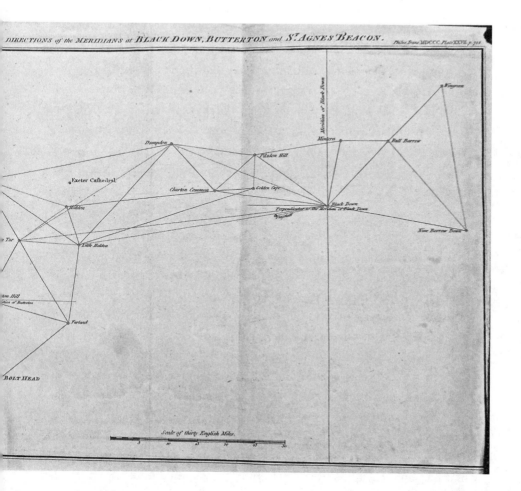

These Acts created a landless working class, many of whom would migrate to cities to work in the new factories and mills once England started industrialising. Enclosure was a hugely contentious subject. Many sympathised with the rural labourers whose mental maps of their surroundings were forced to change overnight. The agricultural reformer Arthur Young imagined a poor man's account of Enclosure: 'All I know is, I had a cow and Parliament took it from me.' But Young also despised what he saw as the dreadful waste of unenclosed common land. Referring to King's Sedgemoor before its enclosure, he lamented: 'what a disgrace to the whole nation it is, to have 11,520 such acres to lie waste in a kingdom that is quarrelling about high prices of provisions!' Young claimed that the land was 'so rich, that some sensible farmers assured me it wanted nothing but draining to be made well worth from 20 shillings to 25 shillings an acre on average'. But Young's priorities were not the same as Mudge's. When King's Sedgemoor was eventually enclosed in the 1790s, the 'improvers' and drainers who scuttled over its surface rendered the tract of land utterly impracticable for a baseline.

Another site whose 'wastefulness' Young bemoaned turned out to be just the place Mudge was looking for. In the late 1760s, Young had complained that 'such a vast tract of uncultivated land' as Salisbury Plain is 'a public nuisance'. Salisbury Plain is a chalk plateau that covers 300 square miles of Wiltshire and Hampshire, and Young suggested: 'what an amazing improvement would it be to cut this vast plain into farms, by inclosure of quick hedges, with portions planted with such trees as best suit the soil! A very different aspect the country would present from what it does at present, without a hedge, tree or hut; and inhabited by only a few shepherds and their flocks.' But Young's vision had not been realised and Salisbury's Plain's unenclosed stretches of flat heathland, and sparse population, offered an ideal site for the Ordnance Survey's 'base of verification'. So in July 1793, Mudge and Dalby travelled to Salisbury Plain to select the ends of their new baseline. For its southern extremity, they lighted on a spot near Old Sarum, an Iron Age hill fort marking the earliest settlement of Salisbury. The baseline then stretched over seven miles north-north-east to Beacon Hill, whose summit provided a 'commanding view' of almost the whole baseline.

During that same summer, the poet William Wordsworth spent two days wandering on foot over Salisbury Plain and he described how, 'though cultivation was then widely spread through parts of it', the land 'had upon the whole a still more impressive appearance'. Returning from Portsmouth where Britain's naval fleets were busy preparing for battle with revolutionary France, Wordsworth was full of 'melancholy forebodings' and predicted the war's 'long continuance' and its inevitable 'distress and misery beyond all possible calculation'. Perceiving Britain's decision to go to war with France as a betrayal of the Glorious Revolution's legacy and the French Revolution's republican spirit, Wordsworth began to feel that the landscape itself was reflecting his sense of gloom and alienation from his fellow Britons in the current climate of martial bonhomie. In his ensuing poem, 'Guilt and Sorrow, or Incidents upon Salisbury Plain', Wordsworth imagined a sailor returning from war, reduced to vagrancy and dressed in a tattered red uniform. He pictured this man, as friendless and homeless as the poet felt himself to be, wandering disconsolately over the plateau. A distant church spire initially fixed the sailor's eye like an emotional trig point, but as 'gathering clouds grew red with storming fire', it soon disappeared 'in the blank sky'. And 'a naked guide-post's double head', which the sailor glimpsed in a flash of lightning, provided only a momentary 'gleam of pleasure'. Thanks to the Ordnance Survey, Salisbury Plain would become integral to the history of Britons' attempts to orient themselves, but for Wordsworth it stood for alienation and disorientation.

Mudge and Dalby's actual measurement of the Salisbury Plain baseline did not begin until the end of the following June and it was completed in around seven weeks. The map-makers did not have the luxury of delay. It was imperative that the baseline should be finished before early autumn, 'when the cultivated ground a mile to the northward of Old Sarum would be ploughed'. Like Hounslow Heath, the Salisbury Plain baseline was an eerie setting for their work. The men shared the land with what Wordsworth described as 'an antique castle spreading wide', 'Pile of Stonehenge!' This giant stone circle, an icon of British history, was just over four miles from the northern end of the Ordnance Survey's baseline. Thought by some to be a celestial map 'by which the Druids covertly expressed/ Their knowledge of the heavens, and imaged forth/ The constellations', Stonehenge could be seen as an ancient counterpart to Mudge and Dalby's own endeavour. Wordsworth also described how the Plain was etched over with giant crop circles and chalk shapes, 'strange lines' left 'on the earth . . . by gigantic arms', which also may have reminded the map-makers of their own triangulation. But for Wordsworth the strange, inexplicable stone circle seemed to symbolise the nation's profound indifference to his isolated and disappointed wanderer and the poet complained how this 'proud' phenomenon seemed 'to hint yet keep [its] secrets'. As Wordsworth's sailor 'plodded on' past Stonehenge, he heard the 'sullen clang' of 'a sound of chains' and 'saw upon a gibbet high/ A human body that in irons swang'.

Salisbury Plain is now inhospitable in a different way. Since 1898 the Plain has been used for Army training. The Ministry of Defence owns 150 square miles, over half of the old Plain. It houses an aerodrome, the Royal School of Artillery, a host of camps and barracks, and even a deserted village called Imber, used for training in built-up-area warfare, especially during the Northern Irish troubles. Live firing takes place on 340 days a year and around thirty-nine square miles are permanently closed to the public. Access to the rest is severely limited. But the Plain's enforced hostility to humans has made it peculiarly hospitable to wildlife. Its chalk grassland has not been lost to agricultural cultivation, and Salisbury Plain now boasts 'the largest known expanse of unimproved chalk downland in north-west Europe, and represents 41 per cent of Britain's remaining area of this rich wildlife habitat'. It has been classified as a Site of Special Scientific Interest and supports around eighty species of internationally rare plants and invertebrates. It is wonderful for birds – in 2003 the Great Bustard was reintroduced to Britain on Salisbury Plain. Many of the features that William Mudge saw in the old Plain, such as its relatively unpopulated and unenclosed landscape, its flatness and expanse, were the same things that later attracted the military to the place and also lie behind its now thriving ecosystem.

BY THE END of the summer of 1794 the map-makers had 'laid down' the coast 'from Fairlight Head to Portland' and some of the inland territory that joined the coast to the capital. Portland is a spit of land in Dorset, just south of Weymouth, and it is separated from Fairlight Head in East Sussex by a distance of about 136 miles. In just three surveying seasons, William Mudge and Isaac

Dalby, with a little help from Edward Williams, had measured two baselines, triangulated through five counties (including parts of the capital), and visited three more to conduct a reconnaissance that would determine the future direction of the Trigonometrical Survey. It was a remarkable achievement.

CHAPTER SIX

The First Map

THE MAP-MAKERS did not pass through the British landscape like ghosts. The sight of a small team of men encamped at the crest of a nearby hill, bustling around a mysterious, squat piece of machinery, and scanning the horizon for rhythmic flashes of light inevitably caught the attention of locals. There was a healthy appetite for public displays of map-making prowess during the Ordnance Survey's first decade. In 1800 Londoners flocked to a demonstration of 'Geometrical and Trigonometrical Operations' at the Establishment for Military Education in Knightsbridge, of which *The Times* gave an enthusiastic write-up. And the Ordnance Survey's endeavours, which were more authentic, proved equally attractive. Wealthy landowners volunteered the turrets, pinnacles, follies, obelisks and cupolas that adorned their mansions and estates as secondary trig points.

William Mudge's presence in the landscape remained visible for months after he had left an area. The surveyors constructed cairns – small mounds of rough stones – to indicate the exact spots at which their theodolite had been placed, to allow the measurements to be redone or verified in future. Newspapers pleaded with locals to leave these memorials untouched. Since 1935, when the Ordnance Survey retriangulated the first Trigonometrical Survey, trig points have been designated more permanently by squat, square, concrete pyramids or obelisks that each bear a brass plate marked OSBM (Ordnance Survey Bench Mark) and with the reference number specific to that station. Although now largely obsolete, having been replaced by Global Positioning Systems, aerial photography and digital mapping with lasers, these concrete trig points are still preserved in the landscape, on hills and plains alike. They are emblems of a productive relationship between humanity and the natural world, and are valuable tools of orientation for hikers: one is generally guaranteed a good view from an elevated trig point.

The Ordnance Survey's work attracted attention from the press as well as from the general public. During the triangulation's first summer, *Lloyd's Evening Post* reported that 'a Survey of the Island of Great-Britain, upon a grand and important scale, has been undertaken by his Grace the Duke of Richmond'. Many newspapers were happy that the foundations laid by William Roy's endeavours were being built upon by a new generation. And once the triangulation was under way, some journalists were enthralled by Mudge's operation as a feat of human exertion. When the Ordnance Surveyors encamped at Bath, at least two national newspapers offered a charming account of how 'the novelty of half-a-dozen tents pitched in that part of the country, cannot fail of attracting the wonder of the surrounding multitude, whose curiosity leads them thither by hundreds at a time. The unknown and supposed magical powers of the surveying instruments, with the flag-staffs, which they erect on neighbouring eminences, suggest the ideas of telegraphs and invasion.'

In early 1794, the Ordnance Survey's nominal director, Edward Williams, a man well acquainted with the machinery of self-promotion, sought the advice of Joseph Banks, President of the Royal Society, about capitalising on the popular fascination with the project. Banks mooted the idea of publishing accounts of the Trigonometrical Survey's progress in the journal of the Royal Society, the *Philosophical Transactions*, the publication in which William Roy had described his endeavours with such success that the relevant issues had sold out. Banks became even more enthusiastic when the Board of Ordnance offered to entirely cover any costs accrued by illustrations for those articles and placed a bulk order for 'five hundred Copies' of the volumes.

On 25 June 1795 the first 'Account of the Trigonometrical Survey' was read before an intrigued audience of Royal Society fellows. The article began by delineating the trajectory from which the Ordnance Survey had emerged. It recalled its origins in Roy's official statement in 1766 of the need for a complete national map, and then described the measurement of the base on Hounslow Heath and the triangulation through Kent during the Paris–Greenwich triangulation. And then, in businesslike prose, Williams and Mudge set about recounting their own experiences of remeasuring the Hounslow Heath baseline: the instruments that had been employed; the complex mathematical calculations that had been used to determine their apparatus's 'rate of expansion' in different temperatures; and the iron cannon with which they had marked the base's extremities. They told their audience about 'the Commencement of the Trigonometrical Operation', 'the Improvements in the great Theodolite', and the progress they had made in the triangulation between 1792 and 1794, giving details of the selected trig points and their bearings from one another. Finally, Williams and Mudge described the measurement of the 'Base of Verification on Salisbury Plain with an Hundred Foot Steel Chain, in the Summer of the Year 1794'. They were clearly overjoyed that their calculations had shown that the actual measurement of the baseline accorded very closely with the figure that had been derived through trigonometric formulae from the angles that had been measured with the theodolite. The article was published shortly afterwards in the *Philosophical Transactions* and Mudge proudly presented a bound copy to King George III.

Two years later, in May 1797, Mudge read and published a sequel, a second 'Account' which described the Trigonometrical Survey's progress in 1795 and 1796. Encouraged by the public demand for these two articles, in the eighty-fifth and eighty-seventh volumes of the *Philosophical Transactions*, Mudge and Dalby set about collating their papers and adding a lengthy preface to make a book-length text, which they published in 1799 as volume one of *An Account of the Operations Carried on for Accomplishing a Trigonometrical Survey of England and Wales*. Sold at £1 8s (which represented nine days' wages for a craftsman in the 1790s), the volume was certainly expensive, but not prohibitively so. Mudge wanted his account of the Ordnance Survey's early progress to reach a wider audience than the specialised readers of the *Philosophical Transactions*.

146

In the early days of the Ordnance Survey, Williams and Mudge had accepted that, in the absence of any formal plan to turn their trigonometrical data into military maps, that information would be used by civilian surveyors. They had good reason to believe this, because in 1793 two map-makers called Joseph Lindley and William Crosley had published a map of Surrey that was based on trigonometrical information from William Roy's published accounts of the Paris–Greenwich triangulation. Roy's godson and sometime assistant was a man called Thomas Vincent Reynolds, to whom he had left his most prized possessions, including his military commissions, his Copley Medal, a watch and a miniature portrait of Roy by Maria Cosway. In that same year, after Britain's entry into the French Revolutionary Wars, the Board of Ordnance had appointed Reynolds to use Mudge's Trigonometrical Survey data as the basis for a rapidly executed 'Military Map of Kent, Sussex, Surrey and part of Hampshire', a small-scale survey showing the roads on which troops and artillery might be mobilised in the event of an invasion.

Writers of guidebooks also found the Trigonometrical Survey's data helpful. A genre known as 'the road-book' had properly emerged in 1675, with John Ogilby's *Britannia Depicta*. Over the next century road-books became sophisticated, compendious directories that provided their users with details of local inns, the times and fares of stagecoaches, and market days. In 1803 Daniel Paterson's immensely popular *New and Accurate Description of the Roads in England and Wales, and Part of the Roads of Scotland* went into a thirteenth edition with a print run of 10,000 copies, and its editor consulted Britain's foremost authorities on land-measurement to verify the book's accuracy. He approached the surveyors and directors of the Post Office, the Commissioners of the Stamp Office, the engraver and Royal Geographer William Faden, the landscape gardener Humphry Repton, and the director of the Ordnance Survey, 'the very ingenious Major Mudge'. William Mudge readily contributed the triangulation's data, and *Paterson's Roads*, as it was known, duly based its table of altitudes on 'The Measurements of the Heights of Mountains and other Eminences, so accurately taken in the grand trigonometrical survey of the kingdom'. Mudge also assured its editor that he 'may depend upon every information that the further progress of this great undertaking can supply'.

Map-makers and travel-writers were not the only ones to benefit from the Trigonometrical Survey. Once Williams, Mudge and Dalby started releasing information about their endeavour to the public through their articles for the *Philosophical Transactions* (which maintained the scientific credibility of the work), and through their own published book, its data was hungrily drawn upon by readers with a variety of professional interests. In 1797 a physician called William George Maton published a book, *Observations Relative Chiefly to the Natural History, Picturesque Scenery and Antiquities of the Western Counties of England,* which used material from the Ordnance Survey's published 'Accounts' to pinpoint the altitude of Ninebarrow Down, a long chalk ridge in the Purbeck Hills in Dorset. The Ordnance Survey's levelling was said to have produced 'for the first time . . . fairly good values for the heights of many British hills'. And two years later, in 1799, the economist Henry Beeke emphasised in his *Observations on the Produce of the Income Tax* how necessary it was that the exact areas of Britain's counties should be precisely known when calculating their revenue from taxation. Beeke eulogised the 'very excellent trigonometrical survey', and plundered its data for this purpose. The Trigonometrical Survey was attracting a great deal of public attention and stimulating both popular fascination and professional engagement with its undertaking. For the first time, a clear sense of the precise size and shape of the country was emerging.

AS THE YEAR 1795 began, the Ordnance Survey lost one of its most ardent devotees. Charles Lennox was sacked as Master-General of the Board of Ordnance. The Prime Minister, William Pitt the Younger, finally lost patience with his erratic temper, and Lennox's old connection with the Reform movement and his antagonism to the King were proving liabilities during the politically fraught years following the French Revolution. Such a prominent military figurehead could not be seen to possess any links to republicanism. In his place, Charles Cornwallis, 1st Marquess Cornwallis, was persuaded to take up the Master-Generalship, which gave him an executive role in the Ordnance Survey. Best remembered today as the British general who surrendered at Yorktown in 1781 during the American War of

Independence, in 1795 Cornwallis had just returned from nine years in India, where his task had been 'to give the crown the power of guiding the politics of India with as little means of corrupt influence as possible'. There, this corpulent, competent officer and administrator had restructured the East India Company, reformed the Bengali police, administered a system of land taxation and waged a successful military campaign against Tipu Sultan of Mysore, paving the way for British dominance in southern India. Cornwallis was an ardent royalist and on his return to Britain he was co-opted into the fight against Revolutionary France.

Mudge was initially dismayed by Lennox's dismissal – the Trigonometrical Survey's very existence had owed a lot to his fondness for cartography. But Mudge's gloom soon proved unwarranted. Cornwallis may have lacked Lennox's personal investment in the Survey, but he approached his role as Master-General with professionalism and, as he frequently sought Lennox's advice about the Ordnance Survey, the voice of its founder continued to be heard. Nothing changed for the worse with Cornwallis's arrival and in fact it was in 1795 that the Ordnance Survey became permanently involved in the activity for which it is now famous and loved: *map-making*.

The first year of Cornwallis's tenure saw the formalisation of the previously ad hoc collaboration between the Trigonometrical Survey and the Tower of London draughtsmen. What Mudge termed an official 'union of the parties' was instigated. The Tower draughtsmen were charged with systematically fleshing out the triangulation that was being conducted by Williams, Mudge and Dalby, and this group of civilian map-makers became known as the 'Topographical' or 'Interior' Survey. They were instructed to make detailed maps of piecemeal chunks of the landscape that were separated by major roads or rivers, whose measurements were rigorously checked against those of the triangulation. These jigsaw pieces would then be fitted together into a survey covering every inch of the kingdom, to be engraved and sold 'for public use'. Although the resulting maps would not be referred to by their makers as 'Ordnance Survey maps' until at least 1801, the following pages will adopt this term. The moment when the collaboration was instigated was a crucial landmark in the Ordnance Survey's history and arguably marks the birth of the institution as we now know it.

Cornwallis decreed that the first Ordnance Survey map would be of England's most south-easterly county, Kent, one of the most vulnerable areas to French invasion. When the triangulation had begun in 1792, Mudge had initially bypassed Kent as William Roy had already plotted a series of triangles over that county during the Paris–Greenwich triangulation in 1787 and 1788. But in June 1795 Mudge began to remeasure Roy's Kent triangles in the name of superior accuracy. He hopped in Roy's shadow from Dover Castle to Ringswold Steeple, which sits on a sand hill between Deal and Ramsgate; to Mount Pleasant House, near Thanet; to the rising ground near Wingham; and to churches at Chislet and Upper Hardes. When Mudge went into Kent, he found that the Interior Survey that was under William Gardner's directorship had already started work on fleshing out Roy's initial triangles into maps. The gangly Isaac Dalby had assisted Roy during the Anglo-French collaboration and Mudge sent him to talk Gardner through the original rationale and methodology of Roy's project.

The Interior Survey consisted of two activities. First Gardner devised a secondary triangulation to complement the principal triangulation and to create an accurate skeleton within which measurements of relatively small portions of land could be fitted. For the secondary triangulation of Kent, Gardner was presented with an eighteen-inch theodolite by Jesse Ramsden. This was a much smaller and more portable beast than the three-foot 'Great Theodolite' of the primary triangulation and it could be easily lifted to the roofs of high buildings. Then a small team of Tower draughtsmen – which in the 1790s included, among others, George Pink and Charles Budgen – filled in the primary and secondary triangulations with detailed surveys of the land that were conducted on the large scale of six inches to one mile. The triangulation was used as a framework to maintain the accuracy of these interior maps and to prevent errors from mounting up, but it was often possible to conduct these adjustments *after* the topographical maps had been completed. The Trigonometrical Survey and Interior Survey worked at different paces, and although it was helpful for the surveyors of the latter to base their maps on a finished triangulation, it was not absolutely essential.

To make their maps, the Interior Surveyors may have followed something similar to the 'General Instructions for the Officers of Engineers Employed in Surveying' that Charles Lennox had set out in 1785. Lennox, and Roy too, had commanded their interior map-makers to use 'small Theodolets and chains' to 'proceed around the Contours and Creeks of the shore; along the great Roads and lanes; and also along the course of the Rivers, and Rivulets. The Boundaries of Forests, Woods, Heaths, Commons or Morasses, are to be distinctly Surveyed,' Lennox had emphasised, 'and in the enclosed parts of the Country all the Hedges, and other Boundaries of Fields are to be carefully laid down.' There is little documentary evidence that describes exactly how the Interior Surveyors mapped Kent, but it is likely that they measured the direction and lengths of the roads and rivers using a perambulator (or surveyor's wheel) and compass, and recorded the results of these traverse surveys in fieldbooks. The road and river surveys were then drawn onto paper on which the primary and secondary triangulations were very lightly traced to fix their exact positions. Then, square mile by square mile, the Interior Surveyors gradually filled in the intervening details of this skeleton map, using theodolites to ascertain the bearings of various landmarks from the roads and rivers. It seems they did not use the tried and tested method of plane-table surveying.

Before long, the Interior Surveyors had created what were known as 'fair plans' of every part of the country. The resulting charts and sketches were often intricate, beautiful productions that were brought to life through a vivid vocabulary of colour. As was customary in eighteenth-century European military surveys, the map-makers delineated rivers and lakes in blue, settlements in red and main roads in ochre. Greens and browns identified pasture and arable land. They recreated hills through the technique of hachuring, in fine black ink. Some altitudes are noted in red. Created over half a century, 351 of these intricate drawings are now held in the British Library in London. They range from unfinished skeleton plans and black-and-white outline drawings to meticulous and complete surveys of irregular portions of the British landscape, and they vary wildly in quality, richness of detail and use of symbols. But, in the words of the historian of the Ordnance Survey Yolande Hodson, some are indisputably 'works of art'.

The river Thames was one of the most vulnerable features of England's geography and laid the nation's centre of power open to naval invasion. So the decision had been taken early on by Cornwallis to include the southern reaches of Essex alongside the whole of Kent on the Ordnance Survey's first map. Its map-makers' representation of the Thames estuary was indebted to information from Britain's new Hydrographic Office. Six years prior to the publication of the Ordnance Survey's Kent map, the Admiralty's first Hydrographer, the Scottish geographer Alexander Dalrymple, had been appointed by George III. In 1800, the first Admiralty Chart, of Quiberon Bay in Brittany, had appeared. As the Ordnance Survey mapped Britain's mainland, the Hydrographic Office was working its way around its coast. Both institutions helped to ensure the safety of this island nation in a period of prolonged turmoil. But it was relatively slow going. It took Gardner and his surveyors over three years to finish their maps of Kent 'in a masterly manner'. By the final year of the eighteenth century, both the Trigonometrical and Interior Surveys of that county had been completed, and the compilation of these materials into a complete map was on the horizon.

IN THE YEARS immediately following the Interior Survey's mapping of Kent, the Ordnance Survey's personnel changed in important ways. In January 1798 the *Sun* newspaper reported: 'Died, Last week in London, Colonel Edward Williams, of the Royal Regiment of Artillery, well known in conducting the Trigonometrical Survey of this Kingdom.' Well known he may have been, but he had not been particularly well respected by his cartographical colleagues. Williams had made his indifference to the mapping project under his command clear when he had been invited to present the King with details of its progress in 1796. He had preferred to talk instead about the improvements he had made to Woolwich's artillery stores and it was left to William Mudge to describe the Trigonometrical Survey's endeavours and to present George III with another printed account of its progress.

In the wake of his superior's demise, Mudge received the much-deserved notification from the Master-General that 'you will accordingly take on yourself the charge as it has hitherto been held by Colonel Williams'. Cornwallis added that he was 'assured of its coinciding with the wishes of his Grace the Duke of Richmond'. On 30 April 1798 the *St James's Chronicle* approvingly described how Cornwallis had once again presented Mudge to George III, but this time Mudge 'had the honour to kiss his Majesty's hand on his promotion to succeed the late Colonel Williams, as Superintendant and Director of the Trigonometrical Survey'. This patient, hard-working man was now given full official control over the project to which he had dedicated every waking hour for the previous seven years. For every spring and summer surveying season between 1791 and 1798, and during many of the winter periods too, Mudge had been separated from his young family, still living in Devon. By now he had one daughter and three sons aged between two and nine. Missing out on so much of their childhood was a sacrifice that would increasingly distress him.

In the late summer of 1799, the Trigonometrical Survey also lost Isaac Dalby, not to the Grim Reaper but to the new Military Academy that had been founded in High Wycombe. At fifty-five years of age, Dalby was 'no longer able to endure the fatigues incident to the service' and he decided to return to teaching. In the second volume of the *Account of the Trigonometrical Survey*, which was published under Mudge's name alone in 1801, Dalby's boss paid ardent homage to 'the extent of his service, his unremitted labour, and attention'. But 'whilst I lament the loss of a man so perfectly calculated to assist me in this arduous undertaking,' Mudge continued, 'I derive every consolation from a knowledge, founded from experience, of the talents and abilities of Mr Simon Woolcot, his successor.'

William Mudge's assumption of responsibility for the Ordnance Survey had important repercussions. First, he applied to the Royal Society to obtain the theodolite that had been used during the Paris–Greenwich triangulation of 1787–8, to supplement the one that the Ordnance Survey already possessed. But more importantly, his superintendency affected the way in which the mapping project was perceived in the public domain. The Ordnance Survey had originated from a mixture of motivations: some martial, some scientific, some ideological. And because the project was not purely military in nature, the resulting maps were not as jealously guarded as other pieces of intelligence. The public utility of the endeavour was on the minds of its progenitors from the start, and on his appointment Mudge argued that the maps themselves should be made available to the general populace. In the first volume of the Ordnance Survey's *Accounts*, he had described how Britons were clamouring 'to possess some general Map, published on the same principle with the *Carte de France*, a performance highly celebrated', and he was adamant that 'it has been very justly expected by the Public, that from the present undertaking, they should derive the advantage of an improvement in the geography of their country'. Mudge's reference to the Cassinis' endeavour indicates that he saw the Ordnance Survey first and foremost as the heir to the Enlightenment's ambitions for cartography. With its trained draughtsmen and its copious funding, the Board of Ordnance was a suitable host for that mapping agency, but Mudge had not forgotten that the Ordnance Survey had developed from a scientific endeavour to triangulate between the Paris and Greenwich observatories, under the sponsorship of the Royal Society and the Académie Royale des Sciences. As a child of the Enlightenment and a descendant of a circle that included a theologian, artist, a politician, a doctor, a lexicographer, a horologist, and even an actor, Mudge had an outlook that was arguably wider than most in the military.

Mudge's bold decision to release the Ordnance Survey's maps to the general public and openly inform the populace of its progress also, perhaps consciously, had the effect of mitigating some potential popular hostility to the project during the 1790s. The climate of that decade was peculiarly conservative in Britain, and some map-makers came under fire as dubious presences in the landscape. The French Revolution had provoked intense fear on the north side of the Channel. The foundation of plebeian societies designed to politicise the working classes, rumours of republican uprisings, and demonstrations at the devastating effects of Enclosure and the exorbitant rise in wheat prices all spooked the British administration. Under the supervision of the Prime Minister William Pitt the Younger and William Windham, Secretary at War, a counter-revolutionary operation was established to suppress such dangerous expressions of discontent in Britain. This entailed the manipulation of the press, the foundation of conservative political organisations to combat their radical counterparts, the stage-management of loyalist public rituals, the prohibition of seditious meetings and publications, the suspension of the Habeas Corpus Act, the appointment of London magistrates and the launch of a permanent police force in Scotland.

Pitt and Windham also oversaw the establishment of an 'Alien Office'. Its remit was initially counter-espionage, but the Office soon developed a network of rigorously vetted spies and informers to scrutinise British radicals. Its charismatic coordinator William Wickham described the organisation as a 'System of *Preventitive* [*sic*] Police', 'the most powerful means of Observation & Information' that 'was ever placed in the hands of a Free Government'. It has been argued that prior to the French Revolution, there was a tacit understanding in Britain that in certain places, such as the home or the coffee-house or the gentleman's club, the individual could speak freely as a private citizen. But during the 1790s, in the period known colloquially as 'Pitt's Terror' (a conservative counterpart to Robespierre's Terror in revolutionary France), these assumptions appear to have been overthrown. Informers seemed to be everywhere. Conversations in coffee-houses were overheard and proponents of anti-royalist sentiment were hauled before courts, such as the Berkshire labourer Edward Swift who was accused of being a 'pernicious seditious & evil disposed Person' for speaking 'treasonous and seditious words, viz. Damn and bugger the King and all that belong to him'. The climate of 1790s Britain appeared to manifest what the political philosopher Jeremy Bentham and, after him, the French historian Michel Foucault termed 'panopticism', all-seeing authoritarian surveillance.

Map-makers had a history of unpopularity among the general populace. The presence of a map-maker in an area often indicated changes in taxes, a new Enclosure Act or 'improvement' to the disproportionately large estates of the nobility. As the nineteenth century progressed, map-makers would also become associated with the unstoppable progress of industrialisation and the sacrifice of the landscape to the pursuit of profit. Wordsworth described with 'astonishment' how 'Engineer agents' who were working for the railways 'came and intruded with their measuring instruments' upon his neighbour's garden. Above all, map-makers suffered from British citizens' fierce protection of their personal privacy. A map-maker was often seen as a busybody, poking his nose and theodolite into other people's business and land. It was the same in France. In the early 1740s one of the surveyors on the *Carte de Cassini* was mapping the region around a village called Les Estables, in the Haute-Loire region of south-central France. Apparently for no other reason than an overriding fear and hatred of strangers, he was hacked to death by the villagers.

In the 1790s these negative associations were augmented by a fear that map-makers might be spies, working for revolutionary France or for the newly vigilant and intrusive British government. Surveying was a standard activity of foreign agents, who often jotted down information about vulnerable locations in the form of maps. For this reason, in the 1790s black-velvet facings were added to the blue uniforms of military engineers, to clearly distinguish them from French soldiers, whose livery was the same colour. A man with a telescope and notebook in his hands, scanning the horizon intently, might also resemble a government mole.

In the summer of 1797, as William Mudge was mapping the West Country, two shady amateur surveyors in the same area, near the Quantocks, provoked enormous suspicion. Spotted by a local doctor 'wandering on the hills towards the Channel, and along the shore, with books and papers in his hand, taking charts and maps of the country', one of these men was instantly suspected of being 'surely a French Jacobin', a foreign agent 'to some principal at Bristol'. The doctor's fear was magnified when it became apparent that the two men were associating with the well-known British radical John Thelwall and were overhead uttering French-sounding words. A possibly apocryphal story relates how the government sent their own spy from the Alien Office to monitor these suspected French intelligence officers. The government spy, a man named Walsh, arrived in Somerset on 15 August 1797 and took a room in Stowey, five miles from his suspects. On tailing the strangers he became increasingly convinced that the two men 'are a Sett of violent Democrats'. But he also worried that he had been discovered, which seemed certain when one of the men apparently began talking loudly about 'Spy Nosey'. Walsh was soon forced to face the humiliating discovery that, far from being French spies, his tails were the poets William Wordsworth and Samuel Taylor Coleridge, who had been talking about Spinoza. Their maps were really '*studies*, as the artists call them', that Coleridge was making in preparation for a topographical poem called 'The Brook'. 'Had I finished the [composition],' that poet recounted cheekily, 'it was my purpose in the heat of the moment to have dedicated it to our then committee of public safety as containing the charts and maps, with which I was to have supplied the French Government in aid of their plans of invasion.'

British map-makers were also to be found spying in France. The map-maker Hugh Debbieg, who had worked beside Roy on the Military Survey, spent much of the 1760s on 'a Secret Service to Survey the principle sea-ports of France & Spain & to make Sketches and Drawings and to take Plans thereof with a view to Discover and State the Strength and weakness of those places'. After the Peace of Amiens in 1802, the Catholic map-maker Robert Edward Clifford, who had fled France after the outbreak of the French Revolution and helped supply the British Army with military maps, decided to return. In November and December 1802 Clifford wrote a series of letters back to England indicating that he was searching for books and maps in Paris and warning that an 'immense invasion' was shortly going to be 'attempted from every port in France' as 'the whole ambition of this country I suspect to be the destruction of England'. Under the guise of a young aristocrat with a peculiar penchant for geology, he managed to gain access to materials in the Bibliothèque Nationale. When war was declared again between Britain and France in spring 1803, Clifford fled and arrived back home in England on 30 May 1803 with 'a box 5 feet long 2 feet broad & a foot high' containing '200 weight of maps plans & manuscripts of france & its environs'. General Simcoe wrote to Clifford's brother that 'the Plans of the Vendée found upon your brother . . . at such a place as Calais & at such a moment would have authorized any legal Government, even any subordinate authority to have executed them at the Instant'. Clifford handed over these maps to the British Army and spent August 1803 making a 'grand military expedition through Kent' to report on the quality of the county's military installations. His religion prevented Clifford from possessing formal employment in the Army, but his shadowy role often worked in his favour as a map-making spy.

Whether map-makers were suspected as state spies or foreign agents, both were unpopular. Many British citizens with left-leaning political tendencies despised Pitt's surveillance mechanisms as hallmarks of a state that had overstepped the mark. One commentator complained, exaggeratedly, that the Alien Office was 'a system of TERROR almost as hideous in its features, almost as gigantic in its stature, and infinitely more pernicious in its tendency than France ever knew'. The essayist and minister Vicesimus Knox was adamant that 'the employment of spies and informers is a virtual declaration of hostilities against the people. It argues a want of confidence in them. It argues a fear and jealousy of them. It argues a desire to destroy them by ambuscade. It is, in civil government, what stratagems are in a state of war.' Employing Swiftian frames of reference, some Members of Parliament had accused Charles Lennox's fortifications scheme of perpetrating a similar invasion of private space in the mid 1780s. 'If this Military Projector was not checked in his career,' one MP had warned, 'a Master General, with his Committee of Engineers, like the Laputan philosophers in their flying island[,] might hover over the kingdom in an Ordnance balloon, descend in a moment, and seize on any man's house and domain [to turn] their pleasure grounds into horn works and crown works'. By insisting upon the open publication of the Ordnance Survey's maps and the accounts of its progress, William Mudge may have hoped to avoid any association of his map-makers' endeavour with this surveillance state.

It has also been suggested that when institutions like the Ordnance Survey render themselves open and accountable to the public, they benefit from greater cooperation. Prior to 1841 the surveyors had no legal right to enter private property and depended on the goodwill of landowners. That Mudge's efforts were not entirely successful in persuading British citizens to acquiesce to its needs was indicated by the fact that the Ordnance Surveyors were forced to apply for a Royal Warrant in 1804 to gain safe passage through the nation. Its assistant Simon Woolcot wrote to Mudge confessing that he felt 'a considerable degree of uneasiness' and thought that 'the authority of a warrant is now become . . . necessary to guard me, particularly on the coast, from those insults and interruptions, in the execution of my business, which I have so frequently and so lately experienced'. A warrant would be 'a measure highly expedient to facilitate the execution of Military Surveys', he asserted. But one journalist spoke for many when he described in impressed tones that 'these maps the Board of Ordnance have very liberally determined to publish for the benefit of the public'. Mudge's decision was also made in the belief that it would encourage a better quality of cartography. He hoped that the maps' surveyors and engravers would proceed with acute consciousness of a wide and discerning audience for their work. Last, and importantly, with the full publication of its maps, the Ordnance Survey became a commercial and potentially profitable concern.

ON THE BASIS of Mudge's ardent support for publication, and now that the Interior Survey had finished its mapping of Kent, the Ordnance Survey could look forward to the appearance of its first map. In July 1799, *The Times* advertised that 'An accurate Topographical Survey of the County of Kent' executed by 'Mr William Gardner, chief Draftsman of the Board of Ordnance', and 'done from materials afforded' by the Trigonometrical Survey, would be published 'in the course of the present year'. But a year later, when Mudge read an account of the Survey's progress between 1797 and 1799 to the Royal Society, the map had still not been published.

As the eighteenth century turned into the nineteenth, materials from the Trigonometrical and Interior Surveys of Kent lay in the hands of a publisher. William Faden was a highly trained engraver and map-seller, whose surname was the result of his Scottish father's decision to 'de-Scottify' it in the wake of the 1745 Jacobite rebellion. He had compressed Mackfaden (also spelt McFadden) to Faden to avoid harassment from anti-Jacobite and Scotophobic Londoners. After a successful partnership with the map-maker Thomas Jefferys, William Faden inherited the business after Jefferys' death. By the 1790s he had been appointed to 'the place and quality of Geographer in Ordinary to his Majesty', and had also become the chief supplier of civilian regional maps to the Board of Ordnance. When Mudge brought out the first volume of the *Account of the Trigonometrical Survey* in 1799, it was Faden who had published it. The Board of Ordnance did not have its own engravers, so Faden was the obvious choice.

Aided by a freelance craftsman called Thomas Foot and a number of assistants, Faden first reduced the six-inch Interior Surveys to the scale of one inch to a mile, which had been established as the standard of county surveying by the Society of Arts' competition. Mudge had ruled out any larger scale because the maps needed to be portable, both for the requirements of the military and to conform to Enlightenment ideas about cartography. Small-scale portable surveys of the country were ideal for the development of routes for the transport of heavy artillery and regiments of soldiers. The Ordnance Survey was designed to accompany tacticians into the landscape. And the one-inch scale also made the resulting maps more affordable and accessible to the general public. After thus reducing the fair plans, Faden then compiled them into a complete county map and began the printing of the Ordnance Survey's first cartographic creation.

To begin the publication process, the engraver marked the positions of the triangulation's trig points with fine dots onto a sheet of hammered and finely polished copper. The plate was then heated and covered with wax, which was left to set. Tracings had been taken from the Interior Survey's maps, and outlines from these tracings were cut in reverse into the waxen plates with a 'graver' or 'burin'. This was a steel-cutting tool with a handle shaped like a mushroom, with an angled steel shaft ending in a very sharp face. The handle rested in the engraver's palm while the index and middle fingers guided and steadied the shaft across the face of the waxy copper plate. The burin engraved the course of roads and rivers, and the outlines of coasts, buildings, fields and the perimeters of woods. Some maps in this skeleton state were distributed among army officers who had no time to lose in the formation of defensive strategies. Place names were next to be engraved and then the ornamental symbols that denoted different types of ground cover, such as miniature trees to represent woods. Last, the hachures were engraved, those clusters of tiny parallel lines that varied subtly in density and direction to denote the steepness of hills. This was the most time-consuming part of the process, requiring a hawk-eyed vigilance, and an absolutely steady hand. Each part of the engraving process was usually executed by a different specialist practitioner, and the highest pay was awarded to those responsible for the hills. (Consequently there were rumours of hills being engraved where there were none on the ground.) Even though the Board of Ordnance had agreed to publication of the finished maps, they appear to have refused to let the working materials out of the Tower of London because of the sensitive nature of such geographical information. Almost certainly the engravers had to traipse to the Tower day after day, to do this work *in situ*.

When the map had been etched, in reverse, printing ink was spread over the waxen plate. This was subsequently cleaned very carefully by hand, with a cloth, to make sure that ink was removed from everywhere but the indented lines. A sheet of paper was laid over the plate, and both were passed through a roller press that exerted a pressure of up to forty tons in order to glean every detail from the engraved plates. In order to effect the greatest transfer of ink to the paper, the plates were reheated slightly and the paper itself was damped. This could cause a slight distortion in the finished image, so when particularly accurate maps were needed, 'dry proofs' were created from undamped paper and cool plates, although this did result in a less dramatic contrast between black and white. Because the first Ordnance Survey maps were all monochrome, only one plate was needed per sheet, whereas coloured maps would have required a different plate for each hue. When the printing press was manipulated quickly, it could produce twenty sheets an hour. The copper plates themselves were capable of producing between 2000 and 3000 impressions before becoming too worn down to function properly.

Upon the production of proof copies of the map, Mudge turned to verifying the place names that had been marked on it. This was a thorny subject. Place names are not set in stone. The names of tiny details of the landscape, such as a dell or a beck, as well as large settlements and major roads, are all subject to change. They mutate over time according to the whims of landowners, the arrival of new members into a community, alterations in the landscape, changing fashions of pronunciation and spelling, local legislation, and so on. At any one moment in time, different communities may have different ways of referring to the same landmark. Toponymy derives from the Greek *topos*, place, and *ōnoma*, name, and it presented a mighty headache to early map-makers. Surveyors had to settle on one particular place name to inscribe onto their maps, out of a plethora of historical and current alternatives. Some map-makers simply wrote down the first name they heard spoken aloud by a local mouth. But more rigorous surveyors applied themselves to unearthing the full range of historical and current variations, before making an informed choice.

In Britain, regional map-makers had been thinking hard about toponymy since the mid sixteenth century. Christopher Saxton, whose atlas of England and Wales had appeared in 1579, had asked local officials to provide information about their place names. But some had worried about undue reliance on such informers. The topographer John Norden had been anxious that even 'the most carefull observer [might] bee led in the Mist by vulgare instruction'. The late seventeenth century marked a watershed in the history of research into Britain's place names. When Edmund Gibson had produced a new edition of William Camden's *Britannia*, he had set about correcting Camden's original choice of toponymy by consulting historical sources to unearth variants, also sending copies of the maps to 'knowing Gentlemen' with requests that they might 'supply the defects, rectifie the positions, and correct the false spellings'. Gibson's practice formed a model for map-makers throughout the following century. Sometimes surveyors had exhibited their maps to the public, inviting helpful criticisms and suggestions. Often they had produced proof copies to circulate among the landed gentry or clergy. These methods guided eighteenth-century map-makers towards a preference for place names that were in current usage. This may sound obvious but, as we shall see, there were often good reasons for rejecting contemporary names in favour of their historical alternatives.

William Mudge undoubtedly felt himself under pressure when tackling the matter of Britain's toponymy. There were no other formal bodies tasked with the standardisation of place names, and as a national mapping agency, the Ordnance Survey inevitably acquired an influential role as a toponymic police force and law-maker. In its early years it did not deviate from the standard eighteenth-century pattern. The surveyors checked the names of large towns and important parishes against institutional and, later, census records and then Mudge explained how, 'to make the work as perfect as possible', proof copies of the first maps were sent 'to different persons in the County, for the purposes of ascertaining whether or not the Spelling of Farms, Hamlets, &c. was correct'. Like his predecessors, Mudge relied chiefly on the advice of landowners and clergymen. Amendments to the maps' toponymy were painstakingly etched onto the plates before the final print run. The whole process of publication, from the moment when the engraver first saw the sketch maps to this final verification of toponymy, was incredibly slow and expensive, and it took William Faden around eighteen months to translate all his material into a complete county map.

FINALLY, ON 1 January 1801, the first Ordnance Survey map was released to the public. Constructed in four massive rectangular sheets, each one around thirty inches wide and twenty inches high, requiring four separate copper plates, the map advertised itself as the first offspring of the 'General Survey of England and Wales'. It proudly stated that it was 'An Entirely New & Accurate Survey of the County of Kent, with Part of the County of Essex' that had been made by 'the Surveying Draftsmen of His Majesty's Honourable Board of Ordnance', under the orders of 'Captn. W. Mudge of the Royal Artillery, F[ellow of the] R[oyal] S[ociety]'. Faden was acknowledged as the map's engraver and a smaller advertisement described how the Trigonometrical Survey's measurements underpinned the whole. The map's title resided underneath the Board of Ordnance's coat of arms, which bore three cannon. In the bottom-right corner the map displayed a dedication from 'their most obedient and faithful servant W. Mudge' to 'Charles Marquis Cornwallis', Master-General of the Ordnance, and 'the rest [of] the principal officers of His Majesty's Ordnance'. Around the borders, a scale resembling a piano keyboard described Kent's precise situation in terms of latitude (measured from the zero-degree line of the equator) and longitude (measured from the Greenwich meridian).

The Kent map presented an intricate black-and-white bird's-eye view of the most south-easterly corner of Britain. It stretched from the Straits of Dover in the south-east to Winchelsea in the south-west, up to London's East

End and over to the many tiny land masses, sandbanks, lighthouses and buoys that peppered the Thames estuary in the map's north-east corner. Kent and south Essex formed a remarkably varied nub of landscape, ranging from the wetlands of Romney Marsh to the precipitous white chalk cliffs of Dover and the Essex marshes that were swathed in mist 'like a gauzy and radiant fabric'. The slow erosion of a chalk, clay, greensand and sandstone 'dome' had created a series of ridges and valleys running east to west across Kent, known as the Kent Downs, which diagonally bisected the map. It also depicted the haphazard sprawl of the areas to the east of central London, including Hackney, Stoke Newington, Bethnal Green and Mile End. It is intriguing to imagine the map-makers surveying at this point where the capital city melted into the surrounding countryside, where the butchers of Whitechapel Road met the farms and fields of Hackney and Tottenham.

The Ordnance Survey did not provide a legend for the symbols on the map, but much was self-explanatory. Miniature trees with tiny shadows to the east were positioned in a haphazard arrangement and signified forests. In orchards they were arranged in neat ranks. Parks and pleasure-gardens were often surrounded by intricate palings whose interior was marked by pointillist stipple. Basic field boundaries were shown by lightly drawn lines, and those marked by trees or hedgerows were denoted with an occasional scattering of tiny spherical bushes. Parallel lines with a range of thickness and shading differentiated turnpike roads from minor and unfenced ones. A cross stood for a church, a tiny windmill was complete with four sails, and a lighthouse was recognisable as such. Cartography shares an etymological root with *cartoon* (both derive from the French *carte*, meaning card or paper), and a perusal of the icons that comprise the Ordnance Survey's vocabulary makes that connection seem particularly close.

THE PUBLIC APPEARANCE of the first Ordnance Survey map had been eagerly anticipated. The Austrian general Andreas O'Reilly was surely not the

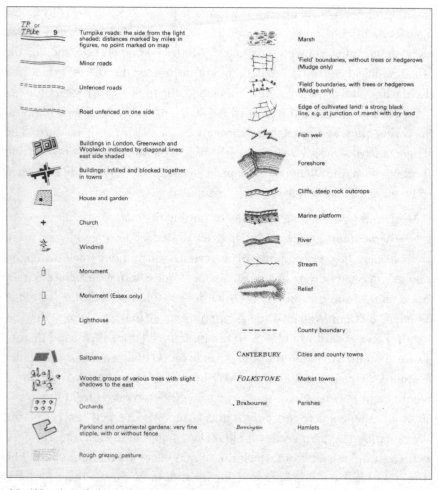

TP or TPike 9	Turnpike roads: the side from the light shaded; distances marked by miles in figures, no point marked on map		Marsh
	Minor roads		'Field' boundaries, without trees or hedgerows (Mudge only)
	Unfenced roads		'Field' boundaries, with trees or hedgerows (Mudge only)
	Road unfenced on one side		Edge of cultivated land: a strong black line, e.g. at junction of marsh with dry land
	Buildings in London, Greenwich and Woolwich indicated by diagonal lines; east side shaded		Fish weir
	Buildings: infilled and blocked together in towns		Foreshore
	House and garden		Cliffs, steep rock outcrops
	Church		Marine platform
	Windmill		River
	Monument		Stream
	Monument (Essex only)		Relief
	Lighthouse		County boundary
	Saltpans	CANTERBURY	Cities and county towns
	Woods: groups of various trees with slight shadows to the east	FOLKSTONE	Market towns
	Orchards	. Brabourne	Parishes
	Parkland and ornamental gardens: very fine stipple, with or without fence	Bonnington	Hamlets
	Rough grazing, pasture		

25. *Although the Ordnance Survey's early maps had no key to the symbols used, it has been possible to compile one.*

only one to exclaim, upon witnessing this remarkable feat of cartography ten years after its first publication: 'the map of Kent was by much the finest piece of Topography in Europe'. On 13 January 1802 Mudge 'presented the Map to his Majesty'. But George III, who suffered from bouts of insanity, was clearly not entirely engaged with the significance of the achievement, and Mudge commented: 'I think [he] still remains to be informed, that it is *an actual map*, and not a written account similar to the last presented.' In 1816, a group of foreign dignitaries requested a private viewing of this iconic achievement. On being informed that the 'Arch Dukes wanted to see everything connected with the Survey', Mudge laid on an exhibition of the Topographical Surveyors' drawings, 'not neglecting the Isle of Wight and Kent plans'. But aside from the triangulation, the Interior Survey's techniques were not innovative, and replicated the standard methods of topographical surveying that had been used throughout the eighteenth century. The Trigonometrical Survey's work was, however, in the vanguard of progress, and for this reason, the attention of the press and scientific community continued to be trained mainly on the primary triangulation.

Mudge always urged the necessity of putting the maps on public view. An advertisement for the surveys in 1818 acknowledged how 'gentlemen' were likely to want 'to procure a map of the country surrounding their own habitations'. The public were able to obtain their long-awaited Ordnance Survey maps either from the Board of Ordnance's headquarters in the Tower of London, or from William Faden's premises at 6 Charing Cross. (After his death, Faden's shop would pass to his apprentice James Wyld, and then to Edward Stanford. Stanford would go on to found the specialist map-retailer Stanford's, which still flourishes today in nearby Long Acre.) As the nineteenth century progressed, the surveys could be procured from more and more map-sellers across Britain. Landowners who desired surveys of the area surrounding their estates could make an appointment at the Tower to peruse an index map from which they could select the relevant chart. The Tower and Faden's shop both sold the first Ordnance Survey maps for three guineas (£3 3s) per county survey: the average weekly wage of one of the map's engravers, or twenty days' wages for a lesser craftsman. From this fee, set by the Board of Ordnance, Faden retained only half a guinea; unsurprisingly, he griped over this meagre commission. By 1816 the Ordnance Survey was charging £6 6s for its map of Devon, which spread over eight sheets: this was around forty days' wages for a craftsman in the building trade.

The price tag mostly restricted access to the early Ordnance Survey maps to the social elite. There were exceptions, such as the Sheffield warehouse apprentice Joseph Hunter, who, in the spring of 1797, borrowed from his local library John Aikin's *Description of the Country from thirty to forty miles around Manchester* and commented intelligently on its 'excellent maps'. And Mudge donated many free copies of the Ordnance Survey's maps to libraries, institutions and associations, through which less privileged members of society could witness the wonderful new cartographical productions. But in the main the consumption of maps in the eighteenth and early nineteenth centuries was the privilege of the affluent. Maps were expensive, and outside military circles it was primarily the upper classes who had been taught how to properly 'read' them. Map-reading is not an intuitive skill, and the language of contours and symbols needs to be acquired. An Ordnance Survey map-maker in the 1830s found that to describe his purpose to a local man in Ireland, he first needed to enquire, 'Do you know what a *map* is?' and to carefully explain that it was 'a representation of the land on paper'.

Wealthy landowners bought the Ordnance Survey's first maps, but at the small scale of one inch to one mile they had little practical utility to most and they are likely to have served instead as rhetorical images of power and ownership or as aesthetic artefacts. The kudos of the Ordnance Survey's early maps was evident from an estate agent's advertisement for the lease of Torbay House, in Paignton, Devon, in which the agent played up the fact that 'the site is laid down in Mudge's map of Devon'. For those who could afford the maps, an explicit mention of one's property was evidently a powerful status symbol. The aristocratic affinity with cartography was borne out by the Scottish nobleman the 8th Earl of Wemyss, who had a pronounced penchant for maps and charts. He scrawled detailed annotations and corrections to the area around his estates all over his copy of John Armstrong Mostyn's *Scotch Atlas*, not unlike map-readers today who make Ordnance Survey maps their own by scribbling routes, viewpoints, travel information, and even pub reviews onto their sheets.

In an era long before the advent of aeroplanes, the Ordnance Survey's maps presented a vision of the nation from the perspective of someone elevated higher than the summits of Kent's rolling hills. The only contemporary comparison was with the viewpoint of a traveller in a hot-air balloon. On 21 November 1783 the first manned hot-air balloon flight was made in a balloon constructed by Joseph and Jacques Montgolfier. The story goes that Joseph devised the conveyance as a possible reinforcement for the

French Army's cavalry and infantry, an embryonic air force if you will. When early models proved successful at raising and carrying animals of various descriptions, Louis XVI agreed to the Montgolfiers' suggestion that a crewed flight might be trialled. Two friends of the Montgolfier brothers, Jean-François Pilâtre de Rozier and François Laurent d'Arlandes, success-fully flew for twenty-five minutes from the Bois de Boulogne in Paris, five and a half miles south-east to the Butte-aux-Cailes, on the capital's outskirts. Sadly Pilâtre de Rozier would become one of the first air fatalities when his balloon crashed during an attempted cross-Channel flight in June 1785.

On 15 September 1784 the aeronautical experiment was exported to Britain when an Italian, Vincenzo Lunardi, demonstrated a balloon flight from the Artillery Company's grounds in London. Lunardi travelled twenty-four miles towards Standon Green End in Hertfordshire. His success spawned a frenzied craze for ballooning in England. Women's skirts were decorated with balloons, bonnets were fashioned in the 'Lunardi' style (bal-loon-shaped, at a height of twenty-four inches), and a whole genre of 'balloon literature' appeared. Fanatics of the phenomenon earned the label 'balloonatics' and, at the outbreak of the French Revolutionary and Napoleonic Wars, some Britons began to worry that the French Army might be developing a balloon division to conduct an invasion of England by air.

Many of the first authors of hot-air-ballooning literature described the delight of witnessing the British landscape from the air for the first time. Mrs Sage was the 'first English female aerial traveller', travelling in Lunardi's bal-loon in June 1785. In her memoir of the landmark voyage, Sage recalled how, 'in crossing over Westminster, we distinctly viewed every part of it; we hung some time over St James's Park, and particularized almost every house we knew in Piccadilly'. This was pure voyeurism. Sage found that she could peek through the windows of the well-to-do, while their owners remained oblivious to her intrusion. When the balloon 'turned on its axis', she recalled how this 'pleased us very much, as it presented the whole face of the coun-try, in various points of view'. Sage's pleasure was not so dissimilar to that of Google Earth and Google Street View users, who can simulate a flight across the earth's digital representation, turning this way and that at will.

One balloonist compared his night-time 'Balloon-Trip' across the capital to the experience of 'looking down on an enormous map of London, with its suburbs to the east, north, and south, as far as the eye could reach, drawn in lines of fire!' (referring to the city's blazing street-lamps). A man called Thomas Baldwin composed 'a narrative of a balloon excursion from Chester, the eighth of September 1785', and he illustrated his text with views of the landscape below, which resembled maps overlain with clouds. If these balloonists imagined themselves to be flying over a map, then the Ordnance Survey's first map-readers may, in turn, have felt themselves suddenly raised above their libraries or desks. Thus suspended miles above Kent, they were treated to a revelation of that county spread out below them, like the views hitherto reserved only for balloonatics.

CHAPTER SEVEN

'A Wild and Most Arduous Service'

THE THREE ELEMENTS of the Ordnance Survey – the Trigonometrical Survey, the Interior Survey and the engravers – worked in separate locations and at different speeds. While the Interior Surveyors busied themselves in the south-east and while the first Ordnance Survey map lay in the hands of its engravers, the Trigonometrical Survey's operations were continuing elsewhere.

After he had finished triangulating Kent in 1795, William Mudge led his team into Dorset and Devon and then into the West Country and Cornwall. There were moments of intense wonderment in this part of the kingdom. While encamped in Britain's south-westerly extremity, 'making observations to determine the distance of the Scilly Isles from the Land's End', Mudge realised with joy that 'the air was so unusually clear' that he was able to distinctly make out through the theodolite's telescope 'soldiers at exercise in St Mary's island'. And the strange effects of terrestrial refraction continued to plague and amuse the surveyors in equal measure. Whilst measuring from Pilsden Hill in Dorset towards Glastonbury Tor, a conical hill in Somerset topped with a roofless tower, Mudge was so proud of the Great Theodolite's capacity to observe long distances 'that, desirous of proving to a gentleman then with me in the observatory tent, the excellence of the telescope, I desired him to apply his eye to it'. This visitor purred his admiration but when Mudge took back his toy and looked again through the sights, he explained how 'the unusual distinctness of this object, led me to keep my eye a long time at the telescope'. 'Whilst my attention was engaged,' Mudge continued, 'I perceived the top of the building *gradually rise* above the micrometer wire, and so continue to do, till it was elevated [far] above its first apparent situation; it then remained stationary,' he concluded, baffled. Terrestrial refraction meant that the air was turned into a prism, by which rays of light were bent to produce these strange hallucinatory effects.

On 20 August 1797 the *Observer* newspaper was able to proudly report that 'Colonel Williams and Captain Mudge of the Artillery, who have for several years past been engaged in a general admeasurement of the kingdom, have compleated [*sic*] a very minute survey of all the Southern and Western Coast to the Bristol Channel'. Britain's entire south coast, from the eastern extremity of Kent all the way to Land's End, and then up the northern coast of Cornwall and on to Bristol, had been measured through sight lines that traced imaginary triangles onto the landscape. But in March 1799 the Master-General of the Ordnance, Charles Cornwallis, commanded Mudge to suspend his own plans for the Trigonometrical Survey and head straight for the unmapped part of Essex, which he designated the Ordnance Survey's highest priority and the subject of its next map. Possessing such a large expanse of coastline and presenting easy access to London, Essex was vulnerable to a French invasion and therefore was thought to urgently need an Ordnance Survey map.

Mudge dutifully spent the spring and summer of 1799 triangulating north-east from London. He first measured a new 'base of verification' to serve a similar purpose to that on Salisbury Plain, but this time between Hampstead Heath to the north and Severndroog Tower on Shooter's Hill, south-east of Greenwich. From this London baseline, Mudge planned to observe triangles proliferating over to Hanger Hill in west London to Bushy Heath near Watford, to Brentwood Church spire, to Gravesend, Langdon Hill in Essex and Tiptree Heath, north of Maldon. But initially it proved difficult. The smog of the late-eighteenth-century city proved impenetrable to the Great Theodolite's sights, even when Mudge used surveying flares bright enough to be seen across the English Channel. It was impossible to observe between Hanger Hill and Severndroog Tower, fifteen miles apart, and Mudge was forced to use the dome of St Paul's Cathedral as an intermediary trig point.

There is something charming about the image of William Mudge and his theodolite perched on St Paul's cross and ball, high above the bustle of the City. Did Mudge, like Wordsworth, reel at the seeming chaos below him? 'What a shock/ For eyes and ears!' the poet cried in 1802. 'What anarchy and din,/ Barbarian and infernal, - a phantasma,/ Monstrous in colour, motion, shape, sight, sound!' Or did Mudge instead revel in it, like another contemporary, the essayist Charles Lamb? Lamb loved

> the lighted shops of the Strand and Fleet Street; the innumerable trades, tradesmen, and customers, coaches, waggons, playhouses; all the bustle and wickedness round about Covent Garden; the very women of the Town; the watchmen, drunken scenes, rattles; life awake, if you awake, at all hours of the night; the impossibility of being dull in Fleet Street; the crowds, the very dirt and mud, the sun shining upon houses and pavements, the print shops, the old bookstalls, parsons cheapening books, coffee-houses, steams of soups from kitchens, the pantomimes – London itself a pantomime and a masquerade . . . I often shed tears in the motley Strand from fulness of joy at so much life.

Mudge admitted that he was forced to measure his London baseline more cursorily than the others, hindered by its abundant buildings and population and the opaque air. But soon the triangulation of Essex was under way, and in the season of 1799 the Trigonometrical Surveyors covered an area of around 5000 square miles as far north as Coventry and as far west as Broadway Beacon in Gloucestershire. Mudge was soon joined by the Interior Surveyors, who began mapping the county in detail alongside the Trigonometrical Survey. On 15 August London's *Courier and Evening Gazette* recounted the progress of the secondary triangulation with interest, describing how the map-makers 'of the drawing room in the Tower' were following 'Captain Mudge with a portable theodolite, for determining the exact situation of every church and remarkable object', before 'fill[ing] up the plans in a style of accuracy and elegance never hitherto attempted'.

26. *Images of the Ordnance Survey's perilous task to measure from the cross and ball on St Paul's Cathedral in London in 1848.*

Mudge, however, had serious reservations about the Interior Surveyors' work. He was not worried by their accuracy or attentiveness to the land-scape – quite the opposite, in fact. He fretted that the enormous six-inch scale on which the map-makers had begun surveying Kent, and the three-inch scale that they subsequently adopted, risked creating an overcrowded map that took too long to complete. 'It does not appear that any advantage will accrue to the Board [of Ordnance], from surveying all the fields of it,' he wrote to a superior. Mudge did not want the Interior Surveyors to spend time mapping any aspect of the landscape that would not find its way onto the engraved map. A superfluity of detail on the fair plans would place an enormous onus of responsibility onto the engravers, who would then have to select what to include and omit on the final maps. Concerned that the engravers were not sufficiently trained in this respect, Mudge emphasised the benefit of 'relinquish[ing] the prosecution of the very minute part of the Survey' to 'attend to what is of real use to the Public at large', another example of his prioritisation of popular concerns over military priorities. He suggested that, if the Interior Surveyors made their maps of Essex on the reduced scale of two inches to a mile, the completed chart would cost 'about one third of the sum which has lately been expended in making the map of Kent' and would consume 'less than a quarter of the time'. He provided the Board with an estimate of 33s per square mile to make a county survey of Essex, which, when combined with the expenses of the triangulation and travelling charges, produced a total cost of £2422. And he predicted that on this basis it would take fifty more years to create maps of the whole of England and Wales. Persuaded by these figures, the Board of Ordnance readily agreed to Mudge's proposition and 'much applaud[ed] the zeal which he has shewed in producing so considerable a saving of expense'.

In 1800, the Interior Survey underwent radical change. At the beginning of that year, William Gardner, chief draughtsman in the Tower of London's Drawing-Room, and the creator of one of the best maps yet of a portion of the English landscape – the 1795 revival of the Great Map of Sussex – unexpectedly died. He was replaced by William Test, who was charged with overseeing the Interior Survey's progress into the next century. And shortly after Gardner's death, all the remaining draughtsmen in the Tower experienced a dramatic alteration of status. Up until then, the Tower map-makers had all been civilians at the heart of a martial establishment. But in December 1800, under the pressure of the Napoleonic Wars, a Royal Warrant reconstituted the Tower draughtsmen, including the Interior Survey's staff, into a 'Corps of Royal Military Surveyors and Draftsmen': they were given military status. Although this would be reversed after the end of the Napoleonic Wars, for the next seventeen years these men found themselves formally subject to the 'rules and disciplines of war' and presented with a new blue uniform to wear, similar to that of the engineers.

The Interior Surveyors duly mapped Essex on the scale of two inches to one mile, for which they were paid 32s 6d per square mile. Each member of the team was instructed to 'represent the Towns, Villages, Woods, Rivers, Hills, omitting only the [boundaries] of the Fields'. Once the problem of scale was overcome, Mudge applied himself to formalising the relationship between the Ordnance Survey and the engravers who had produced the first chart. He hired Thomas Foot, who had worked on the Kent map, and a man called Knight, to work at the Tower of London on a permanent basis for six hours a day, for which they were promised respectively 3½ and 2½ guineas per week: a very reasonable wage. William Faden remained the chief over-seer of the engraving and was given the official title of 'Agent for the sale of Ordnance maps'.

By 1803 the Essex map had been engraved in outline, minus its hills, and proof sheets were ready for distribution, to invite corrections to its place names. Engraving the maps' details was time-consuming and it took two more years until, on 18 April 1805, it was published in four sheets of 23 by 34¾ inches, under the title 'Part the First of the General Survey of England and Wales'. Sheet number one stretched from London Bridge in the west to the River Medway and Rochford in the east, and from Broxbourne in the north to Dartford in the south. Sheet number two sprawled from Osea Island in the west to Foulness and the Thames estuary in the east, and from Goldhanger in the north to Sheerness in the south. The remaining two sheets covered the northern aspects of the county of Essex, and extended from Bishops Stortford in the west to Harwich in the east, and from Manningtree and Saffron Walden in the north to a few miles above Chelmsford in the south.

The Kent survey had been produced as a stand-alone presentation map and had relied on a temporary contract between Faden's engravers and the Board of Ordnance. But the Essex charts were the products of a permanent engraving arrangement that saw the Board of Ordnance turned into a map-publisher, and Mudge conceived of those Essex maps as the first elements of a numbered series of overlapping sheets that would eventually envelop the entire nation. The Ordnance Survey has a number of possible birth-dates, in the Paris–Greenwich triangulation of the 1780s and the date in 1791 when Lennox bought the theodolite and appointed Williams and Mudge to com-mence a national trigonometrical survey. There are also numerous contenders for its 'First Map': Gardner and Gream's map of Sussex of 1795 or the Kent survey of 1801. But the appearance in 1805 of sheet number one (of Essex) marked the birth of the Ordnance Survey as we now know it, an agency whose mapping systematically covers the entire nation. Today the Ordnance Survey's orange *Explorer* maps divide Britain into 403 intersecting squares on a scale of 1:25,000 (or 2½ inches to a mile) and its pink *Landranger* series covers it in 204 segments, on a scale of 1:50,000 (or 1¼ inches to a mile). The publication of the Ordnance Survey's maps of Essex marked the moment when young map-readers could for the first time anticipate possessing a complete and unprecedentedly accurate map of England and Wales within their lifetime. In 1872, when the creation of a second or 'New Series' of Ordnance Survey maps was authorised, this original sequence of maps became known as its 'First Series' or 'Old Series'. But that series would not be completed until way beyond Mudge's estimate of fifty years. Wanting to make the maps as complete an image of all aspects of the nation's face as possible, the Ordnance Survey would not publish the last map of the First Series until 1870.

MANAGING THE Trigonometrical Survey, while simultaneously overseeing the Interior Survey and the engravers, consumed every moment of William Mudge's attention and sent him all over the country. He did not even have a permanent base in London and had been lodging in apartments in his workplace at the Tower of London since the Ordnance Survey's foundation in 1791. It would have been impracticable for Mudge to have moved his family. With his wife Margaret, he now had five children: a daughter, Jenny ('a more meritorious, sensible and affectionate woman does not live,' her doting father wrote), and four sons: Richard Zachariah, John, William and little Zachariah. They would have been cramped in the Tower of London, from which Mudge himself was absent for much of the year. So his family continued to live in Devon, over a day's coach journey away from the map-maker himself. By the end of the first decade of the Ordnance Survey's progress, the weight of responsibility and this separation from his loved ones had begun to weigh on him. It was said that 'he often complained of *depression of spirits*, but found relief from exercise on horseback. It was rather singular, however, that he always got into most unfrequented parts of the country in his equestrian excursions.'

At the turn of the century, Mudge decided that he needed another supervisor to lighten his workload. So in 1801 he travelled to the Royal Military Academy at Woolwich to choose an assistant. One young man instantly caught his attention. Thomas Colby was on the short side, but wiry and fit, with a disarming stare beneath beetling eyebrows and a firm, pursed mouth. This seventeen-year-old showed himself to be 'well grounded in the rudiments of mathematics', and after further examination Mudge decided that he was 'perfectly calculated to be employed in this business'. He arranged for Colby to be given a commission in the Royal Engineers with immediate effect, and from January 1802 the young man was permanently attached to the Ordnance Survey.

Thomas Colby came from a deeply military family. His mother's brothers and her father were artillerymen and Ordnance administrators, and his own father was a Royal Marine. Colby was born in Rochester, but while he was still a young boy his father went to sea and he was sent to live with his paternal aunts in an imposing Gothic mansion called Rhos-y-gilwen, just north of the Preseli Mountains in Pembrokeshire. After being sent to school in Kent, Colby entered the Woolwich Academy. He was a tirelessly energetic young man, who eschewed the temptations of the capital in favour of rigorous physical and intellectual discipline. Caring nothing for superficial accolades and praise, Colby applied himself to the hard, rewarding work of military engineering. He was driven, obsessive and utterly perfectionist, but his spartan demeanour concealed a naturally fiery temper whose 'overheated enthusiasm' occasionally rose to the surface.

Upon his selection of his new assistant, Mudge tasked Colby with overseeing the progress of the Interior Surveying parties while he continued to manage the Trigonometrical Survey. At the end of 1803, Colby joined a party led by a surveyor called Robert Dawson down into Cornwall. Dawson had been intimately involved with the Board of Ordnance since the Trigonometrical Survey's foundation in 1791. An eager trainee map-maker

27. Thomas Colby, by William Brockedon, 1837.

who, like Mudge, hailed from Plymouth, Dawson had been talent-spotted in the 1780s by William Gardner, who had arranged for him to be employed by the Board of Ordnance. In 1794, at the age of twenty-three, Dawson married Jane Budgen, the sister of two brothers who worked for the Interior Survey, Charles and Richard Budgen.

Within five years of Ordnance service, Dawson's professional talents had become wonderfully apparent. He was an exquisitely accomplished draughts-man whose surveys and plans were works of art. In 1800 the newly founded Royal Military Academy at High Wycombe snapped Dawson up as a teacher of topographical surveying. In 1803 the Ordnance Survey became a school for novice engineers: through short placements with the Interior Survey, cadets from Woolwich could acquire skills in many aspects of surveying and draughtsmanship before entering a range of careers in military engineering. Thanks to his previous pedagogical experience and his irrefutable map-making talent, Dawson was given the challenging responsibility of devising a 'Course of Instruction' for these trainee map-makers. He insisted on the critical importance of tutoring his apprentices outdoors, immersed in the landscape.

In December 1803 Colby was observing the work of Dawson and his young charges at Liskeard, in north Cornwall. Although the Trigonometrical Survey's progress was confined to the spring and summer months, this tutor-ing work and some aspects of the Interior Survey, which relied on observations over shorter distances than the triangulation, could be carried out all year round. In Liskeard, a trainee came across 'an old pair of pistols' and raced up to his visiting superior to show them off. Colby held the first pistol by its barrel in his left hand and cocked the second with his right. Perhaps playing up to the cadet, Colby braced himself to shoot. But at that moment, the gun that he clutched in his left hand exploded. In an instant, Colby's hand shattered. Even worse, a piece of the pistol flew into his skull, 'producing a Fracture in his Forehead'. He collapsed, 'almost dying', as a colleague put it.

Dawson hastily gathered up Colby's small frame and carried him back to his accommodation, where he sent for the local surgeon. Shaking his head regretfully, the surgeon nevertheless, on Dawson's request, amputated

Colby's 'violently injured' left hand. He also suggested that the young man's only chance of survival might lie in trepanning: drilling a hole in his skull to release the build-up of pressure caused by the explosion. Shuddering, Dawson refused the operation. Although Colby's chances of survival were slim after such a horrendous accident, Dawson refused to give him up for dead and devoted himself to his care. Mudge raced down from London to Cornwall to see his young colleague, apprehensive that he would arrive merely 'to report his death'. But Colby, nineteen years old, refused to die. Blessed with 'a constitution of unusual strength' and a devoted nurse in Dawson, who sat by his bedside day after day and night after night for a month, Colby eventually rallied and grew stronger. 'At length,' as a friend put it, he 'triumph[ed] over the effects of this appalling accident, and he was mercifully restored once more to health.'

By the beginning of 1804's surveying season, Thomas Colby was back up and running – literally. An eccentric young man before the accident, he disliked walking and preferred to jog everywhere, even through the crowded streets of London. Combined with 'slight peculiarities of dress', Colby cut an instantly memorable figure. Friends remembered bumping into him as he jogged his way to the Tower, yelling after him to 'come back, my boy, and take a beefsteak with me'. But Colby never entirely freed himself from the effects of the explosion. He had lost a hand – a terrible thing for a surveyor – and his forehead for ever bore 'a fearful indent'. Moreover, although Mudge diagnosed that 'the Brain it seems remains free from any injury; Nor any future evil apprehended beyond a year', the accident had a permanent effect on Colby's personality. Within a month of the awful event, the surgeon noted that Colby's 'Spirits [were] raised' and that he had developed an uncharacteristically 'quick manner of speaking which alarmed Mr Dawson'. Before the explosion, Colby had strived to keep an 'over-strained prejudice' under tight restraint. After it, this exertion became even more of a struggle and the young surveyor ricocheted between ascetic self-discipline and uncontrollable outbursts of passion.

WILLIAM MUDGE'S DEVOTION to the Survey did not lessen as he aged, but his health became precarious and his personality developed a fretful streak that qualified his usual bonhomie. By the middle of the first decade of the 1800s, Mudge realised that he needed a break from the Ordnance Survey. Thankfully Colby was energetic, despite his accident, and his presence meant that Mudge could remain in London over some of the surveying seasons, leaving the tiring fieldwork under his assistant's able management. In 1804 Mudge took some time out to allow James Northcote, the protégé of his friend Joshua Reynolds, to paint his portrait. He greatly enjoyed the experience, declaring that the artist 'shall not want the necessary sittings'. But the sheer quantity of work demanded by the mapping project was still so great that Mudge often found himself unable to leave the capital, even over Christmas. When his brother-in-law, Richard Rosdew, wrote in 1804 to invite him for a festive dinner, Mudge declined rather cryptically: 'I should be very glad to sit down to your sirloin; but when I hear that a thousand Artillery are going on the Expedition now fitting out, and also recollect that, but for Triangles and Maps, I might very soon winter elsewhere, I may think myself very well off, although I cannot have that wish [for the sirloin] gratified.' In 1805 he decided to put physical distance between himself and the Survey, and to look for new living quarters away from the Tower. Mudge wrote to Rosdew: 'I am not yet settled in a house, my own, either by year, or lease; very difficult it is to find anything hereabouts to be had at hand, but, having two or three kind friends looking out, I hope soon to be covered with a roof of my own.' But it was to be three years before Mudge moved into his own property, on Holles Street, a small thoroughfare to the west of Oxford Circus, in 1808, when he was forty-six years old.

The Napoleonic Wars continued throughout the first decade of the nineteenth century. After a *coup d'état* in 1799, Napoleon Bonaparte had adopted the title of First Consul of France. In March 1802 the Treaty of Amiens led to an uneasy peace between Britain and France, but war again broke out in May 1803. The following year Bonaparte declared France an empire and was crowned emperor seven months later. These developments worried Mudge intensely, who could foresee no other outcome than a French invasion of Britain. 'How retrograde seem to run the affairs of France from the channel in which we all wish them to steer,' he wrote to Rosdew. 'I believe the great question . . . must be tried on English ground. I shall have my share of it when the period arrives, and so will you; for no man would go further in the defence of his country than yourself.' But as long as the invasion failed to materialise, the Ordnance Survey was allowed to continue mapping Britain.

At the time of Colby's accident, the Trigonometrical Survey had started to concern itself with Wales: a region close to the heart of the young surveyor, who had spent much of his childhood near the Preseli mountains. When he was ready to go back to work, he and Mudge headed west. After measuring a series of triangles that travelled up from the Ordnance Survey's base of verification on Salisbury Plain, through Wiltshire, north Somerset, Bath and south Gloucestershire, the map-makers then made their way through Herefordshire, Worcestershire and Monmouthshire. Before long, they were clustered around the Great Theodolite at Trellech Beacon, just over the border into Wales. Joined to the preceding triangles by sight lines that flew over land and water, over the north of the West Country and across the Bristol Channel, Trellech Beacon was a gateway for the Ordnance Survey's advance deep into *Cymru*.

Trellech Beacon was positioned on a common at the top of the west bank of the River Wye. Removing his eye from the theodolite's sights and stand-ing tall, Mudge looked north and south to see the thickly wooded concertina of the Wye Valley snaking to the horizon, hill bunched against hill, as if an ornamental fan had been laid on top of the earth's crust. The wide, flat expanse of the Wye revealed glimpses of itself far below, between each curve of the towering banks. Contemporary visitors to the area reported how smoke issuing from nearby iron- and charcoal-works sidled between the flanks of the river and rested on top of the water like an ethereal feather boa. Poised at this viewpoint, Mudge bent back over the instrument until his gaze was on a level with its eyepiece. He swung the theodolite's central stack around on its axis and tilted the telescope, until he brought into the centre of the grid of wires that bisected its sights the trig points to the east at which the two men had been standing a few days earlier, at Malvern and on May Hill in Gloucestershire. But Mudge was not alone. Beneath his gaze, a scattered mob of tourists shifted and shuffled around the contours of the Wye, cooing at its varied views.

Over the last half-century, a home-grown tourist industry in Britain had been steadily burgeoning, propelled by a variety of factors. It was significant that the number of turnpike roads had exploded over the second half of the eighteenth century. Between 1750 and 1815, over 1100 separate turnpike trusts were created, enhancing travellers' experiences over some 22,000 miles of road. The quality of Britain's highways was further improved when a Scottish civil engineer called John Loudon McAdam proposed a method of constructing roads by binding stones together with gravel, on a firm base of large rocks, and including a camber at each side to allow rainwater to drain quickly away. (Although the statement is unsupported by historical sources, the writer Thomas De Quincey proclaimed that, thanks to this process of 'Macadamisation', 'all the roads in England within a few years were remodelled, and . . . raised universally to the condition and appearance of gravel walks in private parks'.[1]) Partly thanks to these better roads, there was a rapid increase in stagecoach services: between 1790 and 1836 their number increased eightfold, making journeys faster and cheaper. In 1793 the nation's tourist industry received a further boost when the French Revolutionary Wars and then the Napoleonic Wars placed the Continent largely out of bounds to British travellers. The writer, antiquarian and soldier Joseph Budworth described how 'a once-boasted, though now unfortunate, part of the Continent is become a scene of horror and devastation'. Wealthy families who would once have pursued the Grand Tour and members of an affluent middle class were forced to travel through their own country instead of Europe. But the experience was made more enjoyable by a change in sensibility over the previous half-century, whereby the British landscape had begun to be considered as a source of pleasure and delight.

In 1757 the politician and philosopher Edmund Burke had published *A Philosophical Inquiry into the Origin of Our Ideas of the Sublime and Beautiful*, which

---

1 McAdam's method slowly caught on, and when in the 1880s tar began to be used to bind roads together more firmly, this eventually led to the invention of 'tar Macadam' or 'Tarmac', a staple of roads from the 1920s onwards.

he was said to have written nine years earlier, at the precocious age of nineteen. Meaning literally 'up to the limits', the notion of sublimity increasingly caught the imagination of philosophers, writers and artists in the eighteenth century as a way of describing emotional and imaginative responses to the powerful and destructive aspects of landscape. A glimpse of a yawning chasm, a louring mountain whose dark shadow at dusk appears to follow one 'like a living thing' (as Wordsworth put it), a vertiginous position above a crashing waterfall, or a sudden revelation of a view whose magnitude leaves one speechless are all sublime experiences: moments that take one to the edge of reason. Burke's *Philosophical Inquiry* articulated a growing sense that the value of landscape lay not just in its fertility or functionality, but in its irrational, imaginative or aesthetic qualities too. This was a new mapping of the landscape whereby mountainous and other similarly dramatic scenes, such as those in North Wales, Highland Scotland and the Lake District, became popular visiting sites in the second half of the eighteenth century.

In the summer of 1770 a Cumbrian clergyman and headmaster, the Reverend William Gilpin, who was inspired by this sensibility, began a holiday at Hounslow Heath. Anticipating the footsteps of William Mudge, he made his way to Berkshire, Gloucestershire and the Vale of Severn, and finally arrived in the Wye Valley. Gilpin spent a summer lovingly tracing its contours. It was there that he came up with 'a new object' of tourism: 'that of examining the face of a country *by the rules of picturesque beauty*'. By 'picturesque' Gilpin meant 'that peculiar kind of beauty, which is agreeable in a picture'. When he looked at a natural scene, he sought resemblances with the type of landscape painting executed by the seventeenth-century artists Salvator Rosa, Nicolas Poussin and Claude Lorraine. In the Wye Valley, Gilpin found exactly what he was after. The river's lofty arboreal banks and serpentine course created views 'of the most beautiful kind of perspective; free from the formality of lines'. Almost every vista seemed to be constructed like a theatrical scene, with a foreground, 'two side-screens' composed of the river's wooded sides, and a 'front-screen' made up of the winding Wye. The region's centrepiece, the ruined Cistercian abbey at Tintern, a few miles south of Mudge's viewpoint at Trellech Beacon, provided 'the most beautiful and picturesque view on the river'.

Gilpin published the theories of picturesque beauty that he formulated during his Wye tour in numerous essays in the 1780s and 1790s, in which he defined the types of landscape that most successfully produced the picturesque effects after which he hunted. Preferring the organic to the artificial, and fascinated by natural processes of decay, Gilpin favoured uneven surfaces, ruined buildings, overgrown gardens, and poor and decrepit human figures. Like Burke's notions of the sublime and the beautiful, Gilpin's language of the 'picturesque' was a form of aesthetic mapping or 'cultural cartography', and it inspired multitudes of late-eighteenth-century British tourists. Thousands flocked to the Wye in his footsteps. Many furnished themselves with a dazzling array of accoutrements in their hunt for these picturesque effects, such as a series of lenses and convex mirrors that were designed to alter the hue of a scene and compress and frame it. The Claude Glass, named after the French landscape painter Claude Lorraine, consisted of four or five tinted lenses, each of which was designed to alter the atmosphere of the landscape. 'If the hues are well sorted,' Gilpin explained, 'they give the object of nature a soft mellow tinge, like the coloring of that master.' Clutching their lenses and mirrors, their sketchbooks and sometimes even 'magnifiers for botany, a sixteen-inch tape-measure', barometers and geological hammers, these tourists of the late eighteenth century possessed an enthusiasm for the observation and experience of the natural world not unlike the professional map-making figures of Mudge and Colby. However, whereas the Ordnance Surveyors created an image of the landscape through empirical measurement, Gilpin emphasised that picturesque tourists should eschew reason in favour of spontaneous emotion.

From Trellech Beacon, Mudge and Colby looked down on the Wye Valley and found a very different scene from that perceived by the picturesque tourists that surrounded them. Selecting a viewpoint that allowed him to see as much of the valley as possible, Mudge superimposed in his imagination a schema of geometric shapes, numbers and angles onto the Wye's winding banks and misty water. On that same scene, an Interior Surveyor would later conjure up within Mudge's triangles (which were often traced onto the fair plans) a small number of icons. In place of the orchards that peppered the Wye's banks, each comprised of idiosyncratic trees bearing a variety of fruit, an Interior Surveyor would draw a regimented display of identikit cartoon shrubs. Instead of delighting in the river's coquettish disappearances and returns, the map-maker dragged it out into the open, clearly tracing its course in thick blue ink on paper. The highlight of the Wye Valley tour, Tintern Abbey, became a simple red square, the generic symbol of a building, on the fair plans. The Trigonometrical and Interior Surveyors saw in the Wye Valley another opportunity to exercise their faculties of reason. They mapped the landscape according to the demands of truth and accuracy.

But when 'picturesque' tourists looked at that same scene, they found something quite different: a shifting series of possible paintings. Where Mudge and Colby purposefully attained high viewpoints to fully 'open up' the landscape beneath, Gilpin instead advised travellers to submerge themselves deeply in the valley's nooks and crannies. A footnote added to a work by the poet and scholar Thomas Gray in 1775 emphasised that 'the *Picturesque Point* is always . . . low in all prospects'. With every step they took, these picturesque tourists were presented with different views, each with its own distinct background and foreground. From this sunken stance, the region's latent and overarching geometry and harmony remained obscured, and was replaced by a continually changing vista of exquisite images. Where the Ordnance Survey ironed out the idiosyncrasies of trees, foliage and buildings to a series of uniform icons and lines, picturesque tourists instead revelled in the intricacies of each scene: a piece of moss, growing on the eroded brickwork of the Abbey, or the hunched back of a poverty-stricken labourer. Mudge and Colby prioritised reason, quantification and detailed analysis, but Gilpin described how the sight of 'some grand scene' should affect the observer 'beyond the power of thought' so that 'we rather *feel,* than *survey* it'. The differences between the Ordnance Surveyors and the picturesque tourists were manifested in more mundane ways too. Mudge found the tourists a disturbance and obstruction to his observations, and he fantasised to Colby about 'future almost inaccessible positions' in the landscape, which will 'perfectly free you from all disagreeable intrusions. All those swarms of idle holiday visitors will visit you no more, they will remain at home, like flies from tempests all couched under shade.'

WILLIAM MUDGE AND Thomas Colby completed their observations at Trellech Beacon towards the middle of the surveying season of 1804. They then turned west. Colby was an ancestral Welshman and he knew that from this panoramic trig station the Trigonometrical Surveyors would be able to generate triangles of sight lines reaching into the heart of the Welsh valleys, to Garth Mountain near Pentyrch,[2] over to Ogmore in the Vale of Glamorgan, Llangeinor near Bridgend, up to the mountain of Mynydd Maen (or Mynydd Twyn-Glas), north of Abergavenny, and further and further west. Finally Mudge and Colby would find themselves lodged in the heart of the Preseli Mountains in north Pembrokeshire, revelling in spectacular sea views on almost all sides. This series of triangles would ultimately reach up to Aberystwyth, on the west coast of Wales, which was the hub where multiple series of triangles converged. Here, the triangulation that hugged Wales's south coast would join onto a series that the surveyors planned to measure going east, mountain-hopping all the way through central Wales and back to Trellech Beacon. Aberystwyth was also the junction with an intended chain of triangles that was to travel up through Anglesey to a new baseline at Rhuddlan Marsh, on the north coast of Wales near Rhyl, at the border of Flintshire and Denbighshire.

In September 1806, the Trigonometrical Surveyors found themselves in the small town of Rhuddlan to measure this base of verification, before observing the triangles that would lead them back to Aberystwyth. The discovery of an appropriate baseline of verification was eased by the cooperation of the marsh's owner, who was very happy to hand over to Mudge and Colby a

---

2 Locally renowned as 'the first mountain in Wales', Garth Mountain was made famous in the film *The Englishman who Went Up a Hill but Came Down a Mountain*. Pernickety and culturally insensitive English Ordnance Surveyors, engaged in mapping Wales in the early twentieth century, found the height of the Garth to be less than 1000 metres, the requisite altitude for mountain status. In the film, this prompted the local community to shift enough earth to the top of the mountain to push it over the 1000-metre mark, and provided the context in which one of the English surveyors (played by Hugh Grant) fell in love with Welsh culture, in the shape of a particular local woman (played by Tara Fitzgerald).

private large-scale plan of the territory. For the south-eastern end of the base, Mudge settled on a spot adjacent to Rhuddlan Bridge. This lay in the shadow of a thirteenth-century castle that had been built on the orders of Edward I, the Plantagenet king who subdued the Celts, after much resistance, and conquered Wales. It was here that in 1284 the Statute of Rhuddlan, the document that formally enshrined the English conquest of Wales, was signed. The sight of an English military surveyor mapping Wales at such a historically charged spot may have roused heated feelings among the locals. From this south-eastern end at Rhuddlan Bridge, the baseline extended just over five miles north-west, over a flat, marshy piece of territory, to a point on the coast north of the settlement of Abergele.

The measurement of the Rhuddlan Marsh base allowed Mudge to check to his satisfaction the accuracy of the triangles that joined it with the baseline at Hounslow Heath and another series of triangles that led to a base that had been measured by the Ordnance Survey at Misterton Carr near Doncaster in the summer of 1802. It also allowed him to generate a new chain of triangles along the sublimely mountainous north and north-west regions of Wales, which would eventually meet up at Aberystwyth with the series that had been observed along the south coast. As the triangles left the west end of the Rhuddlan Marsh baseline, they hopped over to Llanelian, near Colwyn Bay, to Great Orme on the Creuddyn Peninsula, to Moelfre on Anglesey and then down into Snowdonia. The Surveyors had no choice but to carry the 200-pound theodolite and its oak-framed 'portable' observatory up the 3560-foot height of Snowdon itself, then over to the rather more gentle peak of Arenig Fawr, and on up the glacial Cadair (or Cader) Idris (in folklore the armchair of the giant Idris).

It was here that the daily drudgery of their task began to weigh on the map-makers. It was, as one put it, 'a wild and most arduous service'. An eye-witness to the Ordnance Surveyors' work at the top of Snowdon recounted with awed admiration how 'the mountain tapered at last into a complete cone, terminating in a summit, on which there was just room to spread out a tent for the protection of some mathematical instruments'. Snowdon's confined pinnacle forced the director to pitch 'several tents, a little lower down, for the accommodation of himself and his men'. But the discomfort at the mountain's peak was offset by its miraculous view. 'Snowdon,' this witness observed, 'lies right in the centre of the British world, and commands from its summit, views at once of England, Scotland, Ireland and Wales, and of the intermediate islands of Anglesey and Man.'

A DECADE LATER, in the mid 1810s, after completing 'the Survey of Norfolk' and Cambridgeshire, the Interior Survey arrived in Wales to flesh out the triangles that Mudge and Colby had measured with the Great Theodolite. Its principal draughtsman, Robert Dawson, relished the mountainous environment. From Carnarvonshire he wrote to the Ordnance Survey's assistant, Simon Woolcot, that: 'the range of mountains is more extensive in this neighbourhood and of superior bulk and eminence and has also in general a nearer approximation to the coast, which, perhaps, may account for a greater attraction of the humid vapours exhaled from the sea'. The Ordnance Survey was distinguished from its precursors by its accurate and detailed – and beautiful – portrayal of the landscape's 'relief', its indentations and protruberances. Hill drawing was the most taxing element of map draughtsmanship. Almost any trainee map-maker could be taught to plot the course of a road from the measurements taken during a traverse survey. But before the Ordnance Survey formally adopted contour lines in the second half of the nineteenth century, the depiction of relief required an eye that was attuned to every fold of the landscape, each declivity and swell. An aptitude for such sophisticated landscape drawing was considered by many military map-makers to be largely innate, and one engineer emphasised that it was vital 'to select those who draw well' for the Corps, as 'it is much more easy to teach an Officer to survey than to draw'. Dawson was partly in Wales to try to teach young map-makers how to represent relief as attractively and accurately as possible in their sketches and fair plans.

The Board of Ordnance invested a great deal of time and money in honing the nascent artistic skills of its cadets. The Royal Military Academy at Woolwich paid illustrious painters to tutor its embryonic map-makers.

These included Paul Sandby, who had begun his career as a draughtsman on the Military Survey of Scotland, and subsequently became renowned as a landscape artist and a founding member of the Royal Academy. Robert Dawson, appointed as an instructor of topographical drawing at the Royal Military Academy in High Wycombe and then the chief tutor of the Ordnance Survey's own trainees, was also employed in the same role by the East India Company in 1810. He put all his students through a rigorous course of hill drawing in the landscape itself, and in the mid 1810s he took some of them to Snowdonia, where he taught them at the same time as overseeing the Ordnance Survey's Interior Survey of Wales. Dawson described how he was 'aim[ing] at a large and striking example of topography' in his own 'new Welsh work'. One witness of Dawson's creations described how they had achieved 'a degree of perfection that had given to his plans a beauty and accuracy of expression which some of our eminent artists had previously supposed unattainable'.

In 1816 Dawson produced a plan of Snowdonia that was bordered by the main road running between Dolgellau and Barmouth, and on which north was shown in topsy-turvy fashion at the bottom of the page. At Cadair Idris, glacial erosion has created a horseshoe of steep walls that plummet from the mountain's top, known as Penygadair ('top of the chair'), down to a lake called Llyn Cau. The Trigonometrical Survey used levelling techniques and found its height to be 2914 feet above the low-water mark. (Altitude is nowadays measured from a 'Mean Sea Level' taken at the Tidal Observatory at Newlyn in Cornwall, which gives Cadair Idris a height of 2930 feet.) On Dawson's map, high points of altitude stand out as exposed patches of relatively blank paper and roads and rivers trace sparse networks of spidery lines across the sheet. Infrequent and tiny dots of red indicate the few small settlements that nestle in the crooks of this mountainous land. But it is Dawson's intricate depiction of relief that makes this map a truly remarkable production. Cadair Idris's spectacular sheer sides are shown in fine black and brown hachures that border Llyn Cau like hairs on the nape of a neck. The shading elsewhere on the sheet is so effective at creating a sense of the landscape's plunges and soaring summits that it makes me want to run my hand over it, expecting to feel the paper pushing up at my palm like a pop-up scene.

In the mid 1810s, Dawson's eighteen-year-old son also applied for a place in the Royal Engineers. In support of his application, Robert Kearsley Dawson submitted a bound sketchbook of experiments 'towards the expression of Ground in Topographical Plans', which included a plan of Snowdonia. Dawson Junior's sketchbook contained a series of paintings of 'mathematical forms applicable to hills', in which he found ghosts of universal geometric forms in the landscape's undulations. The summits of Snowdonia, which his father had represented in exquisite landscape paintings, were delineated by Robert Kearsley Dawson as square pyramids, tetrahedra, and 'conical' and 'hemispherical hill[s]'. This was a standard way of learning how to draw peaks but it created a very different picture of the landscape to his father's maps, which paid loving homage to the minute idiosyncrasies of the slopes, summits and troughs of North Wales.

As he aged, the beauty of Robert Dawson Senior's hills came at an increasingly high price: the neglect of accuracy and detail. In 1832 a colleague was forced to complain to Dawson's superior: 'I feel it my duty to state, that the details such as the boundaries and filling in of the Parks, Woods and Commons are not so defined on these Plans . . . as in my opinion to enable the reducers and engravers to lay them down without further reference.' Dawson remained a revered member of the Ordnance Survey and his boss phrased his rebuke with tact: 'I should be very sorry to see such beautiful Plans as yours mutilated and defaced,' he assured Dawson. 'But we want to be clear and of all things to avoid error, and I should hope you might be able to define, beyond the possibility of mistake, the Boundaries of Parks, Commons, and Woods, as also the termination of streams without injuring the effect of the Plans.' The requirements of art and cartography could pull in opposite directions.

THE INTERIOR SURVEYORS were not just tasked with the mapping of the physical landscape in Wales. They also had the responsibility of surveying its place names, and this would prove much harder than in England.

English map-makers did not have an encouraging history of representing Welsh toponymy. The surveyors that the seventeenth-century map-maker John Ogilby had employed in the construction of his *Britannia Depicta* had tended to jot down the first thing they heard. They duly misunderstood Llandovery as 'Llaniidosfry', Carmarthen as 'Comcarven' and Rydypennau as 'Ruddy pene'. In 1757 a scholar and cartographer called Lewis Morris complained that Wales's place names had been 'murdered by English map-makers'. It is only a slight exaggeration to say that the early Ordnance Survey was no exception. At first it followed the pattern that had been set during its mapping of England: surveyors asked locals for place names, jotted them down in notebooks and checked the principal ones against the census and parish records. Proof copies of the finished maps were later sent to local clergymen and landowners for corrections. Robert Dawson reported that the Ordnance Survey should feel itself 'much indebted' to the Reverend Hugh Davies of Bangor, who had gone to the 'trouble of thoroughly examining the names . . . and supplying a great number of corrections'. Thomas Colby made the most of his family connections and commandeered a 'Mr Colby' in 'the Northern part' of Pembrokeshire to 'put the proofs into the hands of those whom he judged most likely to detect inaccuracies either in the orthography or Surveying'.

But this system was not nearly sufficient to eradicate mistakes from the maps. The language barrier was like the Himalayas. English-speaking Ordnance Surveyors had extremely variable success in noting down the Welsh place names that they heard spoken aloud. In the St David's peninsula, a rugged piece of coastline in Pembrokeshire, they inscribed 'Crosswoodig' instead of Croeswdig, 'Carfry' instead of Caerfai and 'Kingharrod' instead of Kingheriot. In areas of Wales where English and Welsh were spoken side by side, the situation was even more complicated. In Pembrokeshire, Colby attributed his surveyors' difficulties to the fact that 'on one side of it the English language is spoken, on the other side the Welsh, and the orthography of the names of places is continually varying from a Conformity to the usages of the one language to that of the other according to the caprices of the successive persons who possess them'. In such bilingual areas, the same settlement could be referred to by a range of names and spellings. A Welsh antiquary and naturalist called Lewis Weston Dillwyn took Colby to task for the Ordnance Survey's negotiation of the different orthographies of Cardiff. 'You have called it *Caerdiff*,' Dillwyn pointed out accusatorily, 'but both by the Corporation itself and by every body else as well as in all legal proceedings it is now universally spelt *Cardiff*, and in this case I should certainly alter it to the latter, for in fact yours is only a sort of hybrid word half Welsh and half English and the real old Welsh word is *Caerfaff*.'

The Ordnance Survey's map of Pembrokeshire was published in August 1820. By 1821, the Board of Ordnance had received complaints 'that in the Ordnance Map of Part of Pembrokeshire, the names of some of the Places have been omitted, and that some of them are misspelt'. The surveyors engaged in mapping Wales were commanded by the Master-General 'that the utmost care may be taken to avoid such errors in future, and to correct those that may have occurred if not too late'. They admitted that 'we have long been aware of the difficulty, and I might say, impossibility of spelling the Welsh names of places so as to be satisfactory even to those who are considered the best authorities; and it has not infrequently happened that to guard against errors of the kind, we have had to erase from the Copper Plate, and enter again in its original state the same name more than once'. 'But,' the Interior Surveyors added reassuringly, 'we have lately adopted a system which answers the end perhaps as completely as may be possible.'

In the 1820s Wales became a laboratory for toponymic innovation. The Interior Surveyors were given greatly enhanced responsibility for researching the place names that appeared on their plans. One surveyor elaborated how the map-maker 'who is employed in the County . . . together with the drawing which he transmits for the Engravers for publication likewise sends a tabular list of *all* the names on the map, shewing at one view, opposite to *each* name, all the authorities that have been consulted'. Instead of simply relying on corrections made to proof copies by the landed gentry, the Interior Surveyors were now instructed to consult a host of printed and manuscript sources to glean evidence for the existence of different versions of place names in history. The Ordnance Survey built up its own library of reference works, including William Camden's *Britannia*, Nicholas Carlisle's *Topographical Dictionary of England*, and John Adams's *Index Villaris*, a gazetteer of England and Wales's 'Cities, Market-Towns, Parishes, Villages . . . or other Division[s] of each County'. In addition to the records that they had previously perused when researching English names, the map-makers also referred to Acts of Parliament, topographical dictionaries, documents pertaining to local enclosures and parish records. They were also encouraged to make use of the advice of native Welsh speakers. The Interior Surveyors were hunting for evidence of any variation that had existed in the spelling or form of each place name, and they carefully noted down every source they found to justify one spelling over another in specially designated 'name books' and tabular place-name sheets, which were then submitted to the Ordnance Survey's headquarters at the Tower of London. Colby came to formulate a policy of adopting whichever version was recommended by the greatest number of sources. Like Mudge, he preferred forms that were in current usage and spoken by the majority, but he also took into account the historical picture of Wales's toponymy that was painted by the surveyors' researches.

These investigations placed an enormous and in many cases inappropriate onus of responsibility on the military engineers. Many struggled. But some were quick learners and acquitted themselves well. A civilian Welsh map-maker and engineer called Alfred Thomas noted rather angrily that on the Ordnance Survey maps 'the words Lan and Llan are often confounded; one signifies an *Eminence*, the other a *church*'. In his reply, an English Ordnance Surveyor called Haslam demonstrated a sophisticated knowledge of Welsh. 'I cannot help feeling a good deal hurt that the orthography of the Caermarthenshire Sheet should be so unjustly criticised,' he complained. Haslam pointed out that the 'words Lan and Llan I would just hint to Mr A. Thomas do *not* always signify an Eminence and a church. The word Llan has three different meanings in the Directory, and in the Map the word was never written without the approbation of the best authority for its proper application.' But these arguments would not go away. As we shall see, when the Ordnance Survey turned its attentions to Ireland, questions of toponymy would literally split that institution in two.

MUDGE AND COLBY and their assistants on the Trigonometrical Survey left Wales in 1811 when the triangulation of that nation had been completed, and turned their attention back to England. On 16 January 1818 *The Times* was proud to report that 'the Public are hereby informed, that 25 Plates' of the Ordnance Survey's First Series 'are already finished, and that impressions from them are on sale at the Drawing-room in the Tower of London: these plates form a complete Map of the Coast from Folkestone, in Kent, to the Land's End, in Cornwall, and thence to Week St Lawrence, near Bristol, and also from the Thames to Orfordness', in Suffolk. The First Series' state of completion could be imagined as a rising level on a map, and over the next two years this level would ascend as more maps were released of north-east Somersetshire, central Wiltshire and part of Hampshire. In 1820 the line reached south Wales, and a map of Pembrokeshire and Lundy Island was put on sale. By this point, although many of the maps were yet to be published, the Interior Survey had achieved the feat of mapping all of England and Wales south of a line from Aberystwyth to Birmingham, past Madingley to the north-west of Cambridge, to a point just south of Lowestoft on the Suffolk coast. They had even visited some areas north of this line. Engraving was under way for maps of Glamorganshire and the Gower in South Wales; but, as we shall see, the discovery of serious errors on the charts would prompt an extensive process of revision that meant that these particular maps did not make their way into shops until 1833. Consequently the Welsh sheets of the Ordnance Survey's First Series show a wide variety of style. But from 1833 onwards, the public was presented with a succession of beautiful and relatively accurate depictions of surveys of Wales until, in the mid 1840s, they could possess a complete Ordnance Survey map of that nation.

CHAPTER EIGHT

Mapping the Imagination

EARLY IN THE mapping season of 1808 Thomas Colby and William Mudge found themselves in Kettlewell, a tranquil limestone village in the hills on the border of Wharfedale and Nidderdale in the Yorkshire Dales. Wales had mostly been triangulated and the men were pursuing triangles across Staffordshire, Cheshire, Lancashire, Derbyshire, and West and North Yorkshire. From their base at Kettlewell, Mudge, Colby and a small team of artillerymen made their way on foot east out of the village. At first, they followed a track that led to a farm called Hay Tongue; then, on a rougher path, to Hag Dyke cottage and barn; finally, they picked their way carefully through the tussocky ground that led steeply up a small mountain called Great Whernside. The weight of the surveying equipment they carried on their backs made the surveyors unstable and liable to lose their footing, and the tangled heather and bog moss clutched at the men's boots. After hauling themselves to the summit of the 2310-foot peak, Mudge recollected how the surveyors found 'a great number of huge rocks, scattered about in all directions'. From this haphazard miscellany of stone, he chose one rock, flatter than the others, as a base for the theodolite.

On a clear day, a few steps east of Great Whernside's summit give you a view down into the basin of Upper Nidderdale, whose wide expanses of sheep's fescue and bent grass glow golden in late-afternoon sun – truly 'God's own country', as Yorkshiremen say. Today a single drystone wall cascades down from one of the mountain's lesser summits, between two springs, one of them the head of the River Nidd, to meet the southern tip of Angram Reservoir. On the other side of a dark and imposing neo-Gothic dam lies the larger and deeper Scar House Reservoir. These reservoirs were built between the two world wars to supply the Bradford woollen industry with water. Stone was quarried from the surrounding hills to make the two dams, and a temporary village, now almost disappeared, was erected in the remote dale to house, heal and entertain the workforce, complete with a hospital, cinema and concert hall. But in Mudge's day, the River Nidd was allowed to run uninterrupted through this great bowl of moorland, gathering force as it made its way down the valley.

I have been coming to Upper Nidderdale regularly since I was six months old and it has seemed a rare occasion that the long mass of Great Whernside has been visible for more than a few moments. Its horseshoe form wraps around the valley beneath and pours the precipitation that gathers in dark clouds at its summit into the two reservoirs. Many a walk has ended with a dash away from the mountain, with sulky rain clouds or a ferocious blizzard in pursuit. It is likely that Mudge and Colby had a fair wait at the top of Great Whernside for a day clear enough to get the amazing views it concealed for most of the year. If so, the tedium of day after day peering outside their tents to be greeted by mizzle and muted light must have been excruciating. But a clear day was a valuable prize. The summit of Great Whernside offers extraordinarily long views: to the trig point at Boulsworth Hill above Burnley in the south Pennines, twenty-five miles south; nineteen miles in the same direction to the triangulation station at Rombalds Moor, near Ilkley; twenty-two miles south-east to Water Crag, between Arkendale and Harrogate; huge sight lines thirty-two and thirty-five miles east across the Vale of York to Black Hambleton, the highest point on the western edge of the Cleveland Hills, and Hambleton Down, east of Thirsk; twenty-five miles north-west to a gracefully shaped mountain called The Calf, in the corner of the Yorkshire Dales; and sixteen miles almost due west to Ingleborough. The second-highest mountain in the Dales, Ingleborough's distinctive flat summit, like a headless sphinx, even allowed Mudge and Colby to see as far as Snowdonia, a hundred miles south. Alongside The Calf and Great Whernside,

Ingleborough acted as a nexus where series of triangles that came up from the south, from Lancashire, met those coming in from the east, from North and West Yorkshire. These three peaks were also gateways into the Lake District. From Ingleborough, The Calf and Great Whernside, Mudge and Colby's sight lines fanned west, to take in the impressive mountains of Helvellyn, Coniston, Scafell and Black Combe.

Mudge and Colby's team arrived in person in the south-west corner of the Lake District around the summer of 1808. They began their measurements at the top of Black Combe, a mountain on the south-west coast of Cumbria. In preparation for their ascent, the men first made their way to Bootle, reputedly the smallest market centre in England and two and a half miles north-west of Black Combe's summit. The mountain loured over the village. We can imagine that, after a quick meal on arrival at the King's Head Inn on the High Street, the map-makers laced up their boots, pulled on their blue coats and, with their backs to the glinting sea, set off for the peak. A small lane left Bootle at right angles to the main road, and soon petered out into a good, wide path. After an easy ascent the road began again and the men met a guide-post, probably a product of the 1773 Highways Act that stipulated that parish road surveyors should ensure the presence of guide-posts and milestones on the nation's thoroughfares. These made England's landscape easier to navigate, effectively turning the ground into a rudimentary map of itself. After the guide-post Mudge and Colby remembered to keep an eye out for a drystone wall branching off towards Black Combe on the right, which they followed steadily uphill for two miles until treacherous scree slopes appeared on their left. Here they headed south-west and walked in a straight line along the top of the scree until they met an old shepherding path that led directly to the brow of Black Combe.

When the summit was finally gained – if the day was clear – the view would have been phenomenal. Black Combe looks out over the Irish Sea and, at 1969 feet, this position reputedly gives it the most extensive views of any peak in England. The Isle of Man, with its two trig points on the hills of North Barrule and Snea Fell, is sharply outlined in the west. One can imagine Thomas Colby facing north-west, grabbing William Mudge with his right hand and excitedly pointing out Scotland. He could have gestured towards the Mull of Galloway, Scotland's most southerly point and, to its north-east, the peaks of Merrick, Corserine and Bengairn. In front of southern Scotland, St Bee's Head was visible, where the Lake District meets the sea: the western end of the modern Coast-to-Coast Walk. And then, spreading out in a semi-circle from north-west to north-east, like a purple sea of rock and gorse and bracken, were the Lake District's magnificent peaks: Dent Hill, Lank Rigg, The Pillar, Kirk Fell, Skiddaw and Great Gable in the background; in the foreground Buck Barrow and Whitfell; then over to Scafell and Scafell Pike, Esk Pike, Helvellyn, Dollywagon Pike, Swirl How, Dunnerdale Fell, The Old Man of Coniston, High Street and Kentmere Pike. To the east and south the scenery flattens slightly, over towards Burton Fell, Whinfell Beacon, Great Burney and finally to the beginning of the Yorkshire Dales.

Turning south-easterly, Mudge and Colby saw Pendle Hill and Fair Snape Fell in Lancashire, Darwen and Winter Hill in the West Pennine Moors. In the southern background, they made out the young holiday resort of Blackpool and, clear in the foreground, the nearby south Cumberland hamlet of Barrow-in-Furness. Within seventy years, it would alter beyond all recognition into a major shipbuilding centre and home to the world's largest steelworks, but at that point in time Barrow-in-Furness nestled quietly on a peninsula beneath Duddon Sands. Far in the background, south to south-west, were the beaches and peaks of North Wales: Prestatyn, Gorsedd Bran, Llandudno, Moel Eilio, Bwlch Mawr, Moel Penamnen, Snowdon, Yr Eifl, a roll call of beauty that stretched all the way to Anglesey and its peak, Mynydd Eilian. And far, far in the distance, was that Ireland's east coast and Ulster's highest mountain, Slieve Donard? Or just a wisp of cloud? William Mudge had spent the last seventeen years looking at old maps of Britain and making new ones to better them. From the top of Black Combe it was as if all those maps had risen before him simultaneously in glorious three-dimensional technicolour.

THREE YEARS LATER, in the summer of 1811, Bootle received another visitor. When William Wordsworth arrived in the village, he was forty-one.

His poetry seemed to have temporarily dried up and his personal life was in tatters. He had fallen out with his great friend Samuel Taylor Coleridge the previous summer and there was no sign of a reconciliation. His family had recently moved house and at the same time his two children had caught severe bouts of whooping cough. The symptoms persisted, so William's wife Mary had suggested a summer holiday by the sea, where the fresh air might clear the children's lungs. The family found themselves in Bootle.

The holiday was initially a disaster. Eight years previously, Wordsworth's brother John had been given command of an East India Company ship, *The Abergavenny*, on a voyage to Bengal and China. The ship had struck a reef and foundered only two hours after leaving the Isle of Portland, on the Dorset coast, killing 250 men on board, including John. The whole Wordsworth family had invested a great deal in the voyage in hopes that the profits would enable William to continue writing. The tragedy had left him with a huge sense of guilt, ambivalence towards his own career and a debilitating fear of the sea. But Bootle's proximity to water was unavoidable. To the settlement's north-west, west, south-west and south, there was nothing to set eyes on but the Irish Sea, nothing to hear but 'Ocean's ceaseless roar', as Wordsworth dismally put it. To the north and east of this 'bleakest point of Cumbria's shore', his eyes met a wall of black rock: Corney Fell, Waberthwaite Fell and 'grim neighbour! huge Black Combe'. South-west Cumberland is an exposed, dramatic, but stark and rather unpicturesque area of the Lake District. It was a world apart from the Wordsworths' home in Grasmere, among the region's warmer, greener peaks: 'the loveliest spot that man hath ever found'. Wordsworth felt the change of scene cruelly. And his mood was not helped by the terrible weather that summer, which battered the region with biting winds and rain and planted a permanent dark cloud over the summit of Black Combe.

Wordsworth spent much of his time in Bootle indoors. He wrote a peevish letter in poetry to his friend and patron Sir George Beaumont, complaining that, despite being on holiday, 'rough is the time'. He stared out of his window at the huge black bulk of the mountain that thwarted the sun. And he worked fitfully on his intended magnum opus, *The Excursion*. One morning, however, he was cheered by a visit from Bootle's rector, the Reverend Dr James Satterthwaite, whose vicarage was just around the corner from the

Wordsworths' melancholy cottage. Satterthwaite happened to be a friend of William's brother Christopher, and he was a fellow of a college in the University of Cambridge, Wordsworth's own alma mater. Satterthwaite was a welcome presence and full of local anecdotes. He thought Wordsworth might be interested to hear about the rarely visible view from the top of Black Combe and the map-makers who had visited Bootle three years earlier and revelled in its panorama.

Satterthwaite was right. Wordsworth was well acquainted with the progress of the Ordnance Survey through England and Wales in the 1790s, in whose footsteps he had stepped during many of his own wanderings. He had already referred to Mudge in print as 'the best authority' on the Lake District's complex, dramatic scenery. But there was something in Satterthwaite's story that particularly enthralled the poet and inspired him to sit down and compose two poems that told the tale of Mudge's ascent up Black Combe. 'Inscription: Written with a Slate Pencil on a Stone, on the Side of the Mountain of Black Combe' and 'View from the Top of Black Combe' were both published in a two-volume collection of poetry that appeared in 1815, four years after his visit to Bootle. Wordsworth's imagination was transfixed by the thought of the wonderful panorama from Black Combe's top. In one poem, he wrote of how,

> from the summit of Black Combe (dread name
> Derived from clouds and storms!) the amplest range
> Of unobstructed prospect may be seen
> That British ground commands.

He described how the observer's eyesight leapt from 'low dusky tracts, Where Trent is nursed, far southward!' to

> . . . Cambrian hills
> To the south-west, a multitudinous show;
> And, in a line of eye-sight linked with these,
> The hoary peaks of Scotland . . .

And finally Wordsworth imagined Mudge's speechless delight upon discovering that not only 'Mona's Isle', the Isle of Man, was visible from Black Combe's summit but even 'the line of Erin's coast', Ireland, too. (Mona's Isle sometimes refers to Ynys Môn, Anglesey, but was used here by Wordsworth to designate the Isle of Man.) In his guidebook to the Lake District, Wordsworth elaborated on this poetic description by recounting how 'that experienced observer' William Mudge declared that 'the solitary Mountain Black Combe ... commands a more extensive view than any point in Britain. Ireland, he saw more than once, but not when the sun was above the horizon.'

The panoramic vista from the top of Black Combe was not just a breathtaking and sublime experience: it provided a revelation of the entire United Kingdom – Scotland, England, Wales and Ireland – all at once. The mountains of Scotland merged with the Lake District's peaks, the West Pennines and Lancashire coast dissolved into the mountains and beaches of North Wales, and Ireland was a phantasmagoric haze melting seamlessly into the Isle of Man. None of the regional tensions that had rocked the nation through the eighteenth and early nineteenth centuries were remotely evident in that vast prospect. There were no traces of the Jacobite rebellions that had followed the Anglo-Scottish Union of 1707 or of the intense disturbances around the Act of Union that had joined Britain and Ireland together in 1801. In his Black Combe poems Wordsworth exploited the political significance of the view from the mountain to the full. He described the panorama as 'a revelation infinite', a 'grand terraqueous spectacle,/ From centre to circumference, unveiled!' And his poem rose to a nationalist crescendo: the view from the top of Black Combe was a utopian vision of a happily united kingdom, a 'display august of man's inheritance,/ Of Britain's calm felicity and power!'

WORDSWORTH'S POETIC RESPONSE to Mudge and Colby's triangulation from Black Combe is a revealing example of the way in which cartography excited the cultural imagination in the early years of the

Ordnance Survey. The maps inspired emotional and imaginative responses in their readers, then as now. Mentions of maps and surveys in British novels, poems and plays increased dramatically during the first thirty years of the Ordnance Survey's existence. Britons had become 'map-minded' over the eighteenth century, and maps were popular subjects for embroidery and board games. In 1787 'A New Geographical Pastime for England and Wales . . . a very amusing Game to play with a Teetotum, Ivory Pillars and Counters' was released, in which players made their way across a map of England and Wales, travelling on lines that ricocheted from town to town and closely resembled a nationwide triangulation. Such public fascination with cartography continued through the following decades. The popularity of maps in British culture during the early years of the Ordnance Survey had a variety of causes, notably Britain's entry into a prolonged war that was fought on an international stage and was heavily dependent on maps both as tactical military tools and as illustrative aids used to describe the conflict to the reading public. The upsurge in 'literary maps' in this period was also indebted to the Enlightenment's celebration of cartography as the language of reason and political equality. The fact that knowledge of the Ordnance Survey's activities was easily available to the public in newspapers and journals, which covered them in enthusiastic detail, and in its map-makers' own publications and surveys helped to bring cartography to the attention of those engaged in artistic and literary creation.

This is not to deny that maps had a noteworthy place in English culture before the Enlightenment. But in the eighteenth and early nineteenth centuries, they acquired specific contemporary associations. First and foremost, writers associated maps with the military and particular martial events. In James Fennell's comic play *Lindor and Clara: or, The British Officer* (1791) the soldier Firelock was introduced 'reading, with a Map before him, a Musket lying near him'. In his long poem *The Excursion* (1814) Wordsworth described the enrolment of a young shepherd called Oswald in the British Army during the Napoleonic Wars, imagining how 'at some leisure hour' he 'stretched on the grass . . . / Among his fellows, while an ample map/ Before their eyes lay carefully outspread'. Thomas Dermody's poem 'The Invalid' (1800) told the other side of the story and pictured a badly injured soldier returning from the

Continental theatre of war, who referred to a 'shatter'd map' to explain the cause of his 'maim'd stump'. Some writers in this period used maps as the insignia of 'lost lovers' (as the twenty-first-century novelist Nadeem Aslam put it). Thomas Tickell's poem *An Epistle* (1810) told the story of a woman who became 'a mere geographer by love' while estranged from her soldier boyfriend, and 'taught her finger' to 'stray . . . o'er the map' in order to 'span the distance that between us lies'.

The military associations of maps also made their way into literature in more subtle ways. In the late seventeenth and eighteenth centuries, it became popular in poetry and painting to describe the landscape as if from an elevated viewpoint. This trend was known as 'prospect' poetry or painting, and its proponents often linked the physical elevation offered by a hill with moral superiority. For example, in 1726 John Dyer composed a poem called 'Grongar Hill' in which the speaker ascended that peak and, at the same time as enjoying its panoramic view, offered the moralising reflection that 'Hope's deluding glass' can cause one to 'mistake the Future's face'. The idea that a figure becomes more morally or physically powerful by attaining a 'commanding height' (a standard phrase in prospect poetry) arguably derived from military reconnaissance and surveying, whereby map-makers' elevation above the ground gave them a very real power over the subject landscape and its population. Eighteenth-century poetry is full of descriptions of the panoramic views that are available from various summits across Europe, views which are often explicitly compared to maps. In 1735 the poet John Hughes described a vista over '*Hertford's* ancient town' and its surrounding 'flow'ry Vales, and moistened Meads' that 'far around in beauteous Prospect spreads/ Her Map of Plenty all below'. This notion made its way into the vocabulary of the reading public, and when one young nobleman went walking in the Lake District in 1770, he described in his journal the view from the summit of Skiddaw. He recounted that he could see 'all the coast of Galloway – the Solway Firth, Carlisle, Cross fell in Yorkshire – Workington – Cocker mouth and Wigton – Here under your feet is the Lake upon which you were – it is like a map'.

The military associations of surveying were reflected in the cultural sphere, and so too was the idea that cartography was the language of Enlightenment.

Many writers in the eighteenth century used maps as emblems of rationality and republicanism. In her novel *The Young Philosopher* Charlotte Smith pictured a group of characters consulting a world atlas, from which they derived the message that 'true philanthropy does not consist in loving John, and Thomas, and George, and James, because they are our brothers, our cousins, our neighbours, our countrymen, but in benevolence to the whole human race'. A year after the publication of the Ordnance Survey's first map in 1801, a play called *Americana* was published by an anonymous author. In it, one character functioned as an overt emblem of 'genuine liberty' and 'the ever venerated voice of freedom' and she spoke in the language of Enlightenment cartography, describing how 'Europe's continent' was 'spread out like some extensive map beneath'. (However, many writers associated maps with royalism and privilege. The dramatist William Henry Ireland wrote a play called *Henry II: An Historical Drama* (1799), in which the King pointed out on a 'poor dwindled map' the outline of England 'when Harry First, my grandsire, reign'd'.)

As visual images, there were often close correspondences between maps and landscape paintings. Most importantly, both could depend on geometry, which was most commonly manifested in art as perspective. Based on methods devised by the Renaissance painter Leon Battista Alberti, the encyclopaedist Ephraim Chambers defined perspective as 'the art of delineating visible objects on a plain surface' and the rules by which 'the points *a*, *b*, *c*, &c. may be found geometrically'. The technique was used by many artists in history to create accurate representations of space and it often closely resembled mapping, especially the techniques of plane-tabling and triangulation. Perspective drawings of the landscape and these surveying methods both created geometric abstractions, and perspective was embraced by many proponents of the Enlightenment. Perspective was also central to the eighteenth-century school of neoclassical art, which rejected the flourishes of Gothic, baroque, and rococo design in favour of the purity and idealism of ancient classical models. William Mudge's family friend Joshua Reynolds was a childhood enthusiast of perspective and as an adult he came to formulate a neoclassical 'grand style' of painting that demonstrated such ideals alongside an Enlightenment penchant for abstraction. Reynolds wanted to 'reduce the idea of beauty to general principles', and his friend Edmund Burke described how Reynolds's habit 'of reducing every thing to one system' derived from his 'early acquaintance' with Mudge's grandfather, '[Zachariah] Mudge of Exeter, a very learned and thinking man, and much inclined to philosophize in the spirit of the Platonists'. In theory the Ordnance Survey therefore had much in common with neo-classicism, one of the most influential styles of art in the eighteenth century. Mudge and Reynolds were not only personally linked, they both loved and depended on geometry and dedicated their lives to producing abstractions of the world, in maps and paintings.

The late-eighteenth- and early-nineteenth-century movement known as Romanticism was ambivalent and sometimes downright hostile to the Enlightenment's overriding emphasis on reason and its ambition to discover the natural world's universal laws. It is noteworthy that many writers voiced their discontent with attacks on maps. Despite his interest in the Ordnance Survey, in his long poem *The Prelude* Wordsworth was highly critical of the 'rational education' that was forced on children of the Enlightenment and which idolised 'telescopes, and crucibles, and maps'. Coleridge too felt that an upbringing that emphasised reason above all else led to the death of the imagination. Such children were 'marked by a microscopic acuteness', he explained. 'But when they looked at great things, all became a blank and they saw nothing.' Even Wordsworth's description of the Ordnance Survey's map-making in his Black Combe poems was ultimately ambivalent. At the end of his 'Inscription: Written with a Slate Pencil on a Stone, on the Side of the Mountain of Black Combe', Wordsworth imagined Mudge's project being thwarted by a sudden 'unthreatened, unproclaimed' onset of thick fog, which blinded the surveyor 'as if the golden day itself had been/ Extinguished in a moment' – the Enlightenment's ambitions foiled by a simple cloud.

The most outspoken attack on Enlightenment cartography in the Romantic period came from the radical poet, painter and printmaker William Blake. For Blake, the *'esprit géométrique'* that defined the Enlightenment had the odious result of enslaving the human mind to reason and universal laws. Blake's poetry and engravings prominently featured the malevolent character of Urizen, who was enchained to such rational authority, and who was repeatedly described with surveying instruments in his hands. Blake's watercolour and relief etching *The Ancient of Days* (1794) showed Urizen leaning down from the heavens and wielding a pair of dividers in threatening fashion. Blake also confronted the very hero of the British Enlightenment, Isaac Newton. Repulsed by his rational, mathematical, mechanical view of the cosmos, Blake created an image (*Newton*, 1795/*c.*1805) in which the minute idiosyncrasies of the rocks on which Newton sits resist the homogenising effect of his universal laws.

Blake also directed much of his anger about the enslavement of the imagination to reason onto the painter Joshua Reynolds. He annotated his edition of Reynolds's *Discourses on Art* with savagely critical assaults. When Reynolds praised 'truth' and 'geometry', Blake wrote angrily: 'God forbid that Truth should be Confined to Mathematical Demonstration'. He hated Reynolds's view that the 'disposition to abstractions, to generalizing and classification, is the great glory of the human mind' and scribbled beside it: 'To Generalize is to be an Idiot[.] To Particularize is the Alone [*sic*] Distinction of Merit – General Knowledges are those Knowledges that Idiots possess.' Blake incriminated William Mudge's family, too, in his dislike of Reynolds, inscribing 'Villainy' beside a mention of 'old [Zachariah] Mudge' in the *Discourses*. For Blake, it seems that both men were tarred with the same brush of the tyranny of geometric reason.

But cartography was not entirely written off by the Romantic movement. Not at all. Many writers defended the importance of the imagination and emotions against Enlightenment reason, and some re-appropriated maps as images of these faculties. Despite his loathing for the Ordnance Survey's type of geometric mapping, William Blake was nevertheless willing to assist when a friend called Benjamin Heath Malkin approached him to produce a map to illustrate a heart-wrenching volume of memoirs about his son Thomas's extraordinary life and premature death. Thomas Williams Malkin had been a child prodigy, one of whose accomplishments had been to devise at the age of five an entirely 'visionary country, called Allestone, which was so strongly impressed on his own mind, as to enable him to convey an intelligible and lively transcript of it in description'. Malkin had formulated an imaginary history, geography and economy for Allestone, and he had also composed a map of his fantasy kingdom, 'giving names of his own invention to the principal mountains, rivers, cities, seaports, villages, and trading towns'.

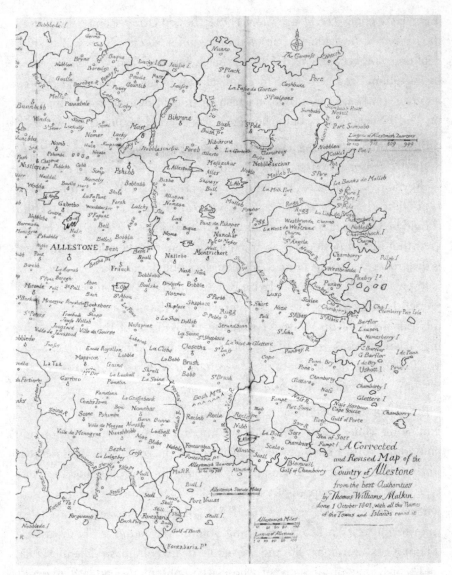

28. 'A Corrected and Revised Map of the Country of Allestone' by William Blake, illustrating the imaginary country of Allestone: detail.

Benjamin Heath Malkin asked Blake to engrave a map of Allestone for his *Father's Memoirs of his Child* (1806). Blake agreed, and his resulting map was described as 'a very remarkable production' that was crowded 'with names, some absurd, but others ingenious and appropriate'. His map was devoid of the type of geometry that characterised maps like the Ordnance Survey's and it was made almost solely of minutely detailed, idiosyncratic place names. It was described as '*an exercise of the mind*', a map of the imagination instead of reason.

ONE SUNDAY MORNING, on 1 August 1802, Samuel Taylor Coleridge 'Quitted [his] house' and set off on a nine-day tour of the Lake District to seek freedom and relief in the landscape from the oppressive atmosphere of his home at Greta Hall in Keswick. He rejected the dominant modes of tourist travel: a coach or horse, and the road. Coleridge was one of Britain's first habitual pedestrian explorers, writing exultantly in his notebook that 'every man [is] his own path-maker – skip & jump – where rushes grew, a man may go'. Motivated variously by the expense of road travel and by Romanticism's praise of solitude, improvisatory wandering and independ- ence, combined with the desire to rebuff the status quo, by the 1790s more and more tourists could be found straying off-road, on foot, over the British landscape. Although the idea of hiking did not properly take off until the late nineteenth century, by the 1810s guidebooks routinely included advice for ramblers, and in 1824 the first footpath preservation society was formed. But when Coleridge set off on his tramp in 1802, there were serious shortcomings in the navigational aids available to him. Human guides had increasingly fallen out of fashion as the eighteenth century progressed; they were expen- sive and an obstacle to ideals of solitude. Many tourists turned to texts for navigational guidance instead, but road-books like Ogilby's and Paterson's trained their users' eyes only on the road ahead, and guidebooks tended to pro- vide travellers with fixed itineraries around a series of principal sights. Other than advice procured from locals, there was little that was designed to allow the rambler to roam free and at will over the landscape without getting lost.

So Coleridge made himself a map. He took a small pocket-book on his hiking tour, and on the first page he drew a chart of the mountainous, craggy countryside over which he planned to wander. (This was probably copied at home from a guidebook or atlas.) Throughout his ramble, Coleridge continued to fill the pages of his notebook with cartographic representations and outline sketches of the western Lake District. On Thursday 5 August, Coleridge walked to Wast Water: 'O What a Lake,' he cried. Describing how at 'the top of the Lake two huge Fells face each other, Scafell on right, Yeabarrow on the left – and between these Great Ga[ble] intervenes, the head & Center-point of the Vale', Coleridge concluded that 'it is impossible to conceive it without a Drawing'. He duly sketched a map. But Coleridge's chart inverted reality. The enormous scree slopes that slant down, in life, to

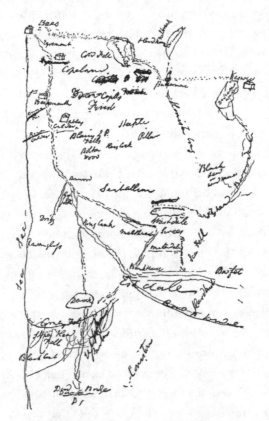

29. A map of the western Lake District that Coleridge drew during his hike there in August 1802.

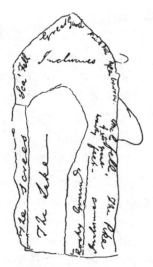

30. Coleridge's map of Wast Water: 'O What a Lake'.

Wast Water's east side – 'steep as the meal out of the Miller's grinding Trough or Spout', in Coleridge's words – appeared in his map on the lake's western bank. Every other detail was likewise turned topsy-turvy. But this mirror-image map was not just a whim or folly: it paralleled the looking-glass that Coleridge found in Wast Water itself. 'When I first came,' he wrote, 'the Lake was like a mirror, & conceive what the reflections must have been . . . all this reflected, turned in Pillars, & a whole new-world of Images, in the water.' His map was a personal record of one man's imaginative experience of and response to a particular landscape, at a specific moment in time.

Coleridge's use of maps during his 1802 exploration was prophetic. He demonstrated how invaluable maps were to travellers who did not want to be coupled to a guide, confined to a set route or chained to the road. A map presented the rambler with an image of the expansive terrain over which he or she could roam. As the nineteenth century progressed, guidebooks increasingly incorporated fold-out maps to assist this new leisure activity of rambling. But Ordnance Survey maps were too expensive and precious to be used in this way by a general readership, and it was not until the early twentieth century that its then director general Charles Close dramatically increased sales among hikers by producing a cheap series of folding maps with stylish, specially designed covers by Ellis Martin. This tapped into the new appreciation of the British landscape in the interwar period. James Walker Tucker's 1936 painting *Hiking*, which shows three young women in berets and shorts, eagerly consulting an Ordnance Survey map, testifies to that institution's central role in the explosion of rambling. But the importance of maps to hiking can be traced all the way back to the earliest decades of the Ordnance Survey's existence, when the British public first started to become truly 'map-minded' and discovered that maps could assist their hunger for freedom and fresh air.

The Enlightenment adopted maps as emblems of reason, but it is the power of maps to give shape to the desire and the imagination that made them so seductive to Coleridge and attracts so many readers today. The love of maps can work itself into the deep recesses of the psyche. Cartography is a language and one can easily start hallucinating the landscape as a map, seeing real hills imprinted with imaginary contours and vast plains traversed by visionary lines of triangulation. And vice versa, from those same icons on the map-sheet itself, one's imagination can conjure up a seemingly living scene. The night before a long walk or run, I love to spread the map out on the table and trace my route with my finger, conjuring in my mind's eye the rises and falls over which the path will ride, and anticipating my fatigue at the top of the bunched contour lines and my relief at the descent on the other side. On my return from overseas holidays, my stepfather will leap up from the sofa and grab his copy of the *Times Atlas of the World* and we'll sit poring over the figures and lines that cartographically recall recently visited places. It was the same in the early years of the Ordnance Survey. Many map-readers would dream over charts of far-off places from their armchairs, and maps particularly stimulated children's fantasy worlds. Coleridge's son Hartley devised the kingdom of 'Ejuxria', to which he disappeared in his mind's eye when pining for his often-absent father, and he drew a detailed fantastical map of the make-believe retreat. The connection between cartography and reason was undeniably powerful, but so too was the capacity of maps to give shape to dreams, and it was not only Wordsworth who expressly thought of himself as 'dreaming o'er the map of things'.

CHAPTER NINE

The French Disconnection

IN SEPTEMBER 1811 William Mudge was mortified to find himself hauled before a Commission of Military Enquiry that had received an anonymous letter with the allegation that the Ordnance Survey's costs were exceeding £10,000 a year. Mudge complained to Colby: 'I cannot tell whether they imagine me to be a rogue or not, but they deal with me exactly as if they thought so. They gave me not a moment's notice scarcely for preparation, taking me literally at the *ground hop* on my return . . . I believe they thought some great secrets were hidden and that torture would be necessary to find them out.' He resolved to 'appear before them rather with the hope of being allowed to assist them in [their] enquiries than as a subject for examination'. Indeed, he had little of which to be ashamed. By the end of 1809, both the primary and secondary triangulations had been completed throughout England and Wales, bar Norfolk, Suffolk and Lincolnshire. The trigonometrical skeleton on which the First Series of Ordnance Survey maps depended was very nearly completed and Mudge had the documentation to prove it was a cost-effective enterprise. Since his appointment to the Ordnance Survey in 1791, this parsimonious man had kept account of every single letter whose postage he had charged to the Board of Ordnance, proving that 'from his entrance into office until the day of his decease, he had never placed the postage of a single private letter to the public account'.

The Commissioners examined Mudge 'on all the essentials of the [Ordnance Survey's] operation, and finally asked the pointed question of its utility and continuation'. Mudge later described to Colby how he 'did [his] best to satisfy them in the first point' and explained how 'little time would be required' to complete the Trigonometrical Survey. He stated his intention to take the triangulation into Scotland the following year, the completion of which he estimated 'would not take more than 5 years', and he presented the Commissioners with 'the *Total* expence of the Trigonometrical Survey' over twenty-one years, which 'was stated at £21,000'. When they demanded more figures, Mudge reported: '*instantly* they had these further sums given', and he described how the total cost of the entire project had reached £54,165 5s 7d, amounting to a average annual expense over twenty years of around £2708. 'Everybody I have shown it to thinks the sum a fleabite,' he confided to Colby, asserting rather defiantly: 'I believe I have built my house upon a rock' and 'I am conscious that, to the best of my abilities, mental and corporeal, I have discharged every trust deputed to me in this undertaking like an honest man.' Mudge was obsessively conscientious and later he would write that 'it is not my desire to have more of the public money in my Hands than is actually wanted'. But in private he was racked with self-doubt and even entertained the prospect of his immediate dismissal. 'I have shortly to look back on the long dream of 20 years,' he wrote mournfully to his family, 'and at the time I exclaim, in truth how has it flown, with the mortification to know that I have toiled to every purpose but that of growing rich.'

In the event, the Commissioners were said to be 'strongly impressed with an idea of the extreme accuracy of the [Ordnance Survey's] work', and few adverse consequences arose from the Enquiry. However, in the name of frugality they did put a stop to the publication of future volumes of the *Account of the Trigonometrical Survey*, despite the fact that the existing books were said to be 'approved of and sell[ing] well' and the third volume was then in preparation. But the Military Enquiry could not have happened at a worse time for Mudge. In that year, he felt embattled and became consumed by a depression that would stay with him, on and off, for almost the next decade. Back in 1809 he had been delighted when the Royal Military Academy in Woolwich, his alma mater, had offered him the position of Lieutenant Governor.

Also in 1809, the East India Company had founded a training academy at Addiscombe in Surrey for cadets specialising in engineering and artillery, and had asked Mudge to take a prominent role as public examiner. He had proudly accepted both invitations, but the added responsibility on top of superintending the Ordnance Survey began to take a severe toll on Mudge's health, spirits and general happiness. By 1811, this industrious map-maker felt himself to be much put upon and wrote dejectedly: 'my labours are great and I am without strength to carry my chains. I can assure you that I am a slave, and not wearing a golden chain.' He elaborated: 'I have more business on my hands than I have strength for, or, if I had strength, even time to perform.'

There were further reasons behind Mudge's depression. In 1810 Henry Phipps, 1st Earl of Mulgrave, had become the new Master-General of the Board of Ordnance. Phipps had been an early member of the Royal Society's dining club and he may have fondly remembered conversations with William Roy about the prospect of instigating a national map back in the 1770s. In the second year of Phipps's appointment, Britain had been at war with France for nearly twenty years and he decided that publication of the Ordnance Survey's maps was too great a security threat, as Britain and France clashed during the Peninsular War. Following a command 'to withhold every map from the public', given on 2 September 1811, the same day as Mudge's appearance before the Commission, the surveys were withdrawn from sale. The maps would not be made available again to the general public until long after the Battle of Waterloo and the end of the Napoleonic Wars, in the spring of 1816.

Mudge was dismayed by Phipps's order. He predicted negative consequences for the quality of the surveys that were to be made during this period of prohibition and he lamented that the public would be deprived of their right to information about their nation's geography. He also disliked an insinuation that he himself might not be entirely trustworthy. Until then early drafts and proofs of the Ordnance Survey's maps had circulated freely among Mudge and Colby, the Interior Surveyors, the engravers and those involved in their revision, but now Mudge was instructed to ensure 'that maps for correction were to be kept with the utmost privacy, and when corrected to be returned to the Tower'. It seemed possible that Mudge himself might even have to apply for permission to consult the maps for whose creation he was largely responsible.

On top of all these worries, Mudge could not even take comfort in the knowledge of the Ordnance Survey's unprecedented excellence. A few years earlier Thomas Hannaford Hurd had been appointed hydrographer to the Admiralty after Alexander Dalrymple's death. Taking to task the quality of coastal charts of Britain, Hurd had discovered significant errors in the Ordnance Survey's calculations of certain longitudinal and latitudinal positions. Mudge passed on Hurd's anxieties to Thomas Colby. 'Ponder the contents of his letter, and when you can speak decidedly on the question, let me hear from you,' he commanded. 'It is of consequence that Captain Hurd should be either proved wrong, or that the errata he has discovered may be acknowledged as such, and these errata put to rights.'

The hassles of 1811 had depressed Mudge. Writing to an officer engaged in the triangulation, he wrote mournfully: 'I believe that I have more difficulties thrown in my way as to the progress of the map-making by Ignorance, Avarice, and Cupidity than you have by the intervention of Mountains [and] Morasses.' The publication of the third volume of the doomed *Account of the Trigonometrical Survey* was also proving unexpectedly problematic, and on 27 July 1811 Mudge had reported how 'ten days have elapsed since I requested the Master General and Board permission to put the Account into the hands of Mr Faden without receiving any answer to my Letter. This silence augurs mischief! . . . I am quite dispirited and dismayed.' When the volume finally arrived on the shelves of bookshops and libraries, even this did not ease Mudge's anguish. It ignited a heated international dispute about the skill and accuracy of his observations that would only trouble him further.

THE THIRD VOLUME of the *Account of the Trigonometrical Survey* described that enterprise's progress between 1800 and 1809. But it also revealed how in that time the Ordnance Survey had not only been concerned with creating the First Series of maps of England and Wales. It had also become involved in a project that, strictly speaking, was a distraction from this central purpose, but which contributed to contemporary questions of geodesy. By taking part in such an endeavour, Mudge returned the Ordnance Survey to its roots in the Paris–Greenwich triangulation of the 1780s and he may have hoped to demonstrate that it was still as much a scientific as a military undertaking.

Since the 1790s, Mudge had been conducting observations of the night sky to ascertain the precise latitudinal and longitudinal locations of a series of key trig points, in order to confirm the calculations made by the Trigonometrical Survey. Longitude could be found by calculating the time difference from Greenwich, for which Mudge relied on one of his uncle Thomas Mudge's famous chronometers. In the earth's northern hemisphere, latitude can be found by establishing the difference in degrees between the zenith (the highest point above the observer's head) and the North or Pole Star, Polaris. Mudge's celestial observations of latitude were initially designed to act as a check on the accuracy of the Trigonometrical Survey's deductions, but towards the end of the 1790s he proposed the extension of these rather rudimentary observations into the measurement of an entire meridian arc through Britain, stretching beyond the length of one degree of latitude, which would be measured through a combination of triangulation and astronomy. This project had got under way in 1800 when Mudge recalled that, back in the early 1790s, Charles Lennox had ordered an instrument called a zenith sector from the notoriously dilatory instrument-maker Jesse Ramsden. Mudge chased up the order and Ramsden had duly 'proceeded with little interruption' on the instrument. In November 1800 Ramsden had died, after a long period of declining health. But Ramsden's principal workman, John Berge, inherited his business and finished the sector for Mudge in such 'a very masterly and accurate manner' that the Superintendent suspected that the instrument 'would not have been superior had the ingenious inventor lived to complete it'.

A zenith sector consists of a dead-straight central frame whose vertical position is established by a plumb line. A telescope pivots on either side of this frame and the difference of its angle from the vertical can be measured

31. The instrument – a zenith sector – used during the Ordnance Survey's meridian arc measurement.

in degrees on a scale at the frame's base. To ascertain his latitudinal position, the astronomer reclined underneath this long, elegant, but cumbersome instrument and trained the telescope's sights on the Pole Star, whereupon he recorded its angular difference from the vertical. A relatively simple calculation produced the angle of latitude from this observation. Upon Berge's completion of the zenith sector, Mudge had taken his new toy to the Royal Observatory at Greenwich. There the Astronomer Royal, Nevil Maskelyne, who was something of an expert in these instruments, had willingly proferred 'advice and instruction' about its use. His instrument completed, Mudge had then been left with the decision of where to position the meridian arc. Although he had desired 'the most extensive arc' possible through England, Mudge also knew that the nation's variable landscape posed significant problems: hills could cause plumb lines to deviate from the vertical, as Maskelyne's Schiehallion expedition had shown. So although an arc measured from Lyme in Dorset up into Scotland would have been very long indeed, the hilliness of the countryside through which it would pass had rendered it unviable. Instead Mudge had alighted on an arc stretching from Dunnose in the Isle of Wight up to Clifton, a small settlement about five miles south-west from the centre of Doncaster, positioned about 1° 13' west in longitude from Greenwich Observatory.

Between 6 June and 28 July 1801, Mudge had overseen the measurement of a baseline at Misterton Carr, at the most north-westerly point of Lincolnshire, near Epworth (where John and Charles Wesley, the founders of Methodism, were brought up). The base verified the accuracy of the triangles that proceeded along the length of the arc, from Dunnose to Highclere, Nuffield, Brill, Epwell, Corley, Bardon Hill, Sutton, Gringley and Clifton, many of which had already been measured for the Trigonometrical Survey. In May and June 1802, the zenith sector had been taken to Dunnose, and in July and August to Clifton too, to astronomically 'fix' the latitudes of the arc's ends. Upon his project's completion, Mudge had taken his zenith sector down to London to the Tower, where he had proudly shown it off to the President of the Royal Society. He had also taken this opportunity to exhibit 'any Topographical Material in his office, which Sir Joseph Banks might be desirous of examining'.

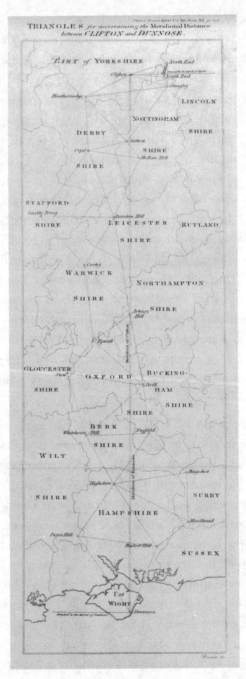

32. *The map of the triangles observed during the Ordnance Survey's project to measure a meridian arc.*

Mudge's procedures during the arc measurement had been seemingly scrupulous. He had even ensured that the temperature was uniform at the top and bottom of the long zenith instrument, by opening 'shutters in the roof, as well as the door of the observatory, a considerable time before the moment of observation'. By the end of the season, Mudge had been able to state 'how thoroughly satisfied [he] was' with the result: 'the length of a degree on the meridian . . . is 60820 fathoms', about sixty-nine miles. And he had been certain that the length 'of the Arc, cannot but be determined with extreme accuracy'. After the completion of the project in 1802, Mudge had written up his methods and results and had read them before the Royal Society on 23 June 1803. His article had been subsequently published in the 1803 issue of the *Philosophical Transactions* and then afterwards as a separate volume. When the third volume of the *Account of the Trigonometrical Survey* was published in 1811, it included this earlier description of the meridian arc measurement that had taken place in 1801 and 1802.

The reappearance of Mudge's article in the public domain provoked new interest in the endeavour, but not all of this attention was complimentary. On 4 June 1812 a paper by the Spanish geodesist Don Joseph Rodriguez, entitled 'Observations on the Measurement of Three Degrees of the Meridian, conducted in England by Lieut. Col. William Mudge', was read before the Royal Society. Rodriguez pointed out that the calculations that Mudge had obtained from his meridian arc measurement were controversial. Instead of confirming the accepted view that the earth was an oblate spheroid (i.e. flatter at the poles than at the equator), Rodriguez said that Mudge was suggesting the opposite, that the earth was an *oblong* spheroid, like an upright egg. Rodriguez voiced 'a suspicion of some incorrectness in the observations themselves, or in the method of calculating them'. He was displeased, too, that Mudge had not made his procedure entirely transparent. He 'had not informed us in his Memoir, what were the formulae which he employed in the computations of the meridian', Rodriguez pointed out, and he compared Mudge unfavourably to his French counterpart, the director of the metre project, Jean Baptiste Joseph Delambre, who had 'published and explained, with admirable perspicuity and elegance, all the formulae and methods' of that undertaking.

Thomas Colby had joined the Ordnance Survey in 1802, in the midst of the meridian arc measurement. When the furore over Mudge's calculations erupted in 1812, he proposed extending the arc from its termination near Doncaster up to the north coast of Scotland. He argued that a longer arc would provide more information about the earth's shape over a larger area and might illuminate the confusion behind the earlier results. And as Mudge had described to the Commissioners in 1811, there were already plans to take the Trigonometrical Survey into Scotland, so the meridian arc extension could be measured simultaneously. In 1813, therefore, Colby set out for Scotland with a small team of assistants, to extend the arc to the northern-most edge of the British mainland.

Scotland presented more problems for the map-makers than England and Wales combined. Robert Kearsley Dawson, the son of the great Interior Surveyor Robert Dawson, was a member of Colby's team and he recalled how 'it was no uncommon occurrence for the camp to be enveloped in clouds for several weeks together, without affording even a glimpse of the sun or of the clear sky during the whole period'. The top of Ben Nevis was 'almost constantly covered with mist, or deluged with rain'. In battering storms, and over sodden and bumpy ground, the soldiers attempted to push carts weighed down by surveying instruments, papers, tents and clothes, 'by the application of guy ropes to support them, and with the men's shoulders to the wheels'. During reconnaissance marches, the map-makers' 'resting places were often miserable hovels, and their only food the porridge of the country'. Even when successfully encamped, the tents frequently blew down during the night. At one camp, the nearest town where food could be procured and letters received and sent was twenty-four miles away. At times, when the rain set in, and the horizon hadn't been seen for days, and the triangulation had come to a halt, it must have seemed as if it was all for nothing.

These hardships did not bother Thomas Colby. The resilient young man had grown up into an extremely tough 35-year-old map-maker. He travelled up to Scotland from London 'on the mail coach' and throughout five continuous days and nights of jolting, swaying, clattering, relentless motion, Colby took a rest of 'only a single day at Edinburgh' before continuing his journey north. For the length of the drive, he shunned the relative warmth of the coach's interior for a seat on the outside. It was said that 'neither rain nor snow, nor any degree of severity in the weather, would induce him to take an inside seat, or to tie a shawl round his throat; but, muffled in a thick box coat, and with his servant Fraser, an old artilleryman, by his side, he would pursue his journey for days and nights together, with but little refreshment, and that of the plainest kind'. Upon arrival at his soldiers' camp, Colby slept fully clothed, resting only 'on a bundle of tent linings'.

Unhappily for his map-makers and assistants, Thomas Colby demanded the same hardiness from them. At first Dawson slept beside Colby, following his example, before realising that a colleague in a different tent had 'put up his camp bedstead, and made himself much more comfortable – a lesson which I did not fail to profit by in my after experience'. But the most gruelling aspect of the surveyors' time in Scotland was the extraordinary distances covered on foot. Relying only on a pocket compass and map and his own intuition, Colby eschewed paths and roads in favour of leading his team at a great rate, and as the crow flies, across 'several beautiful glens, wading the streams which flowed through them, and regardless of all difficulties that were not absolutely insurmountable on foot'. We can imagine the surveyors struggling to keep up with the small, wiry map-maker over ragged grass and clutching gorse. During reconnaissance marches, Colby and his men would average thirty-nine miles a day. Over more prolonged periods of triangulation, he relaxed the pace to a mere twenty-seven miles. In one record-breaking trip through the tricky Highland countryside that lasted twenty-two days, Colby covered a magnificent 586 miles. Even Sundays were not guaranteed days off. On 25 July 1819 Colby led his men in search of a church. Failing to find one, he decided they should use the time to climb the 4049-foot mountain of Aonach Beag, near Ben Nevis. This 'should have been our day of rest', one hapless surveyor bemoaned.

But one particular peak proved too much even for Colby. The Inaccessible Pinnacle that marks the summit of the mountain of Sgurr Dearg in the Cuillins, on the Isle of Skye, is a flat, slippery fin of basalt rock and is only possible for experienced climbers – it presents a notorious challenge for Munro-baggers. Having neglected to bring ladders or ropes, Colby nevertheless tried to reach the summit, but was 'completely foiled in the attempt'.

Indeed, Dawson commented, this was 'probably the only instance in which Captain Colby was ever so foiled'. We can imagine the map-maker's compact body making its way a few feet up the rock, before his hand lost its grip and he slithered back down. The surveyors were left 'admiring for a while the magnificence of the prospect, and the dreary and all but chaotic scene around us', before returning 'to our inn, gratified above measure with what we had seen, though disconcerted with our professional failure'.

There were consolations for the interminably hard work of the Ordnance Survey in Scotland. Sometimes the clouds 'would break away or subside into the valleys, leaving the tips of the mountains clear and bright above an ocean of mist, and the atmosphere calm and steady, so as to admit of the observations for which the party had waited days and weeks'. And working for Colby was an exhilarating and rewarding experience. He took the engineers under his command seriously, and spent time and effort imparting 'a knowledge of Ramsden's great three-foot theodolite, and of its adjustments, and also of the mode of working and entering the computations; explaining still further the position and names of the principal mountains and trigonometrical stations within the range of observation'.

Colby pushed those under his command to their mental and physical limits, and in return they loved him and grew proud of themselves. On the first day of a triangulation excursion, after a thirteen-mile trek on foot in the morning, Colby informed Dawson that 'Garviemore Inn, distant eighteen miles by the road, was to be our next stage'. 'I really thought it was more than I could possibly accomplish that day,' Dawson recalled, 'but Captain Colby said it was not.' After persuading Dawson of his own ability, Colby added a deviation to their route, over 'a rough boggy tract of country', to build a cairn at the top of a nearby 3500-foot mountain. This extended the thirty-one-mile expedition by another nine miles. The ailing Dawson asked to walk along the road, but Colby refused and Dawson recorded: 'I had no alternative but to make the attempt, feeling sure that I should eventually be left upon the ground or carried home upon the men's shoulders.' But, he concluded happily, Colby 'was right. I kept pace with him throughout the remainder of the day, and arrived at the inn at half-past eleven o'clock at night, much more fresh than at the end of our first stage the day before.'

Undoubtedly a slave-driver, Thomas Colby also knew when to relent and allow his team to revel in their achievements. At the end of a surveying season of nearly four months' duration in Scotland, he gave his map-makers 'carte blanc[h]e to provide themselves with a farewell feast'. Dawson explained that 'the chief dish on such occasion was an enormous plum-pudding'. Consisting of one pound each of raisins, currants, suet and flour,

> those quantities were all multiplied by the number of months in camp, and the result was a pudding of nearly a hundred pounds weight. Every camp kettle was in requisition for mixing the ingredients – some breadths of canvas tent-lining were converted into a pudding-cloth, - a large brewing-copper was borrowed to boil it in, - the pudding was suspended by a cord from a cross-beam to prevent its burning, and it was kept boiling for four and twenty hours – a relief of men being appointed to watch the fire and maintain a constant supply of boiling water.

When the giant pudding was ready, 'a long table was spread in three of the marquees' with 'seats being placed also for Colby'. All the map-makers and assistants 'partook of the pudding, which was excellent', and toasted '*Success to the Trig*'.

The press enthusiastically followed Colby and his team's extraordinary endeavour. On 6 July 1815 the *Caledonian Mercury* explained the function and appearance of trig points to a bewildered readership: 'In the course of this very important survey, the points from which the different angles are set off, are marked on the tops of the most prominent hills, by the erection of a pillar of loose stones, staves, or other objects, upon their summits, by which their position and distances from each other, and their height above the level of the sea, are ascertained with the greatest precision.' The journalist sought to persuade his readers of 'the obvious utility of preserving these marks' and 'to intreat that landowners will particularly call the attention of their tenants to the subject, and instruct them to give the necessary orders to their servants'. The Ordnance Survey was gradually being embraced as part of the national heritage.

AFTER THE ORDNANCE SURVEY'S meridian arc measurement through Scotland had been completed, William Mudge received a letter from the French Catalan mathematician François Jean Dominique Arago. By the age of thirty-three, Arago had already been elected an astronomer in the Paris Observatory, a member of the Académie des Sciences, a Commissioner on the Bureau des Longitudes (the French counterpart to the Board of Longitude) and the Chair of Analytical Geometry at the École Polytechnique (and he would even become France's head of state, briefly, in 1848).

In October 1816 Arago wrote to Mudge to propose that the meridian arc that the French geodesists Méchain and Delambre had measured through France and Spain, as part of the project to devise the 'metre' unit of measurement, should now be extended into Britain, to Great Yarmouth in Norfolk. 'This operation,' he wrote, referring to the metre project, is 'equally remarkable for its extent, for its object, and for its precision.' An extension to Great Yarmouth would 'have the more precious advantage of belonging equally to England as to France, and would become one day perhaps the basis of a uniform system of weights and measures' throughout the world: the metric system. Within nine months, Arago's proposition had become even more ambitious. He was excited by the knowledge that the Ordnance Survey had already triangulated to the north of Scotland and eagerly suggested extending the French arc all the way to the Shetland Islands. This was a massive two degrees of latitude further north than the triangulation had thus far extended. Arago felt that 'it was natural to wish that' the Ordnance Survey's meridian arc 'should be joined to that of France', and he wrote persuasively to Mudge that the completion of such a project would contribute 'all the data required for a correct knowledge of a portion of the earth on which we live. And surely,' he added meaningfully, 'to study the earth is far better than to devastate it by conquests, however brilliant and glorious they may appear.' In the wake of the defeat of Napoleon's armies at the Battle of Waterloo in 1815, Arago's words were highly charged.

In spring 1817 the Shetland Islands extension got under way. But Mudge was daunted and gloomy. 'I am overwhelmed with business,' he noted, and described how nevertheless '[I am] going again to turn myself to the stars. 26 years ago I commenced my career with a strong constitution, and with a good supply of bodily health, but I am now perhaps going to close my campaigning service with the performance of as arduous a task as can be given to the execution of any man.' Now fifty-five, Mudge had become depressed by the relentless burden of the Ordnance Survey; after a series of colleagues' deaths, the grim reaper was on his mind and he wrote: 'really I think hereafter the Corps of Engineers should be looked upon as belonging to *Undertakers* as well as to Carpenters and Joiners, and have the Death's Head and bones upon their Buttons by way of a *Memento Mori*'. He had come to depend on Colby entirely; when Colby was absent from the Ordnance Survey's headquarters at the Tower, Mudge wrote frequent plaintive letters, informing his young assistant that 'you are very much wanted' and 'I beg you will *immediately* close your operations and return to London.' But despite Mudge's gloom, newspapers were enthusiastic about the Shetlands extension and *The Times* was overjoyed that 'these joint [Anglo-French] processes, conducted at so high a latitude, may be expected to throw considerable light upon that curious class of problems which regard the figure of the earth'.

In May the French director of the Shetlands extension arrived in Dover. Jean-Baptiste Biot was a physicist, astronomer, mathematician and even an early balloonatic, with large, kindly eyes and a smile that veered from hangdog to playful. Officially exempted from the tedious rigamarole of Customs, Biot was soon in London being wined and dined by the President of the Royal Society, Joseph Banks, now seventy-four and wheelchair-bound from gout. 'No man Could have been more Kindly Received than he was,' Banks later wrote, 'nor did any man I Ever saw appear more sensible of & Penetrated by the universal good Reception he met with.' Mudge, too, appeared to like Biot, although he remained preoccupied by his own health and the risks posed by a surveying season in Britain's most brutal climate. 'What may be in the womb of time, who can tell?' Mudge wrote, pensively. 'I think this will be the wind up of all my campaigns.' But, he added semi-optimistically, 'it is a very great pleasure to me to think that the wind up will be respectable'.

Soon after his arrival, Mudge escorted Biot and his wide array of instruments and belongings up to Edinburgh by coach. The journey was clearly something of a struggle. 'I have been travelling in a chaise with M[onsieur]

Biot,' Mudge recorded, 'who speaks English as imperfectly as I do French.' At first Mudge found the excursion a chore. While waiting for his French collaborator to finish some observations at Leith, he wrote to Colby: 'I am chained up here something like a wild beast against my will.' Nevertheless, he found Biot 'a very able man and a very diligent observer' and he soon cheered up when his own great friend, the mathematician Olinthus Gregory, joined him in Scotland to witness the momentous Shetlands measurement. Mudge had recently been granted an honorary doctorate from Edinburgh University and he took the opportunity to collect this accolade, which pleased him greatly. 'If I choose to sink the Colonel, there will be another Doctor Mudge,' he quipped, referring to his late father Doctor John Mudge. 'Joking however on this point out of the question,' he continued more solemnly, 'this mark of respect from such a University, is a matter extremely pleasing to my feelings.' Biot was also happy in Edinburgh. The French geodesist found the Scottish army officers who were assisting his astronomical observations helpful and courteous. 'If my observations were bad, I had no excuse,' Biot assured his colleagues back home in France.

But soon everything began to go wrong. Before setting foot in the Shetlands, Mudge's vitality rapidly began to fail, as he had feared. 'The state of his health made it impossible that he should go further north,' Biot explained to a colleague. Defeated and dejected, Mudge retreated back to London. To fill his place, Mudge sent his eldest son, Richard Zachariah, a 'very tall and personable' 28-year-old army officer who was already working for the Ordnance Survey. Richard Mudge was 'a very good Frenchman' (in his proud father's words) and he proved an instant hit with Biot, who found him to be 'a young officer full of zeal'. But Richard also brought with him Thomas Colby. From the off, Colby and the French geodesist regarded one another with deep distrust and dislike.

TENSION HAD BEEN brewing under the surface of the Shetlands project since its conception. During the Paris–Greenwich triangulation of the

(A) David Watson by Andrea Soldi, painted after Watson's death in 1762, showing him pointing to his crowning glory: a military survey of Scotland.

(B) A portrait of Robert Dundas (1685–1753), which hangs in the Oak Room at Arniston House in Midlothian.

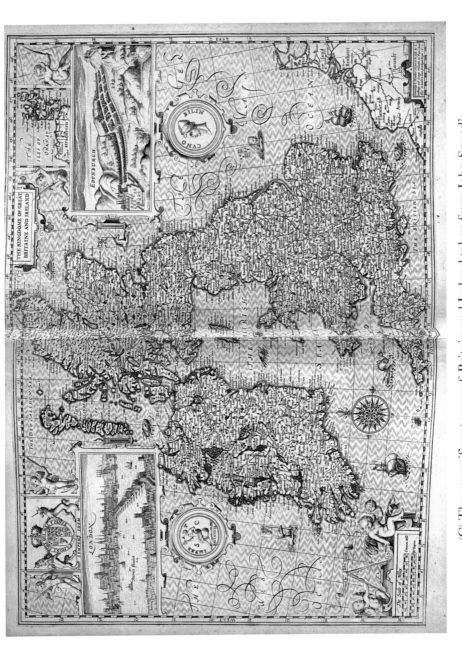

(C) The magnificent map of Britain and Ireland taken from John Speed's *Theatre of the Kingdome of Great Britaine*, 1611.

(D) *A Meeting of the Board of Ordnance*, late 1740s, attributed to Paul Sandby.

(E) Paul Sandby's 'Plan of Castle Tioram in Moidart and Castle Duart on the Island of Mull'. Sandby's landscapes are peopled with tiny redcoated soldiers.

(F) Paul Sandby's *View Near Loch Rannoch*, 1749, shows a surveying party at work in a field on the north side of the River Tummel, a little south of Drumchastle Wood. The man bent over the theodolite is probably the young William Roy.

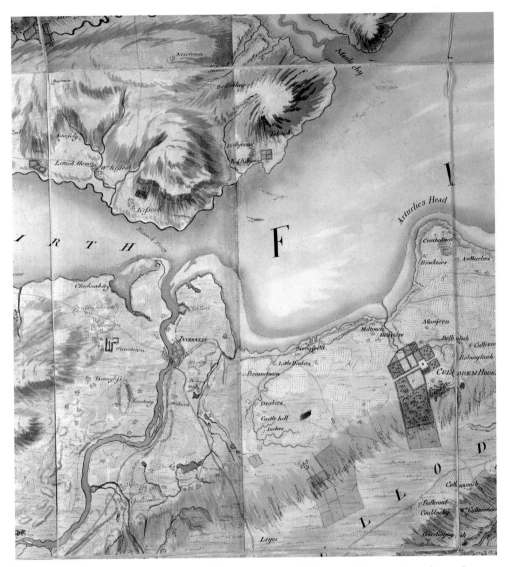

(G) A detail of the area around Inverness and Culloden Moor, taken from the
Military Survey of Scotland.

(H) George Cruickshank's satirical image of the Antiquarian Society from 1812. The figure third from the right, with his pockets stuffed full of papers marked 'ordnanance affairs', may be a posthumous image of William Roy.

(I) Joshua Reynolds's portrait of
Joseph Banks, 1771–3.

(J) Louis van der Puyl's portrait of
Nevil Maskelyne, 1785.

(K) Robert Home's portrait of
Jesse Ramsden, *c.*1790.

(L) J. J. Hall's 1816 portrait of Isaac Dalby,
the Ordnance Survey's first assistant.

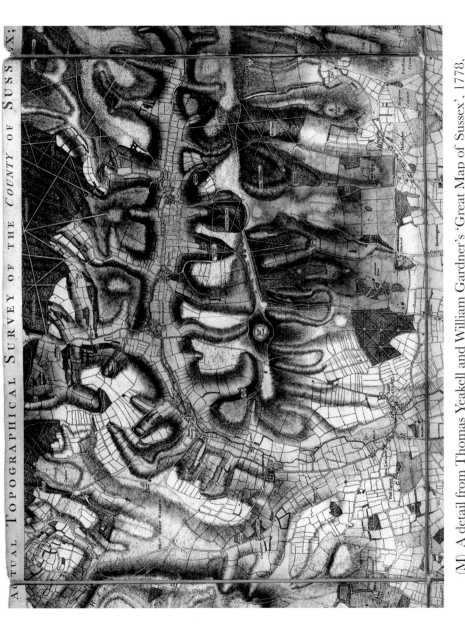

ACTUAL TOPOGRAPHICAL SURVEY OF THE COUNTY OF SUSSEX;

(M) A detail from Thomas Yeakell and William Gardner's 'Great Map of Sussex', 1778, showing Charles Lennox's Goodwood estate in green.

(N) James Walker Tucker's *Hiking*, 1936, was painted as rambling was exploding in popularity as a leisure pursuit. It shows the centrality of the Ordnance Survey to this phenomenon.

(O) The area around the Isle of Dogs, to the east of London, from Sheet 1 of the
Ordnance Survey's first map – of Kent – in 1801.

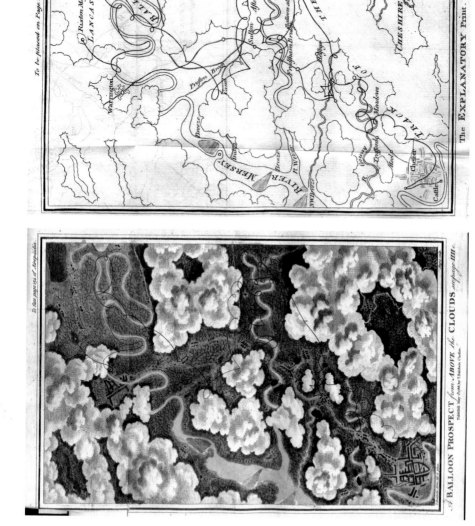

To be placed on Page 155 of Airopaidia.

Chart M

LANCASHIRE

Ruxton Moss

BALLOON

Warrington

Preston Brook

Sutton Affore

Frodsham Bridge

THE

R. WEAVER

TRACK

CHESHIRE

RIVER MERSEY

Chester

Castle

The EXPLANATORY Print. ... see Page IIII d.

To face page 154 of Airopaidia.

A BALLOON PROSPECT from ABOVE the CLOUDS ...see page IIII c.

Published May 5.1785 by T.Baldwin, Chester.

(P)   Two map-like impressions of the view from an early hot-air balloon and its route, taken from
      Thomas Baldwin's *Airopaidia*, published in 1786.

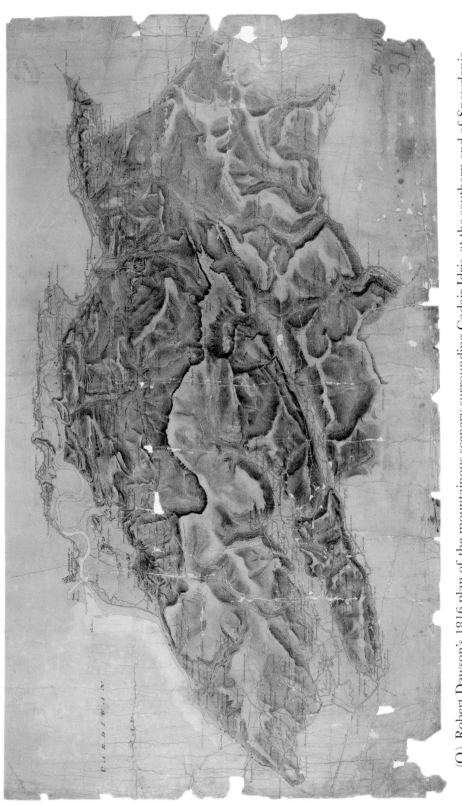

(Q) Robert Dawson's 1816 plan of the mountainous scenery surrounding Cadair Idris, at the southern end of Snowdonia. The summit of Cadair Idris lies near the centre of the map.

(R) Benjamin Robert Haydon's portrait of William Wordsworth, 1842.

(S) Charles Grey's portrait of John O'Donovan.

(T) A sampler 'Map of England', dating from 1780.

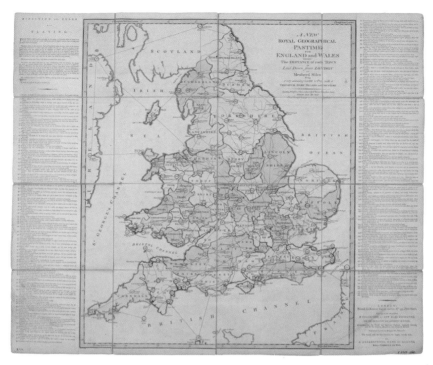

(U) A board game marketed in the late 1780s. Counters move between major geographical landmarks on lines showing 'measured' distances.

(V) William Blake's print of Isaac Newton. Blake opposed the Newtonian view of the universe.

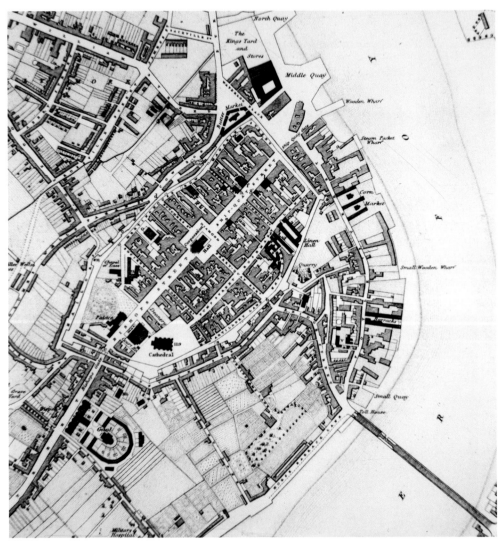

(W) A detail from the Ordnance Survey's six-inch map of Londonderry, 1833.

1780s, the British and French parties had each asserted the excellence of their respective national instruments, Jesse Ramsden's theodolite and Jean Charles de Borda's repeating circle. Thirty years later, the same antagonism affected the Shetland geodesists. From his first glimpse of the zenith sector, Mudge had felt its superiority to the French geodesists' equivalent. 'The French multiplying-circle surely cannot be put on a footing' with the zenith sector, he cried. 'It can do nothing more than . . . approximate to the perfection, which a few Observations, made with such an Instrument as the one I am using, gives.' During the Shetlands measurement each party refused to renounce their own instrument, and the trip therefore acquired added significance as a chance to test the relative merits and accuracy of the British zenith sector and French repeating circle. 'Any opportunity of bringing these two instruments into direct comparison with each other would be hailed with satisfaction by astronomers,' an onlooker recorded. 'Something in that respect was anticipated from the co-operation of the French and English geodesists, during the year 1817.'

But the friction surrounding the instruments paled beside the personal animosity between Colby and Biot. The two men's hostility was fuelled by the presence of Mudge's friend, the mathematician Olinthus Gregory, who 'attached himself to Captain Colby' after Mudge's exit. A histrionic and petulant, but fiendishly hard-working man, Gregory had leapt to the defence of Mudge when Rodriguez had attacked his skill and honesty back in 1812, in his critique of the meridian arc measurement. In melodramatic tones Gregory had accused Rodriguez of corrupting science's impartiality with jingoism, by aiming at 'the depression of English (and perhaps other) ingenuity and exertion, in order to the undue exaltation of the French scientific character'. He had even criticised the editors of the Royal Society's journal for agreeing to print this 'attempt by *a foreigner* to cast discredit upon a great national undertaking'. Thomas Thomson, a chemist and the editor of a scientific journal called *The Annals of Philosophy*, had attempted to mediate between Gregory and Rodriguez, and had reprimanded the former for employing in his censure an unprofessional tone, a 'style quite new to astronomical discussion'. Thomson's rebuke had unleashed from Gregory a torrent of 'low abuse' regarding Thomson's unpatriotic feelings, his lack of scientific credibility, his

33. Olinthus Gregory by Thomson, after Derby, 1823.

parasitic dependence on the Royal Society and his attempted suppression of Gregory's own work. Finally Thomson had struck back: 'Has [Gregory] made any addition to any branch of science whatever?' he demanded. 'I myself never heard of any of his investigations or discoveries.'

When Gregory was introduced to Biot in Scotland in 1817, the meeting sparked memories of the hostile feelings of this earlier squabble. Biot was a close colleague and particularly good friend of Rodriguez, and Colby's apparent intimacy with Gregory, Rodriguez's enemy, earned both Britons the French astronomer's derision. Colby had further reason to dislike Biot as it was rumoured that, while in Scotland, he had penned an article for the *Caledonian Mercury* that expressed opinions highly critical of the Ordnance Survey. Furthermore, Gregory whispered in Colby's ear, *Biot was an atheist*. Richard Zachariah Mudge regretted the antagonism that was splitting the French and British surveying parties in two and conjectured that 'had the ice once been melted between Biot and Colby, they would have mutually valued and respected each other. But it was fated that an eternal frost would separate them.'

In July 1817 these warring factions boarded a boat at Aberdeen and set sail for the Shetlands. Colby was struck by the islands' exoticism and described them as 'the rocks of the ancient Thule'. When their boat finally docked on the Shetlands' mainland, Colby and Gregory instantly set off on foot for Lerwick, at the former's characteristic hurtling pace. Biot found himself left to make his own way and was incensed by the discourtesy. However, the geodesists appear to have temporarily put their animosity behind them, and soon both parties were conducting observations on the island of Unst, the most northerly point of the British Isles. But this semblance of amicability did not last. Colby swiftly and rather arbitrarily decided that Unst was not suitable for his purposes and removed the British surveying team to Balta, an uninhabited island off Unst's east coast, 'where there were just tents sufficient to shelter it and its inhabitants from the weather'. Biot was quite happy with his spot on Unst and declared himself flabbergasted by Colby's erratic and arrogant behaviour. He refused to move.

The surveying parties thus went their separate ways. And with the split came the failure of one of the chief ambitions of the Shetlands measurement, the comparison of the French repeating circle and the British zenith sector. 'One great disadvantage attending your removal from M. Biot is that a comparison cannot be made between the results of the observations . . . with the sector and circle of repetition,' Mudge wrote disapprovingly to Colby. 'This is a very great misfortune.' Rumours of Colby and Biot's spat gradually made their way to London and Joseph Banks intepreted the separation as Biot's admission of the French instrument's inferiority, commenting that he could 'see no Reason' why Biot shouldn't join Colby on Balta, 'but the fear that he Should be obligd to Confess that Ramsdens Zenith Sector is a Capital instrument'.

Biot became happier the moment he was free of the British surveying team. He found Unst a friendly and helpful place, and a young carpenter who expressed interest in Biot's endeavours was persuaded to act as his assistant for two months. Thomas Edmondston was the owner of Buness House, a sixteenth-century whitewashed cottage that is still owned by his descendants, on the island's east coast, overlooking the sea. Edmondston was fascinated by natural philosophy and he gladly opened his home and grounds to the French astronomer. Biot lodged with him for weeks and set up his pendulum apparatus in Edmondston's sheepfold and his repeating circle in the magnificent and exposed garden. Edmondston's 'warm hospitalities made up for a chilly climate,' he wrote. Despite his falling out with Colby, Biot was able to report to his French colleagues: 'I was very much helped in my operations. I found everywhere the most generous attentiveness, and the greatest freedom from those antisocial prejudices, which perhaps often have less force in reality than in the minds of those who imagine them.' Upon his return to Paris, Biot ensured that his host on Unst was formally thanked. Edmondston replied that 'the hope of facilitating even in the most subordinate degree, the advancement of such enlightened, & benevolent views, and the happiness of possessing the friendship of so amiable and excellent a character as Mr Biot, were in themselves far more than sufficient reward for the inconsiderable attention which I could have it in my power to show to him.'

The Shetlands measurement had been intended to mark the restoration of amicable communication between French and British scientists in the wake of Waterloo. *Blackwood's Edinburgh Magazine* had commented that 'it is pleasing to observe the perfect concurrence of two great nations in an operation for the benefit of science'. But the project had quickly descended into xenophobic and childish feuding: a 'French Disconnection' if you will, which formed an unedifying follow-up to the successful Anglo-French collaboration of the 1780s. Although the British and French geodesists both completed their independent observations satisfactorily and the arc was extended to the Shetlands, the proper comparison of the repeating circle and zenith sector never happened. By the end of his stay, Biot could not even bring himself to mention Colby's name. He had never got the hang of its spelling anyway, referring to the hot-tempered map-maker as Kirby or Kolby in correspondence home. Once he returned to France he entirely omitted Colby's name from his published account of his expedition and simply denoted him as 'one of the officers serving under [Mudge]'. Mudge himself was dismayed at the furore that had erupted after his departure. He mourned that, 'as the queen who gave up Calais said that "that word would be found written on her heart", I may say that Shetland will be written on mine, for I have never ceased to deplore, with the keenest recollection, the happiness that I thought before me nipped in the bud, and I sent home, as it were, invalided'.

WILLIAM MUDGE'S HEALTH never fully recovered after his exit from the Shetlands debacle. The following year, he excused himself from the Ordnance Survey to recuperate in France for two months, and shortly after-wards decided to move from central London to Croydon. But the pressure of directing the Ordnance Survey refused to lessen. Although he resolved to spend the Christmas of 1819 with his family, he found himself yet again con-fined to London. It was to be his last Christmas.

In January 1820, while examining at the East India Company's college at Addiscombe, Mudge suffered a severe attack of internal inflammation. He had long been 'subject to the *hyp*, or something like it', referring to hypochondria, but now his worries were materialising in real physical symp-toms. His beloved daughter Jenny rushed to his bedside. The following month Mudge's condition had deteriorated further and she wrote to her aunt's husband, Mudge's great friend Richard Rosdew, asking him to visit immediately. The map-maker rallied for a while, and in March Mudge considered himself well enough to supervise Addiscombe's examinations. But the effort was too much. On 29 March, he became severely unwell. He never recovered and on 16 April 1820 this diligent, kind, rather depressive man died at the relatively young age of fifty-eight. The Interior Surveyor Robert Dawson mourned his death. 'We are left in sorrow and reflection,' he grieved. 'The General's kind and amiable disposition, his mildness of temper, gentleness of command, and many marks of attachment and regard given to me particularly, and evidently always ready for his Friends, have created and nourished a Love for him in my heart, which will ever be the first impulse with which I shall cherish his memory.' *The Times* com-memorated the map-maker as 'the late celebrated and scientific General Mudge'.

William Roy had been arguably responsible for the Ordnance Survey's foundation, but William Mudge had taken charge of it for twenty-nine years. Under his watch, the Trigonometrical Survey had travelled all the way from Land's End in Cornwall almost up to the Shetland Islands. Thirty-seven of the most accurate maps of the British Isles had been published, covering nearly the whole of southern England and the West Country up to a line running between Bath, Oxford, Hertford and Ipswich, and also including Pembrokeshire. A few months after Mudge's death, the *Caledonian Mercury* proclaimed: 'such part of the Survey as has been already published, both for accuracy of delineation and beauty of execution, surpasses any work of the kind ever produced in this or any other country'.

William Mudge had been largely responsible for the early Ordnance Survey's direction and ambition. He had fought the corner of map enthusiasts among the general public, adamant that they deserved access to this astounding feat of cartography. And although his encouragement of the Ordnance Survey's foray into geodesy in the meridian arc measurement and the Shetlands extension had inevitably delayed work on the First Series of maps, those projects signified his unremitting sensitivity to the wider responsibilities of a national mapping agency. It is poignant that, two months before Mudge's death, George III died, and, two months after it, Joseph Banks. The three demises marked the end of an era for British map-making.

CHAPTER TEN

'Ensign of Empire'

WILLIAM MUDGE'S DEATH left a vacancy at the head of what was rapidly becoming a 'national treasure'. Thomas Colby had played a leading role in the Ordnance Survey for the past eighteen years and reasonably expected to be asked to take charge in Mudge's place. After a few months of perplexing silence, he steeled himself to write to the new Master-General of the Ordnance to urge the case for his appointment. Arthur Wellesley, the Duke of Wellington, victor of the Battle of Waterloo that had ended the Napoleonic Wars in 1815, was not entirely persuaded by Colby's description of how he had

> used no interest, I have solicited no one of your Grace's noble friends to paint my character or conduct on the Survey in glowing colors; but I have had a firm but humble reliance that your Grace would, when the press of more important business allowed opportunity, enquire how far my conduct and character would render me deserving of confidence and enable me to conduct the Survey with efficiency and credit to the country.

Wellington replied rather cursorily through a secretary, informing the 34-year-old map-maker: 'His Grace is now making enquiries, and will let you know whether he will appoint you permanently to conduct the Survey or not. In the meantime, the Master General begs you will continue it, in like manner as General Mudge would have done.'

Wellington summoned before him the charismatic chemist Humphry Davy, who gave his opinion that Colby was by far the best and most entitled person to take over the running of the Survey. Wellington also consulted the 'very eminent Professor of Mathematics at the Royal Military Academy', Charles Hutton, who readied himself to offer a panegyric on the map-maker's merits. He began: 'No man more so, My Lord Duke.' But he was immediately interrupted by Wellington, who thanked him, saying, 'that is all I want to know; my time is valuable, and yours, I know, is not less so'. Colby's achievements in the fields of map-making and other scientific endeavours also spoke for themselves. As 'a Gentleman well versed in Mathematics and Natural Philosophy, and employed on the General Survey of Great Britain', Colby had recently been elected a Fellow of the Royal Society. He had also found time to help establish the Astronomical Society and was an associate of the Institution of Civil Engineers, demonstrating 'unceasing attention and liberality, in procuring for and presenting to the [Institution's] Library, the Ordnance Maps, and many other valuable documents, as soon as they were published'.

On 10 July 1820 Colby received the long-awaited letter from the Board of Ordnance that gave him the news that 'His Grace appoints you to succeed the late M[ajor]-General Mudge in the superintendence of the Ordnance Trigonometrical Survey.' Three weeks later, the *Morning Chronicle* happily assured its readers that 'the appointment of an officer who has been long engaged in this laborious undertaking, and who has latterly conducted the scientific part with distinguished ability and zeal, is an assurance that this useful and important work will be conducted in the same admirable style that has hitherto marked its progress'.

Upon taking control of the Ordnance Survey, however, Colby was displeased and disappointed to find that, beneath its veneer of efficiency, the institution had been left by Mudge in a surprisingly poor state. The type of error that the Admiralty had discovered back in 1811 was repeated on many of the maps intended for its First Series. In 1818 the Interior Survey had begun working in Lincolnshire when the gentlemen of that county, spurred by enthusiasm for images of their estates, had requested that the Ordnance Survey map it before the planned date, offering the incentive of a guaranteed purchase of 500 maps. On inspecting the resulting charts of

Lincolnshire after his appointment as director, Colby was dismayed to find them substandard and even 'most slovenly' in their topographical accuracy and representation of place names.

At around the same time, the Admiralty got in touch. One of their hydrographers, who had been mapping the Bristol Channel using the relevant First Series map for reference, had discovered that 'the direction of [Lundy] Island as given by the Ordnance Survey is quite incorrect'. Colby swiftly arranged for the island to be resurveyed and then turned furiously on the Interior Surveyor responsible. 'It was with extreme regret I discovered that your Survey was so erroneous that it was necessary to plane the Island, which had been engraved and published, off the Copper,' he fumed. 'The disgrace and expense of the transaction has been very considerable.' The surveyor defended himself, pointing out that the island had, in fact, been mapped by a trainee under his command, Thomas Compton. Compton had gone on to produce military charts for Charles Lennox, 3rd Duke of Richmond, and then a series of beautiful landscape prints of North Wales, which he designed to function like a guidebook: 'to point out clearly the route which was followed, so that those who may wish to visit the scenes described, will meet with little difficulty in finding them, without any other guide'. But Compton had not distinguished himself in these early days. He had mapped Lundy Island without a triangulation, solely as an exercise, and without ever intending that it should be included on the finished Ordnance Survey sheets. That it had been was a terrible oversight, and although this knowledge did not ameliorate the consequences, Colby was nevertheless somewhat placated. It 'give[s] me great pleasure . . . that [the error] does not attach to any person employed now upon the Survey', he informed Charles Budgen.

Sadly, this type of fault on the Ordnance Survey's early maps was not unique. Colby soon found inaccuracies to be rife, some a result of human error, but others due to changes in the rapidly altering landscape. The latter were admittedly beyond the surveyors' control, but their system was not well designed to cope with these alterations. Britain was industrialising rapidly in the first half of the nineteenth century and some aspects of the landscape changed beyond recognition in the space of ten years. The Trigonometrical Survey and the Interior Survey made their way across the nation at different rates and the time lag before their maps were engraved and released to the public usually extended beyond a decade and sometimes even lasted up to a quarter of a century. It is a truism to say that, by definition, every map is out of date as soon as it is published, but this was especially the case in the 1820s and 1830s. In the time that elapsed between mapping a county and the Ordnance Survey's publication of those maps, railways and factories could be built and a relatively minor town, such as Birmingham, could mushroom into an industrial metropolis.

In the first months of his appointment as Superintendent, Colby tried to tackle this desperate situation. It has been said that 'the idea of tolerable error was foreign to Colby's nature' and this obsessive perfectionist became intensely frustrated by the interference of 'human error' in the ideal of flawless map-making. He instituted a series of reforms of the Ordnance Survey's working practices whereby maps and plans were rigorously checked, second-rate efforts were rejected, mapping executed by trainees or assistants was no longer included on the finished maps, and a systematic process of revision of the already published surveys was set in place. Colby also modified the appearance of the First Series maps, calling for refinements to the depiction of hills, and finer linework throughout, resulting in clearer displays of settlements and smaller, denser writing. Under Colby's care, the Ordnance Survey was transformed into an efficient and systematic national mapping agency.

WILLIAM ROY'S ORIGINAL proposal for a national survey had pointed out the value of covering all of the 'British Islands', but Mudge's principal purpose had been to oversee the First Series of one-inch maps of England and Wales. In its early years, the Ordnance Survey took little interest in a map of Ireland, despite the fact that in 1800 two Acts of Union formally united the nations of Great Britain and Ireland into the United Kingdom. As had been the case with the Anglo-Scottish Union of 1707, the 1800 Act of Union was not smooth. It took place just two years after a republican uprising against British rule in Ireland during which between 15,000 and 30,000 people had died in just three months. Union was proposed as one means to suppress separatists who wanted to pull Ireland and Britain apart, but even in this context in which geographical information might help the British government control its neighbour, the Ordnance Survey showed little interest in an Irish mapping project. When a British engineer had suggested in 1805 that William Mudge should form a party to triangulate that nation, he had shrugged it off with the reply that 'the Irish survey bill has no reference to us'.

In the mid 1810s, the mapping of Ireland became a live issue in Parliament, for reasons of taxation rather than imperialism. The Irish county 'cess' tax was used to fund jails and schools, to repair roads and bridges and to pay the salaries of local officials. Theoretically the levels of tax that citizens were expected to pay were calculated according to individual means, the assessment of which was partly based on the value accorded to their local districts. 'Townlands' are local divisions smaller than parishes, and they formed the basis of the county cess tax valuations. But although the names and rough outlines of Ireland's townlands were generally known, the precise acreages and values of each division remained obscure. By the nineteenth century's second decade, this unsatisfactory situation began to worry Ireland's administrators, who could point to places where the tax was almost ten times as much as in another townland of equivalent value. In 1815 a committee urged that 'some mode should be devised for rendering . . . assessment more equal, the defect appearing . . . to arise in a great degree from the levy being made in reference to old surveys . . . which of course cannot comprehend the great improvements which have taken place in Ireland since the period at which those surveys took place'. At around the same time, the Admiralty was also pressing for an accurate mapping of Ireland's coastline, particularly the St George's Channel, 'for all purposes of navigation'. These calls grew more urgent after 1822 when an English ship was wrecked on an unmapped sandbank off the Wexford coast.

Inspired by this renewed interest in mapping Ireland, Thomas Colby, now Superintendent of the Ordnance Survey, wrote to the Board of Ordnance in

1824, suggesting that the institution under his command was the most apt body to assume such a task. The eldest brother of the Master-General, the Duke of Wellington, was Richard Wellesley, 1st Marquess Wellesley, who had been employed since 1821 as Lord Lieutenant of Ireland, the king's representative and head of the Irish executive. Both were descended from an aristocratic Anglo-Irish family and were reputed to entertain a derogatory attitude to many of their fellow countrymen. Richard Wellesley wholeheartedly supported Colby's idea that the Ordnance Survey should map Ireland and wrote to his brother to that effect. Within a week plans for establishing an 'Irish Ordnance Survey' were under way. Because of the requirement to map Ireland's administrative boundaries in detail for the cess tax, Colby recommended surveying Ireland on the spacious scale of six inches to one mile, and he estimated that such a map would cost £200,000, plus half as much again for a full investigation of its townland boundaries. With these specifications, the project was officially founded in June 1824. It was an evident distraction from the First Series, and one that would ultimately delay proper work on that project by around two decades, but the Ordnance Survey's sojourn in Ireland would be a colourful saga of imperialism, translation, cultural nationalism and local attempts to sabotage the map-makers' measurements. It was also the context in which Colby developed far greater ambitions for the Ordnance Survey of Britain than he had previously entertained.

The mapping of Ireland by a British surveying agency was always going to be contentious, given the active role in the history of Anglo-Irish relations that map-makers had played. In Henry VIII's reign, 'plantation' had begun, the establishment of colonies of English and Scottish settlers on Irish land confiscated from rebels. This redistribution of forfeited estates created a need for new maps of Ireland during the reigns of Henry VIII and his successors, maps that were clear 'ensigns of empire' (in Wordsworth's words in a different context). Then, after a violent uprising in 1641, Oliver Cromwell led his New Model Army into Ireland in 1649 to subdue the country again. Following his brutal success, a 1652 Act authorised the confiscation of lands from Irish Catholics who had opposed Cromwell's troops and the redistribution of those lands among his soldiers. To assist this process, an English natural philosopher called William Petty offered to conduct a new survey of Ireland.

Over thirteen months, Petty sent around one thousand soldiers over the length and breadth of the confiscated lands, clutching simple measuring instruments. It was known as the 'Down Survey' because its results were 'set down' in maps, and Petty's map was accepted in 1656 by a committee which began implementing the redistribution. A century later, a British military engineer called Charles Vallancey was posted to Ireland, where he remained for the rest of his life; in the late 1760s this 'colourful and eccentric' man embarked on a detailed military survey of his new home. Although it was still unfinished when he retired, Vallancey's maps, like Petty's, were images of Britain's military presence in Ireland in the seventeenth and eighteenth centuries. In 1883, Lord Salisbury would express the long-running British view that Ireland's rebelliousness could be partly controlled through cartography: 'the most disagreeable part of the three kingdoms is Ireland, and therefore Ireland has a splendid map'.

Ireland also had its own tradition of maps made under less violent circumstances. By the 1790s, some of the most advanced surveys had been made by the extraordinary Anglo-Irish polymath Richard Lovell Edgeworth, who had mostly grown up in County Longford, in the centre of Ireland. Inventor, engineer, educational theorist, Lunar Society member, politician, landlord and even a dancer, Edgeworth's fascination with science was said to have started at the age of seven, when he was shown an orrery in Dublin, a model showing the positions and motions of the solar system's planets and moons. In his mid-twenties, Edgeworth won a medal from the Society of Arts for his innovative 'waywiser', or surveyor's wheel. Shortly after this accolade, one of many in his lifetime for a host of inventions, Edgeworth's father died and Richard inherited the family's estate at Edgeworthstown, in County Longford. In 1782, Edgeworth, with his children and a new wife, permanently moved to this Irish estate and he became deeply involved in the land's management and improvement, maintaining frequent contact with estate surveyors.

In the same decade, Edgeworth turned his attention to map-making. He became fascinated by William Roy's Paris–Greenwich triangulation and composed an article that described how the event exemplified the need for an international telegraph system, which would have allowed the French and British teams to communicate remotely without having to meet up in person;

and he subsequently devised a scheme to map his home county through a mixture of astronomy and triangulation. Edgeworth used celestial observations to ascertain the latitude and longitude of his estate, and then measured a baseline along a straight road that ran from Edgeworthstown to Longford. Although he left his survey unfinished, a cartographic historian has labelled it 'the first truly indigenous bid [in Ireland] for high exactitude'. And Edgeworth passed on his passion for maps to his son William. After Richard's retirement, William Edgeworth worked for the Bogs Commission, surveying Ireland's principal peat bogs with an eye to their later drainage and reclamation, and he finished his father's incomplete map of County Longford. Then, after commissioning his own ultra-accurate theodolite from a Munich instrument-maker, William applied himself to the mapping of the north of County Roscommon; with his work it has been said that 'native Irish cartography attained a new level'.

However, despite the Edgeworths' remarkable achievements, map-making in Ireland at the beginning of the nineteenth century remained largely unsystematic, incomplete and haphazard. Maps were hard to obtain from Irish print-sellers and bookshops, and were 'as little-known as classical manuscripts', as one disgruntled map-maker remarked.

SO WHEN THOMAS COLBY put the Ordnance Survey forward as the most apt body to map Ireland, he was taking his place in a long history of British military efforts to subdue neighbouring territories through cartography. Early decisions about the staffing of the Irish Survey touched on nerves made acutely sensitive by that history. In 1824 a committee was formed to oversee the foundation of the 'Irish Ordnance Survey'. Chaired by Thomas Spring Rice, Member of Parliament for Limerick, the committee debated whether the mapping project should be conducted by Irish or British surveyors. Spring Rice himself emphasised the importance of Irish involvement, pointing out that it would be helpful if 'private individuals will . . . allow access to such maps and other documents, as can be of service'. But the imperious

Duke of Wellington had other ideas. He declared, '[I] was obliged to hold very strong language, stating my determination to have nothing to say to it if not allowed to perform the service in my own way, and by the qualified officers of the Ordnance. I positively refused to employ any surveyor in Ireland upon this service.' Well-schooled in imperialism after his time as Governor-General in India, his brother Richard supported him and was adamant that an Irish map 'cannot be executed by Irish engineers and Irish agents of any description. Neither science, nor skill, nor diligence, nor discipline, nor integrity, sufficient for such a work can be found in Ireland.' The Wellesleys instructed the Ordnance Survey to staff its Irish endeavour with solely British military engineers, a decision which had a devastating effect on Ireland's own map-makers, who were essentially put out of a job by the appearance of such accurate maps of the territory.

Following these orders, Thomas Colby obtained approval to raise three companies of 'Sappers and Miners', the rank and file of the Corps of Royal Engineers, to work on the Irish Ordnance Survey. These low-grade army personnel were significantly cheaper than civilian assistants but they were also a greater risk, being often poorly educated and with a tendency to drunkenness. Colby tried to take the Sappers' moral health in hand and stipulated that they should be given Bibles, and prevented from growing beards or moustaches (so as to not 'appear extraordinary or ridiculous'), and he demanded that their children be sent to school. In an attempt to sweeten the Ordnance Survey's presence in Ireland, he also made it clear to his superiors that his men were distinguished from the regular Army and would not be prevailed upon to act as an ad hoc police force in cases of local unrest.

The Sappers required a thorough education in military surveying before departing from the mainland. Initially twenty young men were promised for the project, a number which fluctuated wildly throughout the Ordnance Survey's time in Ireland. One newspaper described how they were sent to Cardiff 'for further instruction in land surveying, under Mr Dawson, of the late Corps of Draughtsmen, with whom they will remain about six weeks, and then proceed to Ireland'. Colby also approached a man called Charles Pasley for help in perfecting his assistants' skills. A venerated military engineer, with hooded eyes and a determined mouth, Pasley had served in

Minorca, Malta, Naples, Sicily, Denmark, Spain and the Netherlands during the Napoleonic Wars. After a horrific bayonet wound in his thigh and a bullet through his spine at the Battle of Flushing, Pasley returned to Britain with a silver war medal and a pension. He spent some of his subsequent time writing and in 1810 he published his *Essay on the Military Policy and Institutions of the British Empire*. By November 1812 it had already gone into a fourth edition.

In January 1813, a copy of Pasley's *Essay* was lying on the table of an L-shaped red-brick house, once a farm and inn, in Chawton village, on a busy crossroads in Hampshire. One of the occupants of this house wrote to her sister: 'we are quite run over with books'. '*I*,' she emphasised proudly, 'am reading a Society-Octavo, an Essay on the Military Police [*sic*] & Institutions of the British Empire, by Capt. Pasley of the Engineers.' This was 'a book which I protested against at first,' this correspondent explained, 'but which upon trial I find delightfully written & highly entertaining'. Her esteem for Pasley was passionate. 'I am . . . much in love with the author,' she wrote happily: 'the first soldier I ever sighed for, but he does write with extraordinary force & spirit.'

Jane Austen did not describe exactly what drew her with such ardour to Charles Pasley's *Essay*. But in her novel *Pride and Prejudice*, she jubilantly described 'all the glories of' an army encampment in which 'tents stretched forth in beauteous uniformity of lines, crowded with the young and the gay, and dazzling with scarlet'. Exhibiting such a susceptibility to the attractions of the military, it is perhaps no wonder that she was sympathetic to Pasley's highly militaristic idea that, 'whilst we glory in the freedom, the public spirit, and patriotism of this country, [we must] not give way to the empty delusion, that by them alone we are to be invincible'. His *Essay*'s overriding message was that Britain must become a thoroughly 'military nation' and that all citizens should consider themselves in 'spirit' as amateur soldiers, spies and surveyors in the service of the state. Pasley also stridently asserted the crucial importance of maps to military defence and attack. He emphasised that 'wherever we act, we must have proper plans and information, if we wish to succeed'. Long sections of the *Essay* lamented Britain's failure to possess 'as good maps . . . of most countries, as are in possession of their native governments'.

When Colby contacted him in 1825 to help design an intensive 'Course of Study for Surveying Companies' in Ireland, Pasley was a tutor at an educational establishment at Chatham in Kent, which had been founded in 1812 to teach graduates from the Woolwich Military Academy intent on pursuing elevated careers in the Engineering Corps. At Chatham, Pasley had devised a course of instruction that over 500 days gave his students an advanced education in surveying and reconnaissance techniques, mining, bridging, architecture and siege operations, climaxing in a summer's placement on the Ordnance Survey. On 23 February 1825 Pasley's diary recorded that Colby paid a visit to Chatham, where this distinguished instructor 'show[ed] him the proposed Course' he had devised for the Interior Surveyors in Ireland, which would last about six months. One of Pasley's pointed recommendations was that 'the name Royal Sapper and Miner ought to be abolished' to avoid antagonism with Irish locals who may harbour resentment at the military map-makers' presence. The warning was prescient.

IN THE EARLY summer of 1825, following an initial reconnaissance the previous year, the Ordnance Survey formally began its operations in Ireland. Colby believed that it would take seven years to map Ireland in enough detail to clearly show its townlands' boundaries. After those seven years were up, however, he would be forced to revise the estimated completion date to 1844. So for two decades, the great majority of the Ordnance Survey's personnel were in Ireland, leaving only ten officers of engineers with ten assistant surveyors back in Britain. Very little new surveying would be done in Britain during the Ordnance Survey's time in Ireland; its remaining map-makers were instructed to concentrate on implementing Colby's reforms and revising the maps that had already been completed of a large tranche of southern England and Wales. Many of these surveyors found the task of revision dispiriting compared with the creation of new maps, but over the next twenty years they amended published and unpublished surveys of, among other locations, Glamorgan, Lincolnshire and Lundy Island (which had been so disastrously mapped the first time around). The engravers were also hard at work during this period, and the 1820s and 1830s saw a proliferation of new maps reach the public, including those of Pembrokeshire, the Gower and Glamorgan. By 1841, when the Ordnance Survey started to consider transferring some of its personnel back to Britain, maps had been published of the whole of Wales. But the two-decade sojourn in Ireland would drastically delay the completion of the First Series of one-inch Ordnance Survey maps of England and Wales.

On 9 May 1825 the *Derry Journal* noted that five officers 'with a detachment of the royal sappers and miners, have arrived in this city, to commence the survey of Ireland'. Three months later, the Irish *Examiner* related the news that 'Major Colby, Director of the Ordnance Survey, left the Tower [of London] last month, for the inspection of the . . . survey of Ireland.' After establishing 'head-quarters of his detachments, each under the command of a Captain of Engineers' at Antrim, Coleraine, Derry and Dungarvon, Colby proceeded to Divis Mountain, on the outskirts of Belfast. Under the Ministry of Defence's control between 1953 and 2005, when it was handed to the National Trust, Divis provided the British military with a valuable vantage point over Belfast during the Troubles. The same panoramic opportunities made the mountain invaluable to the Ordnance Survey in the nineteenth century. From Divis, Colby could see with his naked eye to the Mourne Mountains in County Down, to Hollywood in County Wicklow, and to the Sperrin Mountains, which stretched from Lough Neagh in County Tyrone to the south of County Londonderry. More importantly for the Ordnance Survey's purposes, Colby could also see as far as Scotland, to the mountains of Merrick, in Galloway, to Goat Fell on the Isle of Arran and Tartevil on Islay, and to South Berule on the Isle of Man. These long cross-sea observations that were theoretically available from Divis and other similarly prominent peaks like Ulster's highest mountain, Slieve Donard, meant that the Ordnance Survey's triangulation in Ireland could be joined on to the Trigonometrical Survey of the British mainland.

But the Irish weather made such long sight lines from Ireland to Britain very difficult to attain. Colby was plagued by 'inveterate haze and fogginess' on Ireland's east coast, which obstructed and sometimes entirely thwarted his observations. A journalist for the *Belfast News Letter* described his attempts to observe from Slieve Donard in 1826: 'the weather has been extremely adverse to the operations of the gentlemen employed in this arduous task for some time past. For nearly a fortnight the mountain has been so constantly enveloped in mist, that it has been impracticable to do any thing.' The writer hoped that the map-makers' perseverance will 'stand proof against the obstacles of the weather', and he reminded his readers that 'while we enjoy calm weather and sunshine in the regions below, they exist in a constant gale, and [are] enveloped in clouds[,] and might, with great propriety, be . . . denominated the Children of the Mist'.

A relatively new surveyor among these 'Children of the Mist' came up with an innovative solution to the relentless cloud and fog. During his time at the University of Edinburgh, Thomas Drummond had developed an aptitude for mathematics, natural philosophy and chemistry such that his tutor felt that 'no young man has ever come under my charge with a happier disposition or more promising talents'. Long conversations with Colby had persuaded the young man that the military might provide fruitful opportunities to develop his scientific interests. Although he disliked the Army's inflexible discipline, in the early 1820s Drummond was glad to find himself a trained military engineer on his mentor's Irish surveying team. He was an obsessively hard-working officer. Refusing to leave hilltop stations during the most fiendish storms, Drummond regularly contracted severe chills, colds and influenza, and frequently had to take sick leave as a result. When his boss's endeavours to observe with the theodolite from Ireland into England and Scotland were foiled by thick fog, Drummond applied his scientific nous to the problem. He took two recent inventions and developed them to meet the Ordnance Survey's needs.

The heliostat or heliotrope, deriving from the Greek *helios*, sun, was a surveyors' device, originally devised by the German mathematician Carl Friedrich Gauss, that used a mirror to reflect the sun's rays to illuminate a distant station. An article about its invention in the *Gentleman's Magazine* reported how the heliostat could be 'of great importance' to map-makers 'in the measuring of large triangles'. Attracted by this potential application, Drummond tinkered with Gauss's invention, adapting the heliostat so it could be attached to a theodolite rather than a telescope, and making it smaller, lighter, more portable and much easier to adjust. But it was the development of the limelight that made Drummond's name. Invented in the 1820s by the chemist Sir Goldsworthy Gurney, this worked by placing a small globule of lime carbonate 'shaped into the form of a boy's marble, and about three-eighths of an inch in diameter' in a spirit lamp, where it was ignited and burned with magnesium in the midst of 'a stream of oxygen gas'. The resulting bright white flame was harnessed in a reflector that produced rays vastly more dazzling than existing flares. One witness reported how 'the brilliance of the light thus produced is so intense, that it is quite impossible to survey it, and like the sun it produces that painful sensation upon the retina, which renders it necessary to avert the eye almost instantly'.

The limelight was sometimes called the 'pea light' after the small, green appearance of the tiny globe of lime. When Drummond witnessed a demonstration by the chemist Michael Faraday, he became instantly excited. Drummond foresaw that such a flare could be used to illuminate distant signals through dark or foggy weather and he began promoting the limelight, publishing articles in the Royal Society's journal and arranging public demonstrations. One such display took place in a darkened 300-foot-long hall in the Tower of London, where the limelight was lit beside two competitors, the Argand burner and the Fresnel lamp, which had both been used previously by the Ordnance Survey in Britain. A spectator reported that Drummond's invention was so 'overpowering, and as it were, annihilating both its predecessors, which appeared by its side, . . . [that] a shout of triumph and admiration burst from all present'. Another commented that the Fresnel lamp 'is like the moon, compared with the brilliancy of the sun, when placed beside the pea light'. By the 1830s, the instrument was popularly known as the 'Drummond Light' and in 1837 it made its first theatrical appearance, illuminating the stage of the Covent Garden Theatre.

The Ordnance Survey used these products of Drummond's remarkable imagination to penetrate Ireland's 'inveterate haze and fogginess'. The limelight was first used to conduct observations over the sixty-seven-mile distance that separated Divis from Slieve Snaght, in County Donegal. On 28 October 1825, Drummond ascended the pyramidal form of the latter, which reached above 2000 feet, and set up his tent and a surrounding wall 'so that we may consider ourselves safe against any storm'. But the weather was atrocious. Fog quickly gave way to 'a storm of snow' and Drummond wrote on 4 November that 'my tent is blown [down] and I now write from a kind of Cave, formed on the lee side of the Hill'. Nevertheless, he valiantly continued to light the limelight and manipulate the heliostat at the times that he had agreed with Colby, who was perched on the summit of Divis, training his theodolite on Slieve Snaght and scanning the horizon for the telltale flare. In those days prior to telephones and radios, Drummond had no idea whether his limelight was visible from Divis and personally 'despair[ed] of success; there must be a hill in the way'. But on 12 November an officer on Divis gave Drummond

> a very hurried intimation of our having seen the reflector. I have still greater pleasure in communicating the result of the light, it was most brilliant. In the evening, when preparing for our observations, one of the watch called out that there was a much stronger light than the one at Randalstown and a little above it; we immediately turned to that direction and there we saw the light; it was most brilliant, exceeding in intensity any of the Light Houses . . . These two days' observations have completely established the advantage of this admirable invention . . . the light . . . can be very easily intersected from its steadiness and brilliancy.

In August 1826, Colby attempted what was then the longest trigonometrical observation ever made, between Slieve Donard and the mountain of Scafell in the Lake District, the highest peak in England. Poised at Slieve Donard's summit and 2785 feet above sea level, Colby swept the easterly horizon with the theodolite's telescope. Alighting on a hazy shadow behind the northern point of the Isle of Man, he sharpened the telescope's focus, adjusting its vertical position with his right hand and agonisingly slowly pushing it around on its horizontal axis with his maimed left arm. As he focused the lens even further, Scafell, which was a staggering 111 miles away, gradually materialised at the centre. Shaking with excitement, Colby prepared himself to read the horizontal and vertical angles off the theodolite. But at that moment an assistant officer barged into the tent, demanding Colby's attention. Accidentally striking the map-maker's elbow, he jolted the theodolite out of position. Scafell disappeared from the telescope, like the prey vanishing from the sights of a gun. It was said that only 'a momentary ejaculation of anger escaped [Colby's] lips'. Successfully reining in his hot temper, he quickly regained equanimity and, 'though he could not again succeed, and the object was, therefore, lost, he never afterwards alluded to the subject'.

COLBY DID NOT like Ireland much at first. He proclaimed that he did not want to become 'buried' or 'bottled up' in Dublin and continued to travel back and forth to Britain. He appointed numerous assistants to ease his workload, including a man called Major William Reid as his resident deputy. And he also made a wise choice in the employment of a 24-year-old second lieutenant in the Royal Engineers, who originally hailed from Hampshire, called Thomas Aiskew Larcom. Stocky and rather short, with a wide brow and quizzical expression, Larcom had been selected by Colby from among the Corps' new recruits. He counterbalanced his superior's hot-headedness with a cool and organised mind, and Colby's passion for the outdoors with his own relish for more cerebral things like numbers, calculations and statistics. The two men made a wonderfully complementary team.

The scale of the Irish map was six times larger and more detailed than that of the one-inch British surveys, and such an expansive scale demanded exceptional accuracy from the Interior Surveyors. The Ordnance Survey in Britain had set a pattern whereby an initial triangulation was followed by the detailed interior mapping, but an English officer pointed out that in Ireland there were tangible advantages in commencing the Interior Survey *before* the triangulation had got under way. He suggested that if the Interior Surveyors were unable to fudge their measurements to conform to those already discovered by the Trigonometrical Survey, their integrity and skill would be truly revealed, and the maps' accuracy would be brought up to the level required by such a large scale. The Irish map-maker William Edgeworth stepped in and suggested that the triangulation should be commenced

34. Photographic portrait of Thomas Aiskew Larcom, by Camille Silvy, 1866.

immediately, but that its results could be withheld from the Interior Surveyors to achieve the same effect. These discussions, in which Colby was closely involved, significantly delayed the triangulation's progress, so the Interior Surveyors had in practice no choice but to proceed with their work without the trigonometrical results.

Both the Interior and the Trigonometrical Surveyors were faced with hostility from local residents as they made their way across Ireland. Dressed in military uniform, the men were unmistakable emblems of British occupation. A surveyor in the parish of Drumachose in County Derry reported:

> many of the parishioners were greatly alarmed when the engineers first came into the country to make piles at the top of the hills for the purpose of triangulation. They viewed with dismay the strange-looking men erecting (as they thought) redoubts [forts] on the most commanding situations, and actually imagined them to be emissaries of some formidable national enemy, coming to 'spy out the nakedness of the land' and to mark the point from which the country below could be most easily commanded by artillery . . . they conceived that the trigonometrical station on the top of Keady was the spot from whence a fire was to be opened on the lowlands in that neighbourhood.

Other Ordnance Surveyors in Ireland were forced to place trig stations and instruments under armed guard to protect them from sabotage. Surveying poles were moved or entirely removed. Brian Friel's 1980 play *Translations*, which dramatises the Ordnance Survey's endeavours in nineteenth-century Ireland, imagines a young man from County Donegal boasting: 'every time they'd stick one of these poles into the ground and move across the bog, I'd creep up and shift it twenty or thirty paces to the side . . . Then they'd come back and stare at it and look at their calculations and stare at it again and scratch their heads. And Cripes, d'you know what they ended up doing? . . . They took the bloody machine [the theodolite] apart!' Some Boundary Surveyors and engineers were assaulted in Ireland, 'perhaps from a fear that the government had invented a secret weapon', and on Bantry Common in County Wexford the police were called out to protect a party of Interior Surveyors who were being attacked by tenants who were in conflict with their landlord. The map-makers were occasionally suspected of being state mercenaries hired to suppress agrarian uprisings, or perhaps employees of unscrupulous estate owners. Their reputation in Ireland was not improved when one team of Ordnance Surveyors used a cairn revered by pilgrims in Tummock, in County Derry, as a trig point.

As well as out-and-out violence, hostility for the Ordnance Survey's presence in Ireland was often manifested through reticence and ridicule. A map-maker charged with researching place names found the inhabitants of Dungiven to be very shy 'in giving their [own] names because they are afraid that they might be wanted for the "service of war" or some other plan of the Government'. In Drogheda, many 'people [were] not willing to give information, suspecting it may be connected with Tithe affairs'. In Lisburn, in County Down, one surveyor found that 'some people are afraid of any one going about lest he might be a spy, and the subject of tythes is so much agitated that people are afraid of any one sent from the Government'. Elsewhere in that county, a map-maker was accosted by an angry resident who complained that 'you have a great deal of blockheads going about annoying people'. And local residents in Connemara devised a song that, in a brief aside, mocked the map-makers' 'drudgery and labour'.

But not all the surveyors' stories were full of woe. When they were not living out of tents, they brought welcome trade to inns and hostels. And in 1828 the *Dublin Evening Post* drew a charming picture of collaboration between the Ordnance Survey and the local populace. The journalist described how the residents of Glenomara, in County Clare, helped the Ordnance Surveyors to build a trig station. Perhaps they were motivated by simple curiosity in the endeavour, or by good relations with some of the individual map-makers that they had met while the latter were staying in local accommodation, or even by the hope that a cartographical image of Ireland might bolster national identity. Either way, a large crowd ascended the mountain, borne up by music from flutes, pipes and violins, and accompanied by young women carrying laurel leaves. However, even this contained a subversive element. The Glenomara residents insisted on naming the trig station 'O'Connell's Tower', after 'The Liberator' Daniel O'Connell, the Irish political leader who campaigned for the repeal of the Act of Union and for Catholic Emancipation, the right of Catholics to become Members of

Parliament – an entitlement they had been denied by British penal laws from the seventeenth century.

THE FACT THAT the Ordnance Survey's Irish triangulation could be joined on to that of Britain by sight lines extending over the Irish Sea and St George's Channel meant that measuring a baseline in Ireland was not entirely necessary. But Colby decided to do so, to make his map 'as precise and up-to-date as possible', as he put it, and to verify the Irish triangles. So on 6 September 1827 a team of surveyors began measuring a long base by the side of Lough Foyle, the estuary of the River Foyle, where it leaves Derry. Thomas Colby's perfectionism reached new heights during the Lough Foyle measurement. He had become increasingly dissatisfied with the measuring rods and chains that had been used by the Ordnance Survey in Britain. Even the glass rods continued to respond to heat in annoying and unpredictable ways, and although microscopes had been fixed to monitor the expansion of the metals in some places, there were no means of ascertaining the susceptibility to changes in temperature of every single component of the instrument. Colby fantasised about a device that would entirely eliminate expansion and contraction due to changes in heat, so he and Thomas Drummond set about experimenting with different metals. The two men finally came up with a complex amalgamation of brass and iron bars,

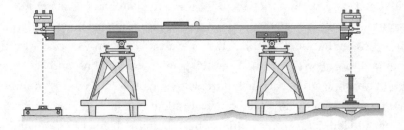

35. *Colby and Drummond's compensation bars.*

which were attached together at the centre but were allowed to expand or contract freely. When exposed to heat, the materials expanded at different rates, but they were fitted together in such a way that their different expansions cancelled one another out, leaving the total length of the instrument unchanged.

The Lough Foyle baseline was measured with what became known as the 'Compensation Bars' in September and October 1827. After a break for the winter, the endeavour began again in July 1828. It was delayed by the impossibility of measuring during the harvest and was finally finished on 20 November. But the final result 'annoyed and vexed' Colby. 'Where a difference of *inches* is scarcely admissable,' he discovered a shocking discrepancy of 'a *few feet*' between the actual length and that predicted by trigonometric equations from the triangulation. Colby was perplexed, disappointed and extremely worried. He could not account for this result, which threw the veracity of the entire British and Irish Trigonometrical Surveys into question. But after a few months, the answer struck him. In 1824 a new Weights and Measures Acts had been passed by Parliament, which officially established the Imperial System in Britain. This defined a foot as twelve inches long, and a one-foot length was duly moulded in brass to act as a standard to which anyone involved in mensuration might refer. A foot had also been defined as twelve inches prior to 1824, but before that date a different brass unit had acted as the standard. The Ordnance Survey's measurements in Britain were based on the earlier standard, and their Irish mensuration on the later one. In theory, they should have been identical, but on a comparison of the two standards, Colby found that there was indeed a minute difference, 'not perceptible to the eye, or in a carpenter's rule', but sufficient when multiplied over long distances to knock the Ordnance Survey's measurements off-kilter. When he adapted his calculations to account for this discrepancy, Colby found the baseline to be 41,640.8873 feet long, just under eight miles. It now accorded with the length produced by calculations from the Trigonometrical Survey to within five inches. 'You may conceive how delighted Colonel Colby was at this result,' the *Belfast News Letter* reported, 'and what a triumph it is for the combined exertions of art and science.'

The measurement of the Lough Foyle baseline, like that in Britain on Hounslow Heath, attracted many interested parties, including the British

36. A sketch showing the Ordnance Survey's measurement of the Lough Foyle baseline with the compensation bars.

astronomer John Herschel, the eccentric mathematician Charles Babbage and the director of the Armagh Observatory, Thomas Romney Robinson. If the surveyors sometimes felt embattled during their time mapping the Irish landscape, Colby and Larcom's experiences at Lough Foyle reassured them that, among the Protestant scientific community at least, they were extremely welcome in Ireland. One man, a 23-year-old with brooding eyes and an intelligent brow, interrogated Colby about the Ordnance Survey's work with particular enthusiasm and persistence.

WILLIAM ROWAN HAMILTON was a wildly gifted mathematician, whose talents were sufficient to garner him the post of Andrews Professor of Astronomy at Trinity College, Dublin in June 1827, at only twenty-one, when he was still an undergraduate. This role also made him the Astronomer Royal of Ireland and director of the Dunsink Observatory in Castleknock, five miles south-west of Dublin. He was appointed because of his prodigious talent, but some people complained that he was not a natural choice for the job of astronomer as his passion was pure mathematics, which operated in abstraction from the material world, rather than astronomy, which was immersed in it. Even Hamilton's friend, the poet Aubrey de Vere, said that he 'did not look through his telescopes more than once or twice a year!' He was 'so much occupied with the purely abstract part of science', de Vere explained, 'that its material phenomena interested him only so far as they revealed laws'. Hamilton's aptitude for pure science was exemplified when in 1832 he used laws of 'the geometry of light' alone, accompanied by no actual experiments, to propose that under certain circumstances a ray of light could be refracted into a cone within a biaxial crystal. Hamilton's 'Law of Conical Refraction', as it became known, was subsequently proved in experiments conducted by the physicist Humphrey Lloyd. Hamilton replied to news of this proof with, 'I told you so', or words to that effect.

Hamilton was fascinated by geometry. He had long felt that it was 'an essential requisite' for all mathematicians and scientists and the keystone of all other disciplines. In the 1850s, he would attempt to sell a game based on geometry and algebra on the commercial market, to a largely uncomprehending audience. (One instruction read: 'the only operations employed in the game are those marked λ and μ; but another operation, $\omega = \lambda\,\mu\,\lambda\,\mu\,\lambda = \mu\,\lambda\,\mu\,\lambda\,\mu$, having the property that $\omega^2 = 1$; was also mentioned'. No wonder one purchaser of Hamilton's game reported back to its creator that 'none of her family would play it'!) Hamilton was entranced by one geometric form in particular: maps. The leaves of his notebooks were scrawled over in places with sketch maps, somewhat like Coleridge's, with whom he would enjoy a correspondence. In one notebook entry, Hamilton recalled 'how much pleasure' he derived from helping his younger sister Eliza 'to find out on the maps the several counties of England & Wales'. This enjoyment grew even more profound when, on reading a short story that was set in Merionethshire, in Wales, he 'recognis[ed] it as a county which I had lately seen marked in the Map. From this little circumstance,' Hamilton continued, 'I was led to reflect on the constant source of pleasure which would be opened to me by an extensive knowledge of Geography.'

In 1824, when the Ordnance Survey of Ireland was first being mooted, Hamilton was introduced to the 57-year-old novelist Maria Edgeworth, the daughter of the amateur map-maker Richard Lovell Edgeworth, who had died seven years previously. The meeting between the two was a great success, despite an age gap of nearly forty years. Maria wrote to her sister Honora soon after this first acquaintance, describing Hamilton as 'an "Admirable Crichton" of eighteen', a highly favourable comparison to the extraordinary Scottish polymath James Crichton. Hamilton, Maria continued, was 'a real prodigy of talents, *who, Dr [John] Brinkley [the astronomer] says, may be a second Newton* – quite gentle and simple'. The approval was mutual. Hamilton also wrote to his sister Grace about '[she] who forms the great and transcendent interest of the place'. 'Miss Edgeworth!' he cried, ecstatically. 'She far surpasses all that I heard or expected of her.' William Rowan Hamilton and Maria Edgeworth formed a friendship that endured until the latter's death in May 1849, and many of their discussions were concerned with maps. Hamilton described his delight upon arriving at Maria's to be 'carried off' to consult 'William [Edgeworth]'s maps and plans, which were just about to be rolled up for departure, as Mr W.E. was to go the next morning on an engineering expedition'. They also talked about geometry and together designed an educational tour of the Lake District to help Maria's younger brother Francis brush up on his knowledge of Euclid's *Elements*. (It is interesting that they considered the Lake District the most apposite location for a geometrical education. Wordsworth and Mudge were clearly not the only ones to imagine that region imprinted over with trigonometrical shapes. In fact, in the late nineteenth century, the map-publisher Bartholomews would produce a series of educational maps for schools, which were based on the Lake District's latent geometry.)

When the Ordnance Survey arrived in Ireland in the mid-1820s, Hamilton welcomed the project with delight. In April 1828 he secured 'a situation among the calculators of the Trigonometrical Survey' for a young assistant who was working under him at Dunsink Observatory, 'thro' the interest of Captain [Richard Zachariah] Mudge'. And when the Ordnance Survey came to Lough Foyle, Hamilton was eager to witness the map-makers in action. He travelled to the site, more 'with the intention of reconnoitring the ground, than with much hope of seeing the base and the officers', as he later wrote to a friend. But Hamilton was lucky enough to find some engineers there, among them 'Lieutenant Drummond at home, and after eating in a tent, for the first time in my life, I took a walk with them along the base line to Roe, and then back again by the shore of Lough Foyle; on our return,' he continued gleefully, 'we found Colonel Colby, and had a pleasant evening, closed by my sleeping under canvas, a novelty which I enjoyed extremely.'

In the summer of the following year, Hamilton received a famous house guest who was connected to the Ordnance Survey: 'Captain Everest, who has been superintending a great triangulation in India, and is going out again for that purpose.' The upright, mutton-chopped and rather humourless George Everest had been appointed assistant to the Great Trigonometrical Survey of India in 1817, and acted as its superintendent from 1823 to 1843. (His surname was given to the Himalayan mountain, but it should be pronounced 'Eve-rest', not 'Ever-est': he would probably not have taken the wrong inflection in good spirits.) The backbone of the Indian Trigonometrical Survey was the longest meridian arc to be measured anywhere in the world, and this 'Great Arc of India' stretched over 1600 miles of the Indian subcontinent. Begun under the aegis of the East India Company, and later taken over by the Crown, the triangulation was designed as the foundation for topographical surveys that would delineate British territories in India, and it was also a vital contribution to geodetic research into the shape of the earth. Everest's dedication to his task was such that contemporaries referred to him as 'Never-rest' (a joke that relied upon the mispronunciation of his name).

George Everest came to Ireland in the summer of 1829 on sick leave from India, to observe the Ordnance Survey's work and to talk with Thomas Colby, whilst staying with Hamilton. The two map-makers had much to discuss, not least the experience of surveying in a colonial territory. Everest noted the huge scale – six inches to the mile – on which the Irish map was being conducted, as well as its wealth of employees and its highly sophisticated instruments. After conversations with Colby, Everest resolved to request more personnel and funding for his own endeavour, as well as a double set of Compensation Bars, and he would also demand that the Indian Survey should be allowed to prioritise the Trigonometrical Survey above the interior maps, just as the Ordnance Survey did. Hamilton had long declared himself fascinated by the 'work on the Meridional Arc of India', and we can imagine that during Everest's visit the two engaged in animated discussion about geometry, triangulation and map-making.

Over the following months and years, Hamilton maintained a fascination with the Ordnance Survey's endeavour. He dined regularly with Larcom and Colby and during a brief interlude, in which Hamilton found himself too busy to visit the engineers' camp, Colby lamented to the mathematician that 'you are so wrapt up in your luminous pursuits that I can never get you to come and see our humble efforts to represent the surface of Ireland as it is. But,' he continued, 'I can assure you my assistant L[ieutenant] Larcom is managing a work here which would repay you for the time of coming to see it.' Hamilton never doubted the merits of Larcom, whom he considered to be 'somewhat of a universalist', and in 1833, he recommended that map-maker as a member of the Royal Irish Academy (of which he, Hamilton, would be president from 1837 to 1845), from his 'own personal conviction of his merits'. And the warmth was mutual, Larcom allowing Hamilton to share in the Ordnance Survey's adventures. Hamilton and Thomas Romney Robinson collaborated in discovering the longitudinal difference between the Dunsink and Armagh Observatories, to assist the Ordnance Survey. A few years later, Larcom provided Hamilton with compass bearings and instructions that enabled him to put his 'telescope correctly on the spot' at home or at the Observatory, and to mimic the map-makers' own observations. 'There can be no harm in your having an eye in the direction every night,' Larcom assured Hamilton, and he even lent the Ordnance Survey's theodolite to the curious mathematician. A Victorian biographer of William Rowan Hamilton described Larcom as 'one of the friends who took most interest in his pursuits both scientific and philosophical'.

From the late 1820s, around the beginning of his friendship with the Ordnance Surveyors, Hamilton's letters and notebooks show that he started thinking much harder about the precise nature and significance of geometry than he had done previously. The Scottish 'common sense' philosopher Dugald Stewart had specifically mentioned the Ordnance Survey's triangulation of Britain in the second volume of his *Elements of the Philosophy of the Human Mind* (1814), describing the Trigonometrical Survey as almost the only example in 'the real world' of the perfect logic that is normally reserved for pure mathematics and abstract geometry. The Ordnance Survey's instruments were so sophisticated, Stewart felt, that they ensured 'we are not liable to be disturbed by those physical *accidents*, which, in the other applications of mathematical science, necessarily render the result, more or less, at variance with the theory'.

Hamilton likewise decided that geometry was a branch 'of the philosophy of mind', a product of the extraordinary powers of independent creation that were possessed by the intellect and which operated quite separately from the world of materiality and sensation. For Hamilton, geometry was 'a language of pure space', and the young mathematician enjoyed imagining the exact moment when the first triangle was 'invented', when 'a new light flashed upon the mind of the first man . . . who demonstrated the properties of the isosceles triangle'. 'The true method' of devising geometrical figures, Hamilton felt, was to 'bring out' an idea that was already lodged in the intuition, rather than following long lists of mathematical rules. The drawing of a triangle should be like a divine act of creation of 'something simple, perfect, and *one*'. Hamilton had a penchant for idealist philosophy, especially the works of Immanuel Kant, and he embraced Kant's suggestions that geometry was an image of the internal, not the external, landscape. This was a very different view to the conviction of many Enlightenment *philosophes* that geometry was a manifestation of empirical reason.

In 1829, shortly after George Everest's departure from Hamilton's residence, another famous friend came to stay with the mathematician. William Wordsworth had first made Hamilton's acquaintance in the summer of 1827, when the young mathematician had been taken to the Lake District to meet Wordsworth by the writer Caesar Otway, author of *Sketches in Ireland*. After climbing Helvellyn, Otway and Hamilton visited the poet at home. Hamilton and Wordsworth instantly liked one another and when they came to part that evening, Hamilton described how the poet 'walked back with our party as far as their lodge'. But, wanting to prolong their conversation further, Hamilton suggested that after dropping off his friends he would 'walk back with [Wordsworth], while my party proceeded to the hotel. This offer he accepted,' Hamilton continued, 'and our conversation had become so interesting that when we arrived at his house, a distance of about a mile, he proposed to walk back with me on my way to Ambleside, a proposal which you may be sure that I did not reject; so far from it, that when he came to turn once more towards his home, I also turned once more along with him.' 'It was very late when I reached the hotel after all this walking,' Hamilton added.

In the late summer of 1829, Wordsworth paid his only trip to Ireland, to stay with Hamilton. He was introduced to Maria Edgeworth, who found the poet 'very prosing – as if he were always speaking *ex cathedra* for the instruction of the rising generation and never forgetting that he is MR WORDSWORTH the author and one of the poets of the lake'. But Wordsworth got on much better with the Ordnance Surveyors in Ireland, many of whom he was introduced to by Hamilton. The mathematician presented Wordsworth to Thomas Drummond, and the poet thereafter referred warmly to his acquaintance with 'Mr Secretary Drummond' of 'calculating celebrity'. He also met Thomas Spring Rice, chairman of the committee that had founded the Irish Ordnance Survey, and was still in touch with him ten years later. And, during his 'short tour' of Ireland, Wordsworth threw himself into political debate with the Irish map-maker and sometime adviser to the Ordnance Survey, William Edgeworth. This Irish visit clearly reinvigorated Wordsworth's earlier interest in the Ordnance Survey's endeavours in the Lake District. When in 1837 Colby and Larcom published a memoir of the *Ordnance Survey of the County of Londonderry*, Wordsworth made sure to buy one of the 1500 copies printed.

During his Irish trip, Wordsworth enjoyed numerous conversations with Hamilton about the nature of geometry and the relationship between literature and science. And it is telling that the poet revised passages in his poetry where he had initially been rather scathing and sceptical about geometry and maps. In his long poem *The Prelude*, Wordsworth erased his denigratory description of the rational child's dependence on telescopes, crucibles and maps. And he significantly expanded a section which encouraged his reader to 'not entirely overlook/ The pleasure gathered from the rudiments of geometric science', but to try to imagine 'the relation those abstractions bear/ To Nature's laws . . . / From star to star, from kindred sphere to sphere,/ From system on to system without end.' It would be nice to think that his contact with Hamilton and the Irish Ordnance Survey may have been partly responsible for Wordsworth's depictions of the joys of geometry in his later poetry and revisions. He clearly came to realise that maps could be interpreted in different ways than purely as exponents of Enlightenment rationality, and was attracted to Hamilton's association of geometry with the independent powers of creation that were possessed by the imagination.

William Rowan Hamilton's contact with Colby and Larcom indicates that their time in Ireland was sociable and intellectually stimulating. But it was not just Anglo-Irish scientists and mathematicians with whom the two map-makers mingled. In the late 1820s, Colby and Larcom started to entertain grand plans for the Irish maps that would bring them into contact with a bevy of geologists, statisticians, anthropologists, linguists, poets and, crucially, orthographers.

CHAPTER ELEVEN

'All the Rhymes and Rags of History'

PHYSICAL CHALLENGES, perilous peaks and endurance exercises excited Thomas Colby, but his assistant was more cerebral. In the summer of 1828, with the Trigonometrical and Interior Surveys of Ireland well launched, Thomas Aiskew Larcom decided to brush up his knowledge of the Irish language. On applying to the Irish Society for tutelage, the surveyor was recommended a Gaelic Irish scholar from south Kilkenny, who introduced himself as John O'Donovan (or Seán Ó Donnabháin). A pleasant-looking 22-year-old, with a wide-eyed open expression and 'peasant garb', O'Donovan had had a peripatetic upbringing. He had been first educated in a hedge school, part of a rural pedagogical system in which students were taught by educated members of the community, usually in a barn or house. There he had learned to read English and Irish, and had been taught in arithmetic and geography. Later O'Donovan attended a private school in Waterford and a Latin school in Dublin, and lessons with a relative gave him facility in Latin and various European languages. But O'Donovan's real intellectual awakening had occurred after his father's death, when he had spent two years with his uncle Patrick, described by his nephew as 'the living repertory of the traditions of the counties of Kilkenny, Carlow, and Wexford'. It was 'from him I first caught that love for ancient Irish and Anglo-Irish history and traditions which have since afforded [me] so much amusement,' O'Donovan fondly recalled.

This revelation of Ireland's ancient history and language changed O'Donovan's life, and it would alter the Irish Ordnance Survey's course too. Although O'Donovan had flirted with a career in the Catholic priesthood, he rejected the idea in order to follow his linguistic infatuations. During a period in Dublin in the mid 1820s, O'Donovan had met a host of scholars prominent in Irish studies, an area in which intellectual interest was rapidly burgeoning. This friendly, eager young man made the acquaintance of James Scurry, author of a paper on 'Grammars, Glossaries, Vocabularies and Dictionaries in the Irish Language' that was read before the Royal Irish Academy in 1826. And Scurry introduced O'Donovan to James Hardiman, Dublin's Commissioner of Public Records. Hardiman was a committed researcher in the history of the Irish language and its literature, and he encouraged interest in Ireland's linguistic history as a way of bolstering the nation's self-confidence. 'After ages of neglect and decay, the ancient literature of Ireland seems destined to emerge from obscurity,' he predicted. 'Those memorials which have hitherto lain so long unexplored, now appear to awaken the attention of the learned and the curiosity of the public; and thus, the literary remains of a people once so distinguished in the annals of learning, may be rescued from the oblivion to which they have been so undeservedly consigned.'

Shortly after their first meeting in Dublin in 1828, Hardiman offered John O'Donovan a job as a copyist of Irish manuscripts and legal documents, which he accepted. And in that same year, Larcom also approached the scholar with an offer of employment as a tutor in Irish, and O'Donovan again happily agreed. For the next couple of years, the two men – one a calm, upright, meticulous military officer, the other an animated and rather irreverent linguist – met three times a week over breakfast, while Larcom developed a proficiency in the language of the nation that he was mapping. But O'Donovan was fearsomely overworked, and after a time he had to put an end to the early-morning tuition.

LARCOM'S INTEREST IN learning Irish was fuelled by ambitious plans that he and Colby entertained for the survey of Ireland. Throughout its mapping of England and Wales, the Ordnance Survey had repeatedly been seduced into diversions from the main priority of the completion of the First Series of one-inch maps, and it was no different in Ireland. There Colby and Larcom both came round to the opinion that 'geography is a noble and practical science only when associated with the history, the commerce, and a knowledge of the productions of a country; and the topographical delineation of a county would be comparatively useless without the information which may lead to, and suggest the proper development of its resources'. In Britain, Colby had been content to produce maps and measurements alone, but he had mostly left their wider applications to the statisticians, industrialists, landowners, agricultural reformers, guidebook writers, travellers, civil servants and others who consumed their data. In Ireland he took a different tack, and Colby and Larcom together formulated a plan of making what the latter called 'a full face portrait of the land'. The two map-makers recognised that their chart was a valuable opportunity to glean information about Ireland's history, culture, economy, geology, religious practices, languages, antiquities, and industrial and agricultural potential. They resolved to make a *survey* of Ireland in the true sense of the word: a bountiful overview of the nation, with an eye to its improvement. As this suggests, Colby had started to feel more engaged with the country. This may have had something to do with the fact that in 1828 he married an Irishwoman called Elizabeth Hesther Boyd, a daughter of the Treasurer of Londonderry. Little is known about Elizabeth, but it seems that his marriage helped to reconcile Colby to his prolonged residence in Ireland.

From the start, the Ordnance Survey's Irish endeavours were more wide-ranging than they had ever been in Britain. Alongside the creation of a Trigonometrical Survey and a topographical map, Colby was also focused on mapping the borders of Ireland's townlands, to assist with the recalculation of the county cess tax. A separate Boundary Commission was formed in 1825 to work alongside the Ordnance Survey under the supervision of a man called Richard Griffith, who ascertained the rough shape of Ireland's townlands from landowners, estate maps, clergymen and various officials. To map them in detail, Griffith then recruited local residents to give the Ordnance Survey's map-makers a tour of the boundaries, for 2s a day. If these impromptu guides were uninformative, or the boundaries indistinct (as was often the case over bogs or mountainous ground), the surveyors tended to simply concoct the course of the borders themselves. The precise area of each townland was needed for the calculation of the tax and at first this was measured on the ground itself by the surveyors, but once the maps had been engraved it became easier to work out these areas on paper.

The Ordnance Survey and the Boundary Commission did not always see eye to eye. Colby and Griffith cordially disliked one another, and each team regularly accused the other of incompetence. When Griffith was given the task of assigning a monetary value to each townland, based on its soil, build-ings, relief, cultivation and population, he realised that the maps that were being produced by the Ordnance Survey in the late 1820s were not up to the job. Tiny errors, such as misplaced buildings, undefined bogs or other omis-sions, hindered Griffith's efforts and he duly instructed Colby to rectify these deficiencies. But the map-makers had much to criticise too. Some accused Griffith of distorting or even ignoring the measurements they had made of the boundaries. After showing proof copies of the charts to a local commu-nity in County Londonderry, one surveyor was forced to report back to Larcom: 'the people deny that many denominations set down as townlands on our maps are townlands'. 'I am inclined to believe,' he continued, 'that Mr Griffeth [*sic*] frequently divided parishes into townlands, more from his own fancy than from the authority of the people.' The situation did not improve with time. Two years later, the same map-maker wrote despairingly: 'I cannot get the people to agree with Griffith's names or Subdivisions. They say that he got Gentlemens' Stewards, who were often not long in the Country, to point out the boundaries, and that these frequently by ignorance, and not seldom by intention, set him astray.' After the Irish Ordnance Survey was over, the *Morning Chronicle* reflected the same opinion: 'Mr Griffith is an excellent and able engineer, but he knows no more of farming than engineers generally do; that is, nothing at all. We are well acquainted with many of the districts that he sets down as cultivable wastes, and we say that practical farmers hold a perfectly opposite opinion about them.'

Colby and Larcom entertained still greater ambitions for their map than the creation of trigonometrical, topographical and boundary surveys. As part of their conviction that the mapping project should chart Ireland's historical face and industrial potential, they also wanted to look *below* the earth's surface. In fact, the idea of establishing a permanent Geological Survey to work in parallel with the Ordnance Survey had been mooted since the early years of the nineteenth century, when Interior Surveyors had conducted rudimentary geological researches to assist with their hill-drawing. And when William Mudge had measured a meridian arc across England, and his results had been unexpected, the geologist John Playfair had pointed out that Mudge's observations may have been distorted by the attraction of the instruments' pendulums to various land masses. 'It would have been of great importance to have added to [the Trigonometrical Survey] a mineralogical survey,' Playfair tutted, 'as the results of the latter might have thrown some light on the anomalies of the former.' The Ordnance Survey had duly remedied this complaint, and between 1814 and 1821 a man named John McCulloch, who was appointed chemist to the Board of Ordnance, turned his attention to geology. McCulloch spent every summer analysing the rocks around the Ordnance Survey's triangulation stations, to ascertain if their levels of attraction might pose a threat to the precision of the map-makers' instruments. An awestruck onlooker commented to McCulloch: 'no person but a man of iron like yourself could have gone over such an extent of rugged country with so much minute accuracy'.

Colby considered the Irish Ordnance Survey as the ideal opportunity to build on these geological beginnings. He issued detailed instructions to his surveyors, and the finished maps of Ireland duly exhibit a plethora of limestone quarries and gravel pits. Colby wanted them to comprise 'the most minute and accurate geological survey ever published'. Nevertheless, it was back in England that an official Geological Survey was set up in this period. McCulloch had remarked on the ease with which such an endeavour could be attached to the Ordnance Survey, and in the spring of 1835 the Master-General of the Ordnance and its Board formed a committee, which included the influential geologists Charles Lyell, William Buckland and Adam Sedgwick, to debate the matter. The result was that 'a grant was obtained from the Treasury to defray the additional expense which will be incurred in colouring geologically the Ordnance county maps'. But before this date a 'gentleman geologist' and owner of a slave plantation in Jamaica, Henry de la Beche, had taken a keen, if amateur, interest in the Ordnance Survey. This man, described by an acquaintance as 'a regular fun-engine', enjoyed colouring in the Ordnance Survey's charts of Devon according to their geological strata. When the income from his Jamaican plantation completely failed in the early 1830s, de la Beche applied to the government for a grant to complete his geological survey and was duly made the Ordnance Survey's official geologist and director of an embryonic geological museum. De la Beche's subsequent career went from strength to strength. He published *Researches in Theoretical Geology* in 1834, an official *Report on the Geology of Cornwall, Devon and West Somerset* four years later, was knighted in 1842, elected President of the Geological Society in 1847, and proudly oversaw the opening of the Museum of Practical Geology on London's Jermyn Street and the foundation of a School of Mines and of Science applied to the Arts. Henry de la Beche remained the bespectacled face of the Ordnance Survey's Geological Survey until 1845.

UNTIL THE COMMENCEMENT of the Irish Ordnance Survey, it was said that Thomas Colby was shy of publication. A 'morbid apprehension of criticism' has been blamed for his failure to provide the public with an account of the Trigonometrical Survey's progress after he had taken over its field work. Mudge and Colby's third volume of the *Account of the Trigonometrical Survey of England and Wales* had appeared in 1811, and then . . . silence. Colby appears not to have subsequently challenged the recommendation of the Commission of Military Enquiry that the *Account* be discontinued, nor to have carried on publishing reports in the *Philosophical Transactions of the Royal Society*. 'It is impossible not to regret this fatal error,' one of Colby's colleagues lamented rather melodramatically, as 'it has materially tended to alienate the Ordnance Survey from the good feelings of the scientific public, and has thrown over a work, with which not merely every scientific, but also every practical man, should be familiar, an air of official mystery and seclusion.'

In Ireland, Colby seems to have revised his opinion about publication. Along with Larcom, he devised a tantalising scheme: the composition of 'memoirs' to accompany the Ordnance Survey's Irish map, which would flesh out its 'paper landscape' into what has been called a 'fully rounded national survey' of its historical and contemporary appearance. In his annual report of 1826, Colby set out his intention that 'a great variety of materials towards the formation of statistical and other reports will be collected'. He commanded his map-makers to carry 'remark books' at all times while surveying Ireland, and dictated that these journals should be filled with 'a great deal of information respecting the means of conveyance, state of agriculture and manufacture, and in short of almost everything that relates to the resources of the country'. Colby was in charge of overseeing the Trigonometrical Survey in Ireland, and it was ultimately to his assistant Thomas Aiskew Larcom that the task of bringing to life the 'memoir project' would fall.

Colby and Larcom's enterprise was not especially unusual. A number of map-makers had marketed memoirs of their cartographic tribulations in the past, mostly to boost sales and attract interest in the finished surveys. In 1809, for example, the cartographer and map-publisher Aaron Arrowsmith the elder had published a *Memoir Relative to the Construction of the Map of Scotland . . . in the Year 1807*, which not only contained one of the earliest uses of the name 'Ordnance Survey' but also described the map-maker's joy at discovering the Military Survey of Scotland stored away in a library in boxes, when he was researching materials for a new map. But the sheer scope of Colby and Larcom's proposal for the memoirs of the Ordnance Survey of Ireland was exceptional.

Larcom intended that, rather than providing an account of the Ordnance Survey's own techniques and experiences, the memoirs should tell the story of the land itself. He hoped that every parish in Ireland would eventually possess its own memoir, which would describe each area's natural and human history, ancient and modern 'artificial state', and 'social economy and productive economy'. He issued his map-makers with hints about the type of questions they might ask the local populace when conducting their researches. They should enquire after 'food; fuel; dress; longevity; usual number in a family; early marriages', Larcom instructed. Are there 'any remarkable instances on either of these heads? What are their amusements and recreations? Patrons and patrons' days; and traditions respecting them? What local customs prevail?' How about 'Peculiar games?' he wondered. 'Any legendary tales of poems recited around the fireside? Any ancient music, as clan marches or funeral cries? They differ in different districts, collect them if you can,' he added eagerly. And what about 'any peculiarity of costume? Nothing more indicates civilisation and intercourse. Emigration. Does emigration prevail?' If so, 'to what extent? And at what season? To what places? Do any return? What number go annually to get harvest or other work in England or other parts of the kingdom? Do they take wives and families with them? Do they rent ground, and sow potatoes for their winter support?' Larcom's motivation appears to have been intellectual rather than commercial: to map Ireland's human face as well as its topographical shape. And there was clearly popular interest in such a project. In County Down, a local clergyman accosted one of the map-makers to ask 'would we publish any book to illustrate the map . . . "If so," says he, "A [*sic*] shall be *locally* interested, and do all in my power to contribute towards its completion."'

Larcom and Colby's memoir project was arguably the epitome of all the digressions in which the Ordnance Survey had dabbled since its foundation. Its superintendents (Edward Williams only excepted) had all been ambitious perfectionists, whose intentions to create an utterly truthful image of the British Isles led them repeatedly into a number of distractions from the central task of map-making, such as the measurement of the meridian arc through England, and the Shetlands venture. Even Colby's decisions to map Ireland and Scotland were diversions from the completion of the First Series of Ordnance Survey maps, which ostensibly formed the project's principal focus. Although one cannot help but contemplate these side projects with awed admiration, it is also true that they slowed down the production of complete maps of England and Wales by decades, and increased the Ordnance Survey's expenditure by eyewatering amounts. As we shall see, this would cause significant problems for Colby.

BY 1830 THOMAS AISKEW LARCOM had been moved from Ireland's 'great outdoors' to the comfort of the Ordnance Survey's headquarters at Mountjoy House, in Phoenix Park, about two miles north-west of central Dublin. He began transforming the building into an efficient map-making factory, successfully juggling the Trigonometrical Survey's measurements with the influx of the Interior Surveyors' plans and the needs of the engravers. He also came up with alternative printing methods, including electrotyping (a way of making faithful duplicate printing plates) to speed up the charts' publication, and he opened Mountjoy House to all who might benefit from the Ordnance Survey's data. Larcom voiced the unnervingly dispassionate ambition to create 'men who work like machines, without thinking or adapting', and by 1836 he was proud to report that Mountjoy was working with 'steam engine rapidity'.

Chief among Larcom's objects of interest in Ireland was its toponymy. Thorough research into its place names was the principal reason for his having sought lessons in Irish in 1828. He also felt that the undeniable necessity of such toponymic research for map-making justified the fecund nature of the memoir project to anyone who might otherwise question its utility. He was adamant that, in order 'to make the Maps a standard of Orthography as well as of Topography', a deep excavation of Ireland's history, culture and language was essential. And 'to trace all the mutations of each name, would be, in fact, to pass in review the local history of the whole country', he explained.

In the first year of the Ordnance Survey's residence in Ireland, Colby had issued a series of 'Instructions for the Interior Survey' regarding research into place names. In this document, he set out his preferred method of settling on toponymy for the map, which reiterated the principles behind the Interior Survey's researches in Wales: that the map-makers should initially discover place names by asking locals and then verify them with reference to printed and manuscript texts and native-language speakers. His mind set on perfection as always, Colby instructed that 'the name of each place is to be inserted as it is commonly spelt, in the first column of the name book: and the various modes of spelling it used in books, writing &c. are to be inserted in the second column, with the authority placed in the third column opposite to each.' However, when the Interior Surveyors' name books began arriving back at the Ordnance Survey's headquarters in Phoenix Park, Colby and Larcom realised the extent of the task they faced. Larcom had hoped that a few lessons in Irish would allow him to reconcile the varying forms of the current and historical place names himself, and come up with a viable standard form for the map. But once he saw just how many variants existed and how wildly they differed, he was forced to admit that his Irish skills were not up to the job and that this crucial aspect of map-making required its own discrete branch.

Any attempt to standardise Ireland's place names depended on the coop-eration of native Irish-speaking linguists. But here Larcom came up against one of the founding principles of the Irish Ordnance Survey: that Ireland's citizens were to be totally excluded from participation. However, the Trigonometrical Survey had already cracked in this respect. The British Sappers had resisted Colby's efforts at moral and intellectual self-improve-ment, and their widespread drunken illiteracy had begun to get on officers' nerves. In April 1825, the employment of Irish 'country labourers' at 1s per day had been authorised on the humblest aspects of the Ordnance Survey, such as laying out the measuring chains. Before long, there were fifty-three Irish labourers employed on the Survey to eighty-seven British Sappers. These numbers rose sharply over the next year, and by 1826 the number of Irishmen easily matched the number of Britons. By the late 1830s, when 2139 Ordnance Surveyors were crawling over the face of Ireland, the number of Irish employees outnumbered the British Sappers by four to one. It is ironic that this institution, whose founders had initially been so derogatory about Irish workers, were forced to turn to them for obedient, skilled personnel when their own British personnel fell short of the mark.

Because of this growing acceptance of Irish participation in the Ordnance Survey, Larcom felt justified in overturning the original decree. So in 1829, he established a branch almost entirely composed of Irish-speakers to research the nation's place names. The so-called Topographical Branch was initially headed by the Dublin-born creator of the first dictionary of the Irish language, Edward O'Reilly, but his health was poor, and in August 1829 he died. Larcom promptly received an enthusiastic letter from his old tutor,

John O'Donovan. Now twenty-four and with an eye to the main chance, O'Donovan proposed that he be given a job on the Topographical Branch. He recorded that Larcom replied 'immediately, offering me a situation at a very small stipend, of which I accepted after some hesitation'.

The Topographical Branch soon became a vibrant centre for toponymic research. After O'Reilly's death, a charming, suave and sociable antiquarian and artist called George Petrie was appointed its director, and he even donated his own home, a large town-house on Dublin's Great Charles Street, as a headquarters for its work. Some of Ireland's most accomplished linguists were recruited for the Topographical Branch: the scholar Eugene O'Curry (Eoghan Ó Comhraidhe), the archaeologist and historian William Wakeman, an assistant called Patrick O'Keeffe and the eccentric translator and poet James Clarence Mangan, 'besides two or three more'. The Branch's employees gave this office the affectionate nickname 'Teepetrie' (*tigh*, pronounced 'tee', signifies house), and it is nice to imagine them huddling under its roof, as intent on surveying Ireland's place names as the Interior and Trigonometrical Surveyors were on mapping its landscape from their canvas tents.

At first the Topographical Branch followed a system whereby the linguists stayed in Dublin and sent enquiries about place names to the officers of the Interior Survey to pursue 'in the field'. But many map-makers lacked the skills or inclination to investigate the often complex questions of the Topographical Branch; Larcom thunderingly berated one officer for his 'strange admission that you can give "no positive information" as to the name of a place surveyed by you! Who is to be looked to but you?' he demanded. 'You might with as much propriety say you can give no positive information as to its area. You are responsible for all that appears on your plan and for making it as perfect as "the scale admits".' The Topographical Branch soon adopted a new routine whereby John O'Donovan went out into the country to follow up the map-makers' preliminary queries about place names with face-to-face conversations with local residents. In name books stuffed with notes, and daily (often twice-daily) letters, this maverick linguist sent his discoveries back to Larcom and to his colleagues in Teepetrie. The linguists then set to, consulting a plethora of printed and manuscript sources to find written authority for historical occurrences of the place names that were being used across mid-nineteenth-century Ireland. 'All sorts of old documents were examined, old spellings of names compared and considered,' William Wakeman recalled.

To glean the fullest picture of Ireland's toponymy, O'Donovan struck up discussions with Irish citizens both young and old, literate and illiterate, from working-class labourers to aristocratic landowners, Catholics and Protestants. As an Irish-speaking Catholic himself who carried a bishop's testimonial to his 'useful, laudable and patriotic pursuit', he received a friendly welcome from many residents who had given the British military engineers the cold shoulder. Between 1834 and 1842 John O'Donovan became 'a kind of one-man local history department'. In his letters, he offered animated verbal caricatures of his interlocutors, and did not hold back when he came across an enemy. After a disagreeable meeting with a Protestant clergyman in Rathfriland, County Down, O'Donovan recorded: 'I was never so disgusted with any little *Cur*, *whelp*, or *pup* in all my life. His petty aristocratic assumptions and ungentlemanly remarks had a very disagreeable effect upon my sensitive nerves.' O'Donovan was entranced by puns and obscenities and reflected to Larcom that 'in the present artificial state of society it is curious to observe that one word is filthy, while another that expresses the same identical idea is honor[able], as a[rs]e, backside, bottom, etc.' His irreverence often clashed with Larcom's moralism. The staid and precise military engineer was not amused by his young employee's digressions, and he complained: 'O'Donovan really writes in a way that vexes me. It is rather too much to suffer him.' Larcom declared himself disgusted by the extent to which O'Donovan's letters were 'defaced (or disgraced) by scurrility' and 'ribaldry', and – with his tongue planted firmly in his cheek – O'Donovan duly promised to his superior that he would, in future, make his letters 'very serious, cold, and un-Irish'.

It should not be imagined that O'Donovan took his work anything but seriously, however. He was convinced that toponymic research was every bit as rigorous a science as surveying, and stated himself to be 'as sceptical an enquirer as any in existence' and 'exceeding (excessive) in love with *truth* to the prejudice of all national feelings'. George Petrie, the Topographical Branch's director, loved the way in which O'Donovan's work exhibited an 'evident approach to the character of *scientific proof*'. O'Donovan's obsession with the scientific accuracy of his linguistic work arguably levelled the playing field on which the Topographical Branch fought for attention with the ostentatiously scientific Trigonometrical Survey, and also demanded that space be made on the British map for *Irish* needs.

O'Donovan's emphasis on the utter seriousness of his researches counteracted a prevalent notion that the history of the Irish language was 'a region of fancy and fable', in Petrie's words. In the mid eighteenth century, Irish literature had suffered from a notorious literary hoax when the Scottish writer James MacPherson claimed to have discovered manuscript fragments of the works of a third-century poet, 'Ossian, the son of Fingal', from which he published 'authentic' translations. MacPherson asserted that his texts offered 'convincing proof' that stories of the legendary Irish figure of Fionn mac Cumhaill were merely versions of a much older Scottish epic about the character Fingal. Many readers were instantly sceptical, mainly because MacPherson could not, or would not, produce his original manuscripts. The poems were eventually written off by most as a fraud that had been composed by MacPherson himself, but in O'Donovan's eyes they had already done a great deal of damage. 'The poems of Ossian have bewildered the minds of the peasantry in Ireland and of the literati in Scotland,' he declared, feeling that they had undermined the status of Irish Gaelic literature. O'Donovan also translated ancient Irish poems and he distinguished his own endeavour from MacPherson's by insisting on publishing the original texts alongside his English versions to give them 'incontestible authority'.

In England and Wales, Thomas Colby had taken a utilitarian stance regarding toponymy and recommended that the versions of each place name that were spoken most widely should be selected for the map. But in Ireland, John O'Donovan fought for more attention to be given to place names' history. He persuaded Larcom that the Ordnance Survey should adopt 'that one among the modern names most consistent with the ancient orthography, not noticing the ancient name merely to interest the antiquary, but approaching as near to it as was practicable without Fancy'. O'Donovan wanted the Ordnance Survey's maps of Ireland to display the forms of contemporary place names that drew closest to the oldest Irish version that the Topographical Branch's researches had unearthed, thus making room on the map for 'all the rhymes and rags of history'. The names' Irish spellings were to be anglicised on the map, and he urged that these translations should be based more on the original Irish names' pronunciations than on attempted transliterations of their meanings. (In respect to house names and private estates, however, the Ordnance Survey continued to defer to the desires of the owner.)

O'Donovan's emphasis on history can be interpreted politically. He hinted that the modern Irish landscape was a tragic Babel in which 'the people do not agree upon the names of [their villages]; the name by which the whole of one mountain is known to one, being that by which a part of it only is known to another'. In O'Donovan's view, the principal cause of such linguistic confusion was the introduction of English settlers into Ireland, and that nation's history of plantation meant that Ireland had developed a 'forked tongue', a bilingual habit that had 'mangled' its old Irish names. He described some of these distortions in disgust: 'To comply with the general custom of sticking *town* as a tail to as many of these names as possible, the ancient name of Queen *Taillteann*, the daughter of *Mamore* was changed to Telltown, as if it were impossible to *tell* what *town* it anciently was (the name of)!' He worried that the Ordnance Survey's military engineers risked perpetuating this history by insensitively anglicising Ireland's place names on their maps. O'Donovan ridiculed the way in which, in a townland near Tobermore, County Londonderry, called Mainister Uí Fhloinn (Monaster O'Lynn, or O'Lynn's Monastery), the surveyor had jotted down the highly anglocentric misspelling *'Money Sterling!'* in his name book. O'Donovan must have been horrified when the misnomer of 'Moneysterlin' actually found its way onto the map.

It seems that O'Donovan thought of himself as a linguistic archaeologist, excavating the 'pure' old Irish names from generations of layers of confusion and change. In a poem, a 'wild rhapsody' that he sent to Larcom in 1834, O'Donovan described how he 'trace[d] his course along the darkened tract' of history, 'Facts shedding light before him as he goes'. He was wary of over-reliance on the spoken evidence given by living Irish residents, and favoured old printed texts and ancient manuscripts. O'Donovan was Adam in the Garden of Eden, doling out names to Ireland's lush pastures, rolling hills and effervescent rills. And his Eden was predominantly *Irish*. In the

Ordnance Survey, O'Donovan appears to have seen an opportunity to resurrect the face of the nation as it had appeared before the arrival of the
English and the ensuing traumatic history of plantation.

At the Topographical Branch, O'Donovan worked alongside a man called
James Clarence Mangan, who was employed as a copyist and scribe and who
also wrote poetry. Mangan cut a strange figure among his colleagues. As a
child, he had contracted a severe influenza that had badly affected his
eyesight and compelled him to don green goggles as protection against the
sun. These looked extraordinary against his vivid yellow hair (possibly a
wig), and 'skin as pale and taut as parchment'. As a poet, he occasionally
adopted the pseudonym 'The Man in the Cloak' and lived up to his name by
wrapping himself in a huge blue cape, hunched beneath a large pointed
hat 'of fantastic shape'. A friend commented that Mangan 'looked like the
spectre of some German romance rather than a living creature'. He was an
unhappy man, who was said to be oppressed by 'the animal spirits and
hopefulness of vigorous young men' – despite the fact that Mangan was only
in his early thirties when he worked for the Topographical Branch – and he
was seen to 'fle[e] from the admiration and sympathy of a stranger as others
do from reproach or insult'. He drank heavily and O'Donovan described
scornfully how his colleage was so well known to Dublin's public houses that
their landlords provided this *'poète maudit'* with 'pens and ink – *gratis*'. Indeed,
O'Donovan related how 'one short poem of his exhibits *seven different* inks,
and seven different varieties of hands, good or bad, according to the number
of glasses of whiskey he had taken at the time of making the copy'.

O'Donovan and Mangan clearly did not see eye to eye, in professional or
personal matters. They lived close to one another, and O'Donovan described
to a friend 'the mad poet who is my next-door neighbour'. '[Mangan] says
that I am his enemy,' the linguist recounted half-mockingly, '[and that I]
watch him through the thickness of the wall which divides our houses. He
threatens in consequence to shoot me. One of us must leave.' In turn, con-
travening Larcom's impression of O'Donovan as an inappropriately jocular
man, Mangan found his colleague to be 'severe, coldly-judging, and testing
ethics by the science of mathematics'. Most crucially, the two men possessed
very different views regarding the method and function of translation. Mangan

278

37. James Clarence Mangan by Frederick William Burton, made after Mangan's death in the Meath Hospital, Dublin, in 1849.

often professed that 'the merit of fidelity' is 'of a very questionable kind in translations', and in 1826 he composed a poem mocking 'bad Etymology and sad Orthography'. For Mangan, the past was irretrievable and he described in a poem how 'Dates, arithmetical tack, and all chronological cumber,/ Shall have been hurried, swept, hurled, as obsolete masses of lumber,/ Into the gulph of gloom, from whence there is no reappearing.' Changes in place names and in language were inevitable consequences of the inexorable march of time, and Mangan saw the task of the translator as that of a parent, fathering new and different offspring from old texts, rather than a linguistic archaeologist excavating the past. He had no compunction about translating from texts composed in languages he did not speak, and it was even rumoured that he did not speak Irish.

As a copyist, Mangan seems to have had little say over the place names that were adopted on the Irish Ordnance Survey maps, and he and O'Donovan clashed most visibly in matters of poetry in translation. In 1852

Mangan published an English translation of a poem by the sixteenth-century Irish poet Aengus O'Daly, which was edited by his antagonistic colleague. O'Donovan surrounded what Mangan designated as his 'Versified Paraphrase, or imitation, of Aengus O'Daly's Satires' with critical footnotes, commenting variously that 'this is incorrect', 'Mangan totally mistook the meaning of this quatrain', 'the translator is here very wide of the meaning', and 'the poet is not very happy here'. The last footnote unwittingly hit the mark: Mangan certainly wasn't very happy. It seems that his distaste for O'Donovan's mission to resurrect old Ireland's place names made collaborations with the linguist painful and his employment on the Topographical Branch an unpleasant experience. Some years after its closure, Mangan described mournfully how his work there had turned him into a ruin of his former self: 'The few broken columns and solitary arches which form the present ruins of what was once Palmyra, present not a fainter or more imperfect picture of that great city as it flourished in the days of its youth and glory than I, as I am now, of what I was before I entered on the career to which I was introduced by my first acquaintance with that lone house [i.e. Teepetrie] in 1831.'

A PLAY FIRST performed in 1980, which is now a stalwart of school syllabuses, has meant that the Irish Ordnance Survey is most popularly thought of as a tool of English imperialism. Brian Friel's *Translations* was the first play to be staged by the Field Day theatre company, which had been established by Friel and the actor Stephen Reagh to take drama to rural Irish communities. 'Field Day' is a pun on Friel-Reagh, in which the phrase's military connotations are overturned in favour of more joyous associations. Friel and Reagh entertained particular hopes for Ireland. Alongside the nation's traditional four provinces of Leinster, Munster, Connacht and Ulster, Field Day wanted to provide 'a fifth province to which artistic and cultural loyalty can be offered'. Friel described how the theatre company 'has grown out of that sense of impermanence, of people who feel themselves native to a province or certainly to an island but in some way feel that a disinheritance is offered to them'.

Translations used the Ordnance Survey's naming activities in 1830s Ireland as an image of this disinheritance. It tells the story of the clash of a fictional rural Irish community with the military might of the British Engineers of the Ordnance Survey, and narrates how the map-makers' brutal translation of Ireland's native place names 'into the King's good English' left its residents 'imprisoned in a linguistic contour which no longer matches the landscape of . . . fact'. In the play, the Ordnance Survey's acts of Anglicisation are nothing more than 'a bloody military operation', perpetrated by brutal and philistine English colonisers on the disenfranchised Irish. In Friel's eyes, John O'Donovan was a perfidious 'quisling' whose actions were an 'eviction of sorts'.

Translations is a play about bad acts of translation, and Brian Friel carried out many of his own wilful mistranslations of history in the play, such as describing how the surveyors carried firearms (they didn't). Most importantly, we have seen how the Irish Ordnance Survey's character was much more complicated than the purely imperialist endeavour that Friel describes. Comprised almost entirely of Irish-speaking Catholics, the Topographical Branch's 'imperative duty' was to promote Irish cultural heritage in the face of the 'collective folly and stupid intellect of the Empire', in Eugene O'Curry's words. O'Donovan's defence of the seriousness of the study of Irish literature played a crucial role in a movement known as the 'Gaelic revival', in which Ireland's ancient culture was seen to contribute meaningfully to the nation's contemporary identity and confidence. The Topographical Branch has even been termed 'the first peripatetic university Ireland had seen since the wanderings of her ancient scholars'.

So *Translations* is hardly a faithful account of the Ordnance Survey's activities in Ireland. But Friel never intended that it should be. His next play, *Making History*, explored how the 'empirical truth' of history might be justifiably transformed into 'a heroic literature' for the purposes of bolstering national self-confidence. *Making History* authorised 'the tiny bruises' that Friel openly admitted had been 'inflicted on history' by his earlier play. *Translations* had turned the Ordnance Survey into a mythical villain on which audiences could project frustration and anger at imperial history. Friel openly admitted in his diary that *Translations* was 'inaccurate history' and he expressed frustration that it 'was offered pieties that I didn't intend for it'.

IN ITS TOPOGRAPHICAL Branch and memoir project, its geological interests and its boundary delineation, the Irish Ordnance Survey showed just how ambitious a map could be. It was Thomas Colby's proudest achievement and his most luminous legacy. He saw the survey as 'a monument for future generations': a testimony to his own achievements and an aid to Irish economic progress. But the same all-embracing character that made the Irish Ordnance Survey such an awe-inspiring achievement was also the greatest hindrance to its principal objective: the completion of six-inch maps of the entire nation. Thanks to Colby's obsessive perfectionism, the project advanced very slowly in the years following its foundation. The Survey's lack of tangible results provoked a governmental enquiry in 1827 and 1828. The chair of this investigation, a rather overbearing military engineer called James Carmichael Smyth, who had been chosen specially by the Duke of Wellington, insisted on the Ordnance Survey's adoption of a number of time-saving and cost-cutting measures. He recommended that its director stay focused on the completion of the Irish map, and then the First Series of England and Wales, with as much expediency as possible, and put a stop to the Ordnance Survey's geological researches. Colby was incensed and he complained so vociferously at Smyth's downright philistinism that a new enquiry was set in place a year or so later. This time, Colby's old compadre Charles Pasley was appointed as its chair. Pasley's diary described confidentially how 'our report went to give the responsibility to Colby again, giving him his own way as far as we could'.

Colby had initially quoted £300,000 for the completion of the Ordnance Survey's national map and boundary survey of Ireland. But by 1832 its various projects had consumed this entire amount and not a single finished map had yet been published. Colby blamed the boundary measurement for much of the expense, arguing that he could not have known either how many townlands Ireland possessed or the complexity of their borders. He also pointedly remarked that the time-saving devices that had been implemented by Smyth in 1828 had only succeeded in producing sub-standard maps, whose revision had significantly delayed the Ordnance Survey's progress:

'You recollect the cry of haste, haste, etc., useless accuracy and so forth against me. Now the valuation is delayed, the survey is delayed, by the unavoidable revision [of] the early plans.' At this point, he estimated that another £420,000 and twelve years would be required to finish the Irish map. The Board of Ordnance and the government had little choice but to acquiesce.

In 1835, the newly founded British Association for the Advancement of Science hosted its annual meeting in Dublin. Its conferences were lively and pioneering affairs: the very word 'scientist', now such a staple of our vocabulary, had been coined by 'some ingenious gentleman' at an early gathering. (It is sometimes said that it was invented by Coleridge who, along with Wordsworth, attended a number of the Association's assemblies.) The 1835 Dublin meeting lived up to the British Association's illustrious reputation. William Rowan Hamilton was secretary, and he presented an address in which he argued that collaboration between different scientists from a host of multiple disciplines and nations was the bedrock of progress. 'Genius is essentially sympathetic,' Hamilton urged. It 'is sensitive to influences from without, and fain would spread itself abroad, and embrace the whole circle of humanity with the strength of a world-grasping love'. He held up an early copy of the first production of the Ordnance Survey's Irish memoir project and circulated it among the British Association's delegates as an epitome of the 'social feelings' that he encouraged. The response to this short draft of the Ordnance Survey's memoir of the parish of Templemore, in County Londonderry, was overwhelmingly positive. The British Association's *Report* of its Dublin meeting heaped praise on the book and its members urged Colby not to relinquish the memoir project until every one of Ireland's parishes possessed such an admirable volume.

Buoyed up by the British Association's praise, Colby and Larcom spent eighteen months greedily cramming their tome with as many products of the surveyors' researches as possible. When the *Ordnance Survey of the county of Londonderry, volume the first: memoir of the city and north western liberties of Londonderry, parish of Templemore* was published in November 1837, it began by describing how 'a map is in its nature but a part of a Survey, and that much of the information connected with it, can only be advantageously embodied in a memoir, to which the map then serves as a graphical index'. Over the next 350 pages, the memoir provided a compendious account of the 'physical features of the ground' of Templemore, its

> aspect, climate, and geological structure, as introductory to several branches of natural history, which in great degree depend upon them. The Second Part, in like manner, based upon the map, describes, in detail, the roads, the buildings, and other works of art, whose positions are shewn upon it . . . From this point, the Third Part commences; its first division, social economy, beginning with the earliest history of the people, the septs, or clans, whose descendants still may inhabit the district, and the various changes or improvements which have gradually led to the present establishments for government, education, benevolence, and justice. This account of the people and their establishments, leads naturally to the productive economy, which closes the work, as resulting from the means the people have been shown to possess for calling into beneficial action the natural state at first described.

Behind a contents page that resembled an Enlightenment tree of knowledge, the Ordnance Survey's memoir tried to produce 'a full face portrait of the land'.

It is possible to view Colby and Larcom's memoir project as the creation of an imperialist archive of knowledge about Ireland, and some historians have done so. That Larcom went on to supervise the 1841 Irish Census also demonstrated how the field of statistics emerged in this period as a means of social control. The two map-makers' hopes that their researches might tend towards the improvement of Ireland's manufactures, communications and industry echoed certain colonial 'experiments' that had been conducted in Ireland by British map-makers in the past. The seventeenth-century surveyor William Petty had openly compared himself to a 'political anatomist' who, just 'as students in medicine practise their inquiries upon cheap and common animals', had 'chosen Ireland as such a political animal . . . in which, if I have done amiss, the fault may be easily mended by another'. And that Larcom later prepared an edition of Petty's 'Down Survey' for publication only reinforces the comparison between his and Petty's endeavours. But however one interprets the political implications of the memoir project, there was no denying that it far exceeded its original remit. Larcom had initially estimated that each parish memoir would occupy only six pages, and the Templemore volume was spread over 350. He had budgeted £500 to produce memoirs for one county, but this single tome, which covered a solitary parish, had cost £1700. The Templemore memoir was relatively popular among readers and sold 1250 copies in the space of six years (Wordsworth bought one), but its exorbitant production costs meant that the memoir project's days were numbered.

Colby and Larcom's brainchild had, in fact, been deeply flawed from the start. The military engineers found it extremely tricky to make time for both surveying and memoir research. At the British Association's Dublin meeting, the mathematician Charles Babbage had optimistically claimed that the memoir work could be conducted 'in the evening' after a full day's map-making, but this was ludicrous: both tasks were full-time jobs in themselves. Even more problematically, the map-makers were not trained in anthropological research methods. The quality of their investigations was inconsistent and their discoveries were occasionally '*manifestly* wrong', as John O'Donovan put it. He acknowledged that 'the officers of the Royal Engineers were most excellent Mathematicians and accurate Surveyors' but he was extremely 'doubtful as to what extent their habits and education had qualified them to judge of the nature and quality of soils or the rotation of crops, &c.' An anxious witness of the surveyors at work on the memoir project warned that 'if such descriptions and remarks . . . should ever be published, they would be laughed at by all those who are acquainted with the soil and productions of the respective townlands'. In the late 1830s, the *Dublin University Magazine* and *Dublin Evening Post* foresaw that, thanks to its excessive costs and imperfect researches, the memoir project was doomed. Despite fierce protestations about its closure from certain Irish nationalist political groups (whose support for the memoir project perhaps gives the lie to those historians who interpret it as an imperial endeavour), by 1840 it was stuttering to a halt.

The memoir project was arguably also affected by a pronounced change in the spirit of the age. William Rowan Hamilton had celebrated Larcom as 'somewhat of a universalist' and the Irish Ordnance Survey's embrace of anthropology, statistics, toponymy, antiquarianism, geology and map-making was held up as an exemplary model of interdisciplinarity. Such an approach can be construed as a product of the Enlightenment's enthusiasm for universalism and a vibrant 'public sphere' of fevered collaborative discussion. With these same ambitions, the Royal Irish Academy had been founded in 1785 to promote the combined 'study of science, polite literature and antiquities', and it openly sought to incorporate '*all* the objects of rational inquiry'. But by the middle of the nineteenth century, the zeitgeist was changing and such 'general knowledges' began to be considered dangerously amateurish. As the century progressed, the literary personalities of the Royal Irish Academy were increasingly separated from the scientists, until two distinct cultures of 'the Arts' and 'the Sciences' had emerged. It is possible that the Ordnance Survey was also affected by this shift from universality to specialisation. In July 1840 Colby was commanded to forget his fascinating sidelines and to 'revert immediately to [his] original object', map-making and nothing more. The memoir project was doomed, and over in Britain, the Geological Survey also suffered under this changing spirit of the age: it was detached from the Ordnance Survey and handed to the Office of Woods and Forests.

In this climate, and with the impending completion of the Irish Ordnance Survey, the continuation of the Topographical Branch could not be justified much longer. But it was politics that finally sounded its death knell. On 2 May 1842 the Chancellor of the Exchequer, Henry Goulburn, received an anonymous letter from one who signed himself simply 'A Protestant Conservative'. Claiming to have worked under George Petrie, this mysterious correspondent warned Goulburn in hushed tones that the Topographical Branch was composed of Irish Catholic nationalists, who formed a hothouse of sedition. The linguists' 'bigotry and politics are carried to all parts of the kingdom', the writer cautioned, and 'persons sent from this office' had a habit of 'taking down the pedigree of some beggar or tinker and establishing him the lineal descendant of some Irish chief, whose ancient estate they most carefully mark out by boundaries'. Here his voice rose to a hysterical crescendo: 'They have actually in several instances, as I have seen by their letters, nominated some desperate characters as the rightful heirs to those territories.' The scaremongering worked, and the Topographical Branch was closed down on the basis that it was 'stimulating national sentiment in a morbid, deplorable, and tendencious manner'.

But all was not woe: the maps themselves tell a happier story. In 1833, the first Irish Ordnance Survey maps had been published, of Derry. On 9 May of that year, its director was proud to present a six-inch map of County Londonderry to King William IV (George III's third son, who succeeded George IV on 26 June 1830). Colby reported to his wife that the King 'looked over the whole atlas . . . very carefully, sheet by sheet, asking many questions about them and the Survey, and expressing his approbation, not only of the maps, but also of his Corps of Engineers and Artillery and of the Ordnance Department under which the work was executed'. Colby was also amused that, when he had been introduced to the King, he had said that '"he ought to have been ashamed of himself for not immediately remembering Colonel Colby" – that he thought I had belonged to the Artillery, and did not recognise me in the Engineers' uniform'. Maps of Antrim and Tyrone were published in 1834, and two more counties the following year. By 1842 every inch of Ireland had been mapped by the Interior Surveyors and the resulting charts were received by the public and the military with almost unalloyed pleasure. 'No one who has seen the six-inch Ordnance Survey maps of Ireland, especially those of the southern counties, can fail to be struck with their great clearness and beauty,' wrote one of Larcom's colleagues. 'They, for the first time, showed that by properly proportioning the relative strengths and sizes of the outline, writing and ornament, even an outline map [without hill shading] might be made a work of great artistic beauty and merit.' A German visitor to the Ordnance Survey's Irish headquarters noted that the maps, which had only been made by 'common workmen', ranked among the best in the world.

What strikes one on looking at the six-inch Irish surveys is how white and sparse they are in comparison to the one-inch maps of England and Wales, whose smaller scale brings a cluster of icons and lines together into darker and more claustrophobic sheets. The Irish maps appear more relaxed some-how, as if the engravers have exhaled deeply while marking the widely spaced symbols and strokes. The presence of neat field boundaries, which Mudge had commanded to be omitted from the First Series maps, gives the Irish surveys an orderly, geometric appearance. Despite the tiny icons of shrubs and forests that nestle in the corners of walls and on the borders of beautifully rippling lakes, the Irish six-inch sheets are perhaps less evocative than their English and Welsh counterparts, whose frenzied hachures and swift transitions from hilly to flat lands inspire a breathlessness in the reader not unlike the experience of traversing the landscape itself.

The Ordnance Survey's Irish maps proved far more immediately popular than their English and Welsh surveys. Thanks to Larcom's experiments in printing methods, they were much cheaper and more accessible, the equivalent price of a hardback bestseller now, and salesmen touted them from door to door. The Ordnance Survey extravagantly donated full sets of the maps to almost every governmental department and major library in Ireland. Some private booksellers blew the sheets up to three or four times their intended size, and flogged them to landowners as superior estate maps (a mistake, as the magnification highlighted minuscule errors). The head of the Ordnance Survey's Boundary Commission, Richard Griffith, reported how 'the principal gentlemen of Ireland . . . all had 6-inch maps on their study walls'.

Colby was gratified to see that these Irish maps were put to many of the practical uses of which he had dreamed: the boundary survey was used to calculate not just the county cess tax, but poor-law rates and income taxes too. Road-builders and railway-engineers found the maps invaluable in their work, as did those involved in the draining of Ireland's bogs. And the Geological Survey continued to base its work on the Ordnance Survey's charts, even after that department had been detached from the Ordnance Survey and given to the Office of Works. When the last map of Ireland was published in 1846, the Ordnance Survey had finally produced a map of a nation. But it was not the nation that it had originally set out to map. The First Series of one-inch maps of England and Wales still remained unfinished.

CHAPTER TWELVE

'A Great National Survey'

ON 30 OCTOBER 1841, at half past ten at night, a sentry guarding the Jewel Office at the Tower of London noticed a bright light at the windows of the Bowyer Tower, on the north side of the complex. He left his post to investigate the strange flare more closely. Transfixed at first by its shifting intensity, it was a few moments before he realised that it was a fire. Sprinting towards the Tower's central quadrangle, the sentry started shouting with all the force he could muster. Soldiers began pouring from every doorway and the air was soon full of the sound of bugles. The Tower possessed its own fire engines, which made their way to the conflagration, where they were joined by those from neighbouring parishes, and the main gates of the complex were closed to the public. But the fire, which had begun in an overheated chimney, had already made its way to the building in front of the Bowyer Tower. The Grand Storehouse, or Armoury, was reputedly the largest room in Europe and contained some 300,000 guns, military carriages, cannon, gunpowder, bombs and various spoils of war. The soldiers who had been running towards this building quickly turned on their heels when they saw the proximity of the blaze to this informal arsenal. The next day, numerous newspapers reported how, once the fire had got hold of the Armoury, flames burst 'from several windows with extraordinary fury'. Around midnight, the fire was reported to be 'gushing forth from every window of the building, which had all the appearance of the crater of some volcano'. The inferno and explosions were discernible across the city, and a journalist reported how 'the reflection on the surrounding houses, and on the shipping in the river, produced a most striking effect'.

The Map Office, which had once been situated in the White Tower, in the middle of the complex, had been moved in 1752 to the residence of the Principal Storekeeper. During the fire, this removal at first seemed fortunate. In the broadwalk that separated the Armoury from the White Tower, the heat was found to be 'so intense, that it was utterly impossible for a human being to stand'. But soon the wind changed direction and, leaving the White Tower unscathed, the flames began to make their way west towards the Jewel Tower. Here the Crown Jewels were protected behind a locked iron grating, whose only key was in the possession of the Lord Chamberlain. With no time to hesitate, a police superintendent used a crowbar to wrench apart the iron railings and squeezed himself inside the cage. As he began passing the gems through the gap, panicked voices outside warned of the fire's swift approach. But the superintendent continued to prise the ornaments from their stands, and after twenty minutes a small group of soldiers, firemen and policemen left the building unharmed, cradling a large pile of precious jewels and coronets.

As the fire passed the Jewel Tower around two o'clock in the morning, it made its way towards the Map Office. We can imagine Thomas Colby arriving at the scene and, with the assistance of a team of soldiers, frantically packing the paper records of fifty years into boxes. Along with its instruments, the Ordnance Survey's truly priceless papers and maps were quickly conveyed to safety in another part of the complex. But in the hurry and the confusion, Jesse Ramsden's ten-foot zenith sector, which had been used for the meridian arc measurement in 1801 and 1802, and for the Shetlands extension in 1817, could not be rescued in time. When the flames finally reached the Map Office, this uniquely precise instrument was engulfed. As dawn began to appear, the headquarters of the Ordnance Survey were left in ruins.

By October 1841, most of the Ordnance Survey's personnel had either returned to Britain from Ireland or were in the process of doing so. Colby himself had arrived back in London in 1838, and by 1842 only fifty Sappers would remain across the water. The 'Great Fire' at the Tower made them all homeless. The Board of Ordnance's officials quickly scouted around for new quarters for its map-makers, and within six days of the fire they had come up with the rather unenticing prospect of an abandoned barracks that had been used for a military school until lack of pupils had led to its closure. This building was in Southampton, about eighty miles south of the capital. Colby had spent much of his time in Ireland fantasising about his return to London, and he must have been devastated by this removal of the Board of Ordnance from the centre of government and from the grand buildings that housed many of the nation's scientific and intellectual societies.

On 31 December 1841 the Ordnance Survey reluctantly made its move south. A member of its party recalled how 'early in the afternoon, the first arrivals asked for admission at the entrance gates – which was refused by the old barrack-sergeant then in charge of the buildings'. But an officer called William Yolland, who had worked on the Survey for three years and was described by a colleague as 'not a person to hold a long parley with', soon gained admission. Over the next hours, surveying parties descended on the Southampton barracks from all over the British Isles. Inside the bare walls, there was nothing to make them comfortable and 'everything at first was confusion'. But the following morning an Ordnance Surveyor recalled that 'a large load of furniture, bedding, and other requisites arrived from Winchester Barracks, which was soon put in order, and by night all was prepared'. A team of 'carpenters, bricklayers, smiths, masons, painters, and every kind of workman which the corps could produce were set to work' and 'in a very short time the establishment was got into working order'. Soon the engravers arrived too, and the Ordnance Survey's Southampton headquarters were ready for business. In 1842 Colby, now fifty-eight, moved his wife and their four sons and three daughters permanently to the city, and after a visit to his comrade's new abode, Thomas Larcom described with a hint of surprise how 'at home, the steady calm reasoner of one moment became the next, almost the giddy boy, when playing joyously and without restraint with his children'. Over time, the Ordnance Survey took to its new residence. The mapping agency would remain at the Southampton barracks for almost a hundred years, until bombing during the Blitz devastated the site and forced the Ordnance Survey to relocate temporarily to Chessington. In 1969 a new home was purpose-built on the outskirts of Southampton, where it is still to be found today.

Thomas Colby had been distracted by the Irish survey for almost twenty years. From his new place of work, he turned his attention back to the First Series of maps of England and Wales. This project had been over forty years in the making and was only now inching its way to completion. While the Ordnance Survey had been in Ireland, a very small team of map-makers had been engaged in revising the finished First Series maps and the engravers had been busy producing charts of Wales. There were also plans afoot to re-examine the height of Britain's peaks, measuring all altitudes from a 'datum', a standard position, which the Ordnance Surveyors eventually took as the 'mean sea level' at the Victoria Dock in Liverpool. What became known as the 'primary levelling of Great Britain' took place between 1840 and 1860, although it would be subsequently recalculated from a different datum, at Newlyn in Cornwall, in the early twentieth century. And the Trigonometrical Survey was also shortly to be recalculated in order to better compensate for the effects of local attraction and the size and shape of the earth; this exercise would be finished in 1852. But mapping a nation is like the well-known analogy of painting the Forth Railway Bridge: almost as soon as the map of a region has been published, a new map is likely to be necessary to keep up with changes in the landscape. And it is not only the scenery that alters. As time passes, fresh demands are made of maps that require adaptations in their content and appearance, and place additional burdens on their makers.

DURING THE YEARS of the Ordnance Survey's Irish adventure, Britain had fully entered the era of the Industrial Revolution. Every year saw dramatic change in which, at first near Leeds, Stockton and Darlington, and then between Liverpool and Manchester, and in South Wales, and finally stretching across England and southern Scotland, lengths of iron and then steel railway track were laid. The clouds of smoke and soot that puffed from

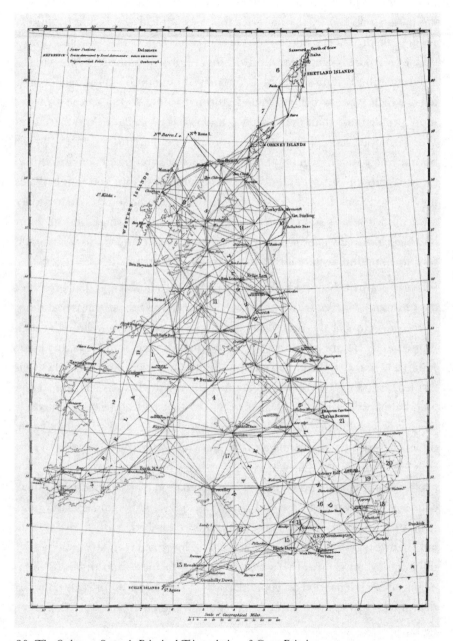

38. *The Ordnance Survey's Principal Triangulation of Great Britain.*

the engines that travelled along these railway lines were vastly multiplied in the nation's skyline by the factories that were quickly replacing traditional small-scale producers, and by the clusters of chimneys of the urban centres that were mushrooming around the new hubs of industry. The skies of industrialising Britain were not a conducive home to the map-makers' sight lines, which were forced to struggle through smog so thick that it proved more problematic than the dense fog and cloud of Ireland. And there were changes under, as well as over, the ground. The demand for coal to power both the new steam engines and Britain's factories required new workforces for the mines of central and northern England, Scotland and South Wales, in which 'unguided, hard-worked, fierce, and miserable sons of Adam' toiled away in the nation's subterranean world. The kingdom whose infrastructure had been gradually coalescing through the previous century was now rapidly transforming into a state integrated through industry.

It was not just the rapid alterations to Britain's landscape that presented the Ordnance Survey with a challenge; the new uses for the maps that many of the projectors of the Industrial Revolution devised did as well. In the years immediately before and after Queen Victoria's accession to the throne in June 1837 many of the nation's administrative and governmental bodies were formed, including the Poor Law Commission and the Board of Health. These found applications for the Ordnance Survey that had never been dreamt of when it was established in 1791, and it quickly became clear to Colby that the project he was now managing and the nation it was mapping were both radically different to those in which he had been first employed in 1802. The explosion in the size and density of the population of cities had made social concerns like poverty, disease and contagion, and sanitation unavoidably pertinent. In 1834 the Poor Law Commission was established, partly to research the links between poverty and disease, and also to investigate sanitary conditions in the large urban centres that grew up around factories and mines. Railway engineers also depended on good maps of the landscape to plot the course of their tracks. It was a new country under a new monarch, and it needed a new map.

His experiences in Ireland had persuaded Colby of the benefits of mapping on a larger scale than the one-inch First Series. His return to Britain at the height of the Industrial Revolution reinforced this view, as the Ordnance Survey became the focus for a number of urgent requests. In the mid 1830s, the Poor Law Commission contacted the Ordnance Survey to ask it to produce maps of Britain's towns on the huge scale of five feet to one mile, to reveal the presence of every existing tap, drain, sewer and water source, and improve urban sanitation. The Tithe Communication Act of 1836 had also created a demand for large-scale maps to accurately delineate property boundaries, in order to calculate the tithes that were owed by landowners to the Church. So in April 1840, Colby wrote to the Inspector-General of Fortifications with a proposition. England's northern counties still needed to be mapped in order to bring the First Series to completion, and a separate survey of Scotland had been begun in the 1810s. Colby suggested that these largely unmapped parts of Britain should now be surveyed, not on the one-inch scale of the First Series, but on the larger six-inch scale that had been used in Ireland. He proposed that the resulting maps of England's northern counties could subsequently be reduced to the one-inch scale, to finish the First Series on a uniform basis.

In October 1840 the Treasury agreed to Colby's scheme. In February 1841 the Master-General of the Board of Ordnance passed a 'Survey Act' through Parliament, which gave the Ordnance Survey's map-makers the legal right to enter any private property on British soil with three days' written notice. It was specifically designed to assist those engaged in mapping the nation on a large scale, who needed access to private estates to accurately depict their extent, internal groundcover, buildings and boundaries. Buoyed up by this new power, in 1842 the Interior Surveyors began mapping Lancashire and Yorkshire on the six-inch scale. Three years later, some of the Ordnance Survey's map-makers were diverted away from their work on the First Series to survey the urban centres of Fleetwood, Clitheroe, Manchester and Lancaster, on the enormous five-foot scale requested by the Poor Law Commission. Mapping on such a large scale created surveys in which individual shrubs might even be identified: it was enormously time-consuming and depleted the surveyors available for the small-scale mapping of England and Wales. In 1845 the number of map-makers remaining at the Ordnance Survey's disposal was further decreased. In that year, known as 'the great railway year', it was said that 'every one seemed to have a mania for new schemes' and 'many lines were proposed'. But whereas the Poor Law Commission was working in collaboration with the Ordnance Survey, the railway companies became a rival when they decided that the First Series' one-inch maps were too small to plan new railway lines. They offered the huge salary of three guineas a day to surveyors to draw up bespoke maps, and 287 employees left all levels of the Ordnance Survey for this lucrative employment.

The following year, amid all these upheavals, the Ordnance Survey lost its director. In 1846, Thomas Colby was promoted to major-general, a much-deserved accolade. But under the Army's rules he was no longer allowed to remain director of the Ordnance Survey with this rank, so in April 1847 he had no choice but to take retirement. Although he may have been initially disoriented and found it hard to adjust to his new relaxed pace of life, Colby soon transferred his perfectionist energies to his seven children, moving the family to Bonn for a while to further their education. Before he left, Colby set the Ordnance Survey's finances and achievements before the House of Commons and the Board of Ordnance. The year 1847 marked exactly a century since David Watson and William Roy had begun surveying Scotland in the wake of the defeat of the 1745 Jacobite Rebellion at Culloden. In the intervening period, British map-makers had wrenched the nation away from its state of cartographic myopia, a feat in which Thomas Colby had played no small part.

Under his watch the Trigonometrical Survey of Britain had been very nearly completed. Ireland had also been enveloped in its own national triangulation and Interior Survey, and maps of its entirety were now available for purchase. Interior Surveys of Scotland and northern England were under way on the six-inch scale, and First Series one-inch maps of England and Wales had been completed and published up to a line between Hull and Preston. After three major price changes, these iconic maps were no longer prohibitive luxuries. Sold in individual sheets of forty by twenty-seven inches, fifty-nine were available for seven or eight shillings each (less than two days' wages for a craftsman in the building trade), and the rest retailed even more cheaply at between three and six shillings. Thanks to increased demand and these reduced prices, over 28,000 sheets of the First Series had been sold in

1845 alone. By 1846 the First Series' sales had brought in £4259. This was, however, only a tiny fraction of the total expenditure on all the Ordnance Survey's endeavours since 1791, of £1,462,522 (around £85 million in 2010's currency). Of that enormous sum, £28,375 had been spent on the Scottish survey and £574,439 on the First Series, and Colby estimated that to complete the latter project, a further £316,492 would be required. Thomas Colby had devoted forty-five of his sixty-three years of life to the Ordnance Survey and he could retire with the gratifying knowledge that he had played a vital role in setting Britain's lands in order.

AS HIS SUCCESSOR, Colby recommended to the Board of Ordnance the officer William Yolland, who had been employed in the Royal Engineers since 1828 and had joined the Ordnance Survey after instruction at Chatham and Woolwich and a four-year stint in Canada. Since its move to Southampton, Yolland had been in charge of the Ordnance Survey's general running. But to Colby's annoyance and disgust, in March 1847 it was announced that Lewis Alexander Hall, a 53-year-old lieutenant-colonel in the Royal Engineers with no experience on the Ordnance Survey, would be appointed its Superintendent. Hall, who possessed a flabby jawline, a small petulant mouth and beady eyes, was Chief Engineer of the London District, and it may have been that the commander of the Royal Engineers and Corps of Royal Sappers or Miners, together with the Master-General of the Ordnance, were both seduced by his experience in military engineering. Colby commented angrily that this decision 'will show most distinctly that neither' of his two superiors 'have any notion that the charge of a great national survey requires any experience of such a duty'. And indeed, over the seven years of Hall's superintendency, the Ordnance Survey fell into deep trouble. Amid desperate indecision, mostly around the scales on which the maps should be conducted and engraved, the First Series was nearly forgotten.

In November 1847 the Metropolitan Sanitary Commission contacted Hall and Yolland (who adopted an active deputy role in the project), to request that the Ordnance Survey make a map of London on the five-foot scale, to help the improvement of drainage in the capital city. The two men optimistically estimated that such a survey would take between six and eight months to complete, at a cost of £104,000, and that it would cover 900 sheets. It actually took two years. Between 1848 and 1850 the Ordnance Survey's map-makers made the most minute and accurate survey of London then in existence. Although there were some objections about the expense of such an undertaking, the *Daily News* reported gleefully in January 1848 that 'the new brooms of the metropolitan sewer commission are at work'. On 9 May, another journalist for the same newspaper offered a breathless account of how 'a company of Royal Sappers and Miners were employed at the summit of St Paul's Cathedral, erecting a temporary observatory immediately beneath the ball and cross, for the purposes of the ordnance survey of the metropolis'. He continued: 'the dexterity with which the men pursued their operations at that dizzy height, their only support the architectural ornaments of the building, excited considerable curiosity'. From this elevated platform, a surveyor called James Steel made some 10,000 observations in the four months between May and August with the smaller eighteen-inch theodolite that was used by the Interior Survey. Although on two occasions large pieces of wood were dropped from this great height, one of which hit the pavement with a noise 'like the booming of a piece of Ordnance', nobody was harmed in the course of the map-making.

As one division of the Ordnance Survey was working on the London map, further teams were surveying thirty-five towns on the five-foot scale for the Poor Law Commission. Between them, these responsibilities meant that the six-inch mapping of northern England was proceeding very slowly, and by 1850 only twenty-three sheets of Lancashire and Yorkshire had been published and none had been reduced to the one-inch scale of the First Series. The irresolution about the optimum scale on which the maps should be conducted soon became farcical, in what is now referred to as the 'Scales Dispute'. When the Board of Health was formed in 1848, it recommended that the five-foot scale of the town maps was, in fact, much too small to cope with the multiple demands of the Industrial Revolution, and advised the adoption of a scale twice as large, at ten feet to the mile. The

Tithe Commissioners also weighed in, and called for the entire nation to be remapped on a scale of 26⅔ inches to a mile. Over the next four years, amid a babel of recommendations from various Select Committees, the only constant was an almost ubiquitous assertion of the need to complete the First Series of one-inch maps of England and Wales as soon as possible, which was still widely accepted as the best scale for a 'general map' of the nation. The Treasury eventually despaired of the Ordnance Survey's indecision and took the matter into its own hands, asking a variety of landowners, authorities and institutions for advice on the matter of scale. A preference emerged for the one-inch scale for general purposes, and a scale of between twenty-four and twenty-six inches to the mile for civil purposes.

In 1854, Lewis Hall left the Ordnance Survey to take up a command in Corfu. His resignation occurred just two years after the sudden death at sixty-nine of Thomas Colby, who had been living in New Brighton, near Birkenhead, with his family. A colleague described how 'with little or scarcely any warning, the Spirit, yet active, was summoned from the Body, yet firm and hale and . . . he passed from the scene of so much bodily and mental exertion to rest and peace eternal'. After his death his widow was awarded a life pension by the government, in recognition of his achievements. There could not have been two more different superintendents than Thomas Colby and Lewis Hall. While the first was obsessively committed, eccentric and skilled, the latter was indecisive, inexperienced and unimaginative. Hall's directorship had been characterised by confusion and false starts, and the comparison with Colby that the latter's death must have provoked was not flattering. An obituary of Thomas Colby in the *Memoirs* of the Institution of Civil Engineers celebrated the Irish Ordnance Survey as 'really the great work of his life', where he had brought 'into harmonious action, the labours of about forty observers and of several hundred surveyors and draughtsmen', not to mention the shared invention of the Compensation Bars which 'would alone suffice to give to their author a claim to a high place in the list of improvers of geodetic science'. It had been Colby who 'was not only fully aware of the direct advantages of an accurate survey of the country, but was the first to point out the collateral benefits, to be derived from combining with it searching investigations into the geology, mineralogy, natural history, statistics, and antiquities of the country'. The obituary ended by reflecting that 'his life was a course of scientific research, and his name will hereafter be inseparably connected with the history of the Ordnance Survey' and 'it might be said of him, that few men were more sincerely regretted'.

Few, on the other hand, were sorry to see Hall go, and the limited nature of his achievements is perhaps signified by the fact that he has no entry in the *Dictionary of National Biography* beside those of his predecessors, Roy, Mudge and Colby. He was replaced by Henry James, who had worked on the Ordnance Survey for twenty-eight years, mostly in Ireland (with a brief interlude of four years away from the Survey, in which he was employed variously by the Admiralty and a commission for investigating the use of iron in railways). At the beginning of James's directorship, his employer, the Board of Ordnance, which was under the Master-Generalship of Lord Raglan, experienced devastating disgrace during the Crimean War. The first Russian winter of the war, 1854–5, was extremely harsh; the Board found itself woefully unprepared and failed to supply adequate provisions. Many of the British Army's modes of transport and supply broke down, and the hospital arrangements were notoriously bad, as a result of which many died from the cold, lack of food and their injuries. Blame was laid at the door of the Board of Ordnance and Lord Raglan, who wrote: 'I am charged with every species of neglect.' In 1855, this 400-year-old institution was abolished by an Act of Parliament. Control over the Artillery and Engineers, and the Ordnance Survey too, was passed to the War Office, a department of the British government that had been responsible for administering the British Army since 1684. The Secretary of State for War replaced the Master-General as the Ordnance Survey's chief.

While adapting to this new regime, James launched himself into the ongoing Scales Dispute. He decided that a 1:2500 scale (roughly twenty-five inches to a mile) would be adopted as the basic scale in Scotland, to replace the six-inch scale on which its maps had hitherto been constructed. The

Treasury agreed but stipulated that the six-inch scale should be retained for uncultivated areas, and it again pleaded with the Ordnance Survey to *please, at last* complete the First Series. Over the next five years, more committees bombarded James with a variety of conflicting recommendations regarding scale, but he continued to survey on his favourite scale of 1:2500, although he depicted some major Scottish towns at 1:500 (about ten feet to a mile). He implemented the 1:2500 scale in the northern counties of England too and recommended that southern England be remapped on this basis. By 1859 the survey of Scotland was nearly finished on both the six-inch and the 1:2500 scales.

At the end of the 1850s James forecast that the First Series might soon be completed. Acutely aware of its iconic status, he had issued instructions immediately after his appointment that, as soon as an unmapped area of England was surveyed, its maps should be reduced to the one-inch scale and engraved as a matter of urgency. The public was starting to gripe at the colossal amount of time it was taking for the First Series maps to appear. On 17 September 1862, *The Times* carried a letter from an anonymous correspondent called 'Surveyor', who described his expectation that 'the Ordnance Survey of England would . . . furnish me with a correct and accurate plan of the country as it exists at the present day, with all the late additions of railways, alterations, and increase in the neighbourhood of the rising towns &c.' This man then proceeded to relate his disappointment 'on receiving' maps of Plymouth, south Devon, Dorchester and Weymouth,

> to find in many sheets railways that have been in existence for years not laid down; towns and cities that have grown to double the size they were half a century ago still appearing on the map as they were then; and that most of these maps (I speak principally of the south of England) were engraved in the early part of the century 1809-27, and remain as then, with the exception of now and then a railway having been added after the lapse of years, but more frequently left out altogether.

Henry James drafted a measured reply to this correspondent, which *The Times* printed on 22 September. Although he pointed out that 'Surveyor' had mistakenly been provided with old unrevised impressions of the maps of Plymouth and Dorset, he admitted that 'no one is more conscious than

I am of the fact that the old Ordnance maps do require a very extensive revision to bring them up to what they ought to be'. But, he added hopefully, 'half a loaf is better than no bread'. James proceeded to explain the rationale behind the decision to resurvey the entire nation on a much larger scale. Even though it would take a little longer than a straight revision of the one-inch maps, mapping on a large scale would still 'give us the means of correcting the 1-inch map without incurring the expense of sending surveyors specially for that single object'. The poet and cultural critic Matthew Arnold wrote a piece in the *London Review* shortly after James's letter to *The Times*, reflecting on the gravity of the Ordnance Survey's undertaking. 'A Government's first and indispensable duty in the way of map-making is to provide a *good* map of its country, not to provide a *cheap* one,' Arnold emphasised. 'Let [Sir Henry James] clearly understand what is expected of him.'

By 1863, only eight sheets of the First Series remained to be published. In his letter to *The Times* the previous year, James had promised that 'this year I shall be able to get the survey of the north of England finished, and complete the 1-inch map of England and Wales'. But many of his readers had heard similar promises before and, like its forerunners, James's assurance failed to materialise in the time he had predicted. The reason for the delay, however, was unprecedented. Between 1864 and 1869, Ordnance Survey map-makers were sent to the Middle East, to Jerusalem and the Sinai Peninsula. The recently completed large-scale map of London that had been made for the Metropolitan Sanitary Commission had so impressed the Dean of Westminster, Arthur Penrhyn Stanley, that he had called for a similar survey to be made of Jerusalem, 'to improve the sanitary state of the city' and protect pilgrims from water-borne diseases. The Palestinian territories were then under the rule of the Ottoman government, who voiced no objection to the presence of British surveying parties in the area. And the War Office may have seen in the proposed survey an opportunity to spy on French engineers' construction of the Suez Canal, which, despite fierce hostility from the British, was under way in Egypt. But it is hard in retrospect not to grow a little exasperated with the Ordnance Survey's easy distraction from the task in hand, even if such digressions were the result of an admirable intention to engage with geography in its fullest sense and piece together a complete image of the world in which we live.

IN SEPTEMBER 1864 a party of six map-makers departed from Southampton to make two maps of Jerusalem, on scales of 1:2000 and 1:10,000. Upon arrival, they found the city's sanitary state every bit as bad as they had been led to expect. 'The city is at present supplied with water from the numerous cisterns under the houses in the city, in which the rainwater is collected,' one map-maker explained. 'But as the water which . . . runs through the filthy streets is also collected in some of these cisterns, the quality of the water may be well imagined, and can only be drunk with safety after it is filtered and freed from the numerous worms and insects which are bred in it.' 'Of the drainage of the city,' he commented brusquely, 'it is sufficient to say, that there is none in our acceptation of the word, for there are no drains of any kind in the city, and the accumulation of filth of every description in the streets is most disgraceful to the authorities.' The map-makers quickly set about selecting and measuring their baselines and conducting a triangulation that stretched across Jerusalem, before beginning the Interior Survey.

It was very hard work. The Survey Act of 1841 had given the men carte blanche to enter private property in Britain, but no such power existed in Jerusalem, and the Turkish governor advised the surveyors to 'proceed with caution' amid the city's 'diversities of religion and population'. Furthermore, the terrain was rough, the air hot and dry, and the conditions dirty. Later the Ordnance Survey would consider that only map-makers who had previously worked among the rugged territory of Highland Scotland were hardy enough to survey in such conditions in the Middle East. The surveyors were forced to ascertain the position and depth of Jerusalem's cisterns with a ruler or tape measure and compass by descending into the holes on rope ladders. One map-maker noted: 'in some of the smaller cisterns the shaft was not large enough for this, and a rope tied around the breast was used, the arms being held well above the head to diminish the width of the shoulders as much as possible'. 'It is no easy matter to work with a candle in one hand and up to the knees in water,' he emphasised.

But there were significant consolations to the Middle Eastern mapping project. Henry James had instructed the men to attend to 'the geological structure of the country, and to bring home specimens of all the rocks, with their fossils'. And he had also provided them with a relatively newfangled toy, a camera, 'to enable Serj. McDonald, who is both a very good surveyor and a very good photographer, to take photographs of the most interesting places in and around Jerusalem'. This sort of activity paved the way for the widespread adoption of photography in surveying in the early twentieth century. After their arrival back in Britain in July 1865, James and another mapmaker, Charles Wilson, wrote up and published an account of the expedition, the *Ordnance Survey of Jerusalem*. The text could not hide the surveyors' excitement and fascination with their task. They may have been initially scathing about Jerusalem's drainage system, but they recounted how:

> when we come to examine the ancient systems for supplying the city with abundance of pure water, we are struck with admiration for we see the remains of works which, for boldness in design and skill in execution, rival even the most approved systems of modern engineers, and which might, under a more enlightened government, be again brought into use.

They described mosques and fortresses in detail, especially Haram-es-Sherif, 'the name now commonly applied to the sacred enclosure of the Moslems at Jerusalem, which, besides containing the buildings of the Dome of the Rock and Aksa, has always been supposed to include within its area the site of the Jewish Temple'. The map-makers were hopeful that their researches allowed scholars to exactly pinpoint biblical locations in the contemporary Middle Eastern landscape.

So it was not with displeasure that the Ordnance Survey received the news that it would be back in the region between 1868 and 1869, this time to map the Sinai Peninsula. The Royal Society and Royal Geographical Society collected public donations to fund the expedition, driven by the idea that detailed knowledge of the area's geology, botany and zoology might elucidate the precise route that had been followed during the biblical Exodus. The resulting account of the mapping adventure explained 'that there is a great need to carry out such a survey [which] must be manifest to all students of Old Testament history; among the most important and interesting questions which are now subjects of inquiry are the locations of the Passage of the Red Sea, the Route and Encampments of the Israelites, and the identification of the Mountain of the Law Giving'. The knowledge that the Indian Navy had recently made a map of Sinai's coast may also have been significant in persuading the War Office of the project's military utility. A map-making party arrived in Sinai in November 1868, with the aim of producing a map of the area on the scale of two inches to a mile. By April 1869 their project was complete and they were back in Britain by the end of May. The maps and an accompanying written memoir were published in three parts in March 1872.

THE SINAI MAP-MAKERS returned to Britain in time for a momentous day. On 1 January 1870 the single remaining map of the First Series, sheet number 108 of south-west Northumberland, was finally published. One can imagine the maps being delicately carried from the printing room at Southampton by military engineers flushed with excitement, and placed in carriages that distributed them to map-sellers across the land. For by now there were around 150 official 'agents for the sale of Ordnance maps' in the United Kingdom's major towns. Six famous publishers in London, Edinburgh and Dublin also sold the charts, including Stanford's (to this day a popular London map-retailer), and we can imagine the crowds clustering around this historic artefact as shopkeepers proudly positioned them in their windows. A journalist for the *Northern Echo* hailed the fact that 'the trigonometrical survey of England and Wales on the scale of one inch to one mile has just been completed'. He added wryly that 'the circumstance that it was eighty years ago since this survey was ordered to be made may be taken as an instance of the speed at which things sometimes move in official quarters'. But he acknowledged the rapidly changing landscape with which the surveyors had contended, pointing out that the 'railway system itself had no existence when first the Ordnance Survey of England and Wales was authorised by Parliament'. Two years before the completion of the First Series, the

Ordnance Survey's work had been put on show at the Paris Exhibition, where it had been commented that 'the Ordnance Survey is a work without precedent and should be taken as a model by every civilised nation'.

The publication of Sheet 108 marked the conclusion to the Ordnance Survey's project to map England and Wales for the first time, from scratch, with utmost accuracy, on a uniform scale of one inch to one mile. (Sheet 100, of the Isle of Man, would not be published until 1873, however.) The First Series was completed 123 years after William Roy had taken his first steps in the Scottish Highlands, there developing the skills that would ultimately lead to this momentous event. Prior to the initial landmark of the Military Survey of Scotland, the inhabitants of the British Isles had only been able to contemplate their respective regions in a shattered looking-glass of fragmentary and defective maps. But by 1870, English and Welsh citizens owned a lifelike cartographical mirror of their countries, whose sheets were then retailing at the modest price of 2s 6d, which was considered to be 'within the reach of all who may require such aid'. Although the Ordnance Survey's sheets were usually experienced individually, it seemed to be important to readers that someone, somewhere could piece them all together into a complete image of a unified nation. The *Daily News* described with satisfaction: 'the maps fit together at the edges without any overlapping or duplicate engraving, so that they form, not merely separate maps, but . . . one map'.

Over the same period that William Roy, William Mudge, Thomas Colby and a host of other talented contributors had dedicated themselves to the creation of this complete map of England and Wales, the United Kingdom itself had come into being as a country that was integrated through, among other factors, regional unions, the development of state-wide networks of transport and communication, and the fostering of a strong sense of nationalism during a long period of warfare. The changing nature of the national state created applications for the Ordnance Survey that meant that the consumers of the First Series of maps ranged from economists, civil servants and aristocrats to estate agents, walkers and industrialists. They were privileged to possess a mapping agency whose outlook was centralised and national, intent on supplying readers from a host of civilian and military backgrounds with maps of every mile of the British Isles. Despite its name, which had well and truly stuck, the Ordnance Survey was clearly no longer just a military map, and this was reflected by the decision in 1870 to transfer the mapping agency to the government department known as the Office of Works, within the Office of Woods, Forests, Land Revenues, Works and Buildings.

When one spreads out an array of First Series maps over the broad tables of map-reading rooms in libraries across the country, and casts one's eye over their sheets, the attention flits from the thin parallel lines that denote where the south coast meets the sea and resemble gentle ripples in a pond, to the dramatic hachures of the rolling hills of Hampshire. The eye can travel upwards, across sheet piled on sheet, roughly following the meridian arc that Mudge himself measured, through dark forests speckled with miniature trees, along invariably empty roads and oil-black rivers, to the exposed summit of Arbury Hill in the Midlands, then through meticulous grids of settlements in Nottinghamshire to the dense clusters of parallel lines that evoke the undulations of the Yorkshire landscape. There the eye races through a flock of carefully engraved place names and scattered icons of churches and tiny ruins, until it finally meets the sea once again, just north of Burleigh, where the solid markings of the land disappear into a blank space on the map.

Despite the years that separate us from their creators, each First Series map still provides enough information to translate its language of lines and symbols into a three-dimensional landscape in the mind's eye. And the sheer pleasure that many find in the act of map-reading today draws us into sympathy with the earliest purchasers of the Ordnance Survey's maps. I, for one, cannot help but be moved when I read of how one such map-lover described in 1862 that he could 'stand an hour at a time' in front of one of its sheets, 'tracing a good run, or, if that wasn't his line, planning rides and drives. Every glance at a name on the map recalled pleasant days, reminded him of an acquaintanceship falling a little in the rear, and suggested pleasant visions.' This besotted enthusiast for the early Ordnance Survey was delighted to find 'how many unexpected bearings of the country are disclosed in a good map, and how many mysteries of the landscape reveal themselves in it!'

Maps of Freedom

A COUPLE OF summers ago, while researching this book, I spent a few days in the Upper Ward of Lanarkshire tracing the paths of William Roy's childhood. At around three o'clock one afternoon on a glorious Saturday in July, I stopped for a swig of water and a quick consultation of the Ordnance Survey *Explorer* map in my backpack. The chart showed that about a mile ahead on the minor country road on which I was walking, a small path would loop away to take in what the map described in archaic font as a 'Non-Roman Tower'. Intrigued, I followed the footpath away from the road, walking along lush fields that sloped down to the Clyde, then through a small dense wood and out the other side. There the path unexpectedly forked into a neat gravelled driveway curling up to a beautiful and somewhat imposing manor house, merely an innocuous rectangle on the map. Continuing along the left side of the fork, I met the Non-Roman Tower, a satisfyingly decrepit pyramid of stones nestling amid bracken and shadowy trees. Now, according to the map, the thin black dotted line of the path should curve around to rejoin the yellow streak of the road. And so it did. But between the two was a high and solid set of electronic gates, which were shut.

Shrugging my bag from my back, I unfolded the map again. As I was scanning it, a low growl came from somewhere below the sheet and I raised the map to reveal a large Kangal Dog, which in other circumstances might have been a rather handsome beast, but was now looking decidedly threatening.

Growling persistently, he nudged at my knee. I took a step back. The dog pushed again, harder this time, and I took a large backwards sidestep that placed me behind a gate, in a field speckled with sheep. The animal seemed pacified and stopped growling, and after a couple of minutes he even lay on the path in front of the gate and closed his eyes. Resolving to retrace my steps and rejoin the road the way I had come, I decided to tiptoe around my sleeping gaoler. But as soon as I pushed the gate open, the dog was on his feet and snarls and barks replaced the earlier growls. Hurriedly retreating into the safety of the field, this pattern continued for a good ten minutes. Dog slept – I tried to escape – dog awoke. I couldn't stay there all day, and after a bit of yelling had failed to rouse anyone, I decided to ring Directory Enquiries on my mobile phone and ask to be put through to the inmates of the large mansion down the drive.

'Name and town?' a bored voice enquired.

'Um, I don't know the owner, but I can give you the name of the house,' I replied hopefully.

'I need the surname of its occupant, madam.'

It seemed faintly absurd that I knew the names of the families who had lived at that house through the 1720s and 1730s, through William Roy's childhood and early adolescence, but not its present owners. And, rightly, I didn't hold out many hopes that this antiquated knowledge would prove much practical use. The phone call to Directory Enquiries failed to free me from my tight spot and I resorted to shouting again. After twenty minutes, a fresh-faced young man in tweeds eventually emerged up the driveway and, patting his thigh, summoned the dog, who gave up his sentinel position. As I blustered apologies and explanations, the two walked me along the path and the man typed in a code to open the electronic gates and restore me to the map's yellow streak.

Back on the road, I was confused. Had I been right to apologise? Or did the electric gates and angry guard-dog contravene Scotland's Land Reform Act of 2003, which enshrines the right to universal access to the land? I am still not sure. But it strikes me now that one of the reasons I find Ordnance Survey maps so seductive is the promise they seem to offer of the unfettered freedom to wander across the British landscape. Although maps also have the important function of demarcating where one *cannot* go, those as detailed as the Ordnance Survey's allow one to stray from the road and improvise routes at will, without a set itinerary. In England and Wales, they appear to bolster the Countryside and Rights of Way ('Right to Roam') Act of 2000, which 'provides a new right of public access on foot to areas of open land comprising mountain, moor, heath, down, and registered common land, and contains provisions for extending the right to coastal land'. Some philosophers feel that maps stimulate free wondering as well as wandering: that they encourage the mind to eschew linear logic and play with randomness and free associations, like the rambler who leaves the confines of major roads to skit over wide expanses of purple heath.

This association between maps and freedom has recently been manifested in campaigns to 'Free Our Data', which seek to 'persuade the government to abandon copyright on essential national [information], making it freely available to anyone, while keeping the crucial task of collecting that data in the hands of taxpayer-funded agencies'. Campaigners have targeted the Ordnance Survey as a significant 'culprit' whose pricey information (in the form of its database of geographical information) was described as 'useful to any business that wanted to provide a service in the UK, yet out of reach of startup companies without deep pockets'. There is a widespread feeling that the Ordnance Survey should foster the same principles of free access in the digital domain as those granted by their maps to ramblers who pass with relative liberty over moors and mountains. And on 1 April 2010 the agency made thirteen data sets freely available to the public, ranging from a small-scale map of the whole of Great Britain, to boundary information and a national gazetteer of road names.

The idea of freedom that many of us associate with the Ordnance Survey was arguably born in its first decades, over the late eighteenth and early nineteenth centuries, in the period described in this book. Not only did a home-grown tourist industry and penchant for hiking begin to take off then in Britain, but travellers gradually transferred their dependence from human guides to guidebooks and published itineraries and finally to maps. And with each shift, they gained a wider view of the landscape through which they journeyed, allowing freer improvisation over its contours. Although thinkers of the Enlightenment and Romantic periods often entertained conflicting interpretations of the significance of maps – one as emblems of reason and the other as images of the imagination – the modern connection of Ordnance Survey maps with freedom in the landscape is indebted to both. It was during the Enlightenment that maps acquired their associations with political liberty and egalitarianism, and it was during the Romantic period that this was supplemented by a deeply felt love for nature and solitary wandering. Proponents of both movements projected these ideas onto the great 'national undertaking', the Ordnance Survey.

As I write, well into the twenty-first century, the Ordnance Survey looks very different to the young mapping agency we have seen come to life. Now over three hundred map-makers trace the shape of the British Isles with the most advanced technology, such as laser-driven theodolites, hand-held pen computers, digital aerial images from satellites, and Geographic Information and Global Positioning Systems. The maps themselves are taking different forms, and the view of the landscape that was once offered by the 'traverse survey' or strip map now finds a modern equivalent in satnav systems that translate the landscape into a series of instructions of 'turn right', 'turn left' and 'turn around'. There are some who lament that these 'new maps' represent a step backwards, into the relative myopia towards Britain's landscape that characterised public geographical awareness before the Ordnance Survey. For example, the *Sun* has worried that Google maps omit important features like Stonehenge from the scenery that separates major roads, and that 'the simple directions from the internet and satellite navigation systems are wiping our historic landmarks and monuments from our national consciousness'. But it seems to me that Britons enjoy a greater geographical acquaintance with the country in which they live than ever before. Satnavs provide only one form of this knowledge, and can be supplemented with a wealth of online maps, hand-held digital navigational devices and mobile-phone applications, not forgetting the more traditional aids of the map and compass. Ordnance Survey maps never had the monopoly on geographical information, and now, as in the earliest years of its endeavour, map-readers continue to be drawn to 'the equal wide survey' (as the poet James Thomson expressed it in his poem *Liberty* in 1735) that they offer of the nation's varied scenery. That we are fortunate enough to live amid such abundance of knowledge can largely be traced back to a small group of men hunched over a theodolite on top of a mountain, in bucketing rain, over two hundred years ago.

Notes

After 1752 Britain began to adopt the revised Gregorian calendar, which was ten to eleven days ahead (in the period covered by this book) of the Julian calendar. I have given dates as they appeared in manuscript sources. Where possible, I have also retained the original spellings used by the documents' authors, with very occasional use of '[*sic*]' to prevent confusion. Manuscript and newspaper sources are given in full in the Notes and are not listed in the Works Cited section. Otherwise, the Notes refer to printed sources described in full in the Works Cited section.

ABBREVIATIONS

BL	British Library
NA	National Archives
NAS	National Archives in Scotland
NLI	National Library of Ireland
NRAS	National Register of Archives in Scotland
PO	Paris Observatory
RA	Royal Archives
RS	Royal Society
SP	State Papers
TCD	Trinity College, Dublin

PROLOGUE: LOST AND FOUND

xiii *On the evening of 16 April 1746* For contemporary descriptions of Lord Lovat's life and flight after Culloden, see [Forbes], Anon [1746] and Arbuthnot. There were also lengthy depictions in *London Evening Post*, 2902, 10 June 1746, and

2910, 28 June 1746, and *George Faulkener the Dublin Journal*, 2012, 24 June 1746. Shorter newspaper accounts appeared in, among others, *London Evening Post*, 2889, 10 May 1746; 2904, 14 June 1746; 2906, 19 June 1746; 2907, 21 June 1746; 2909, 26 June 1746; and in the *Penny London Post*, 480, 23 May 1746; 484, 2 June 1746; and 496, 30 June 1746.

xiii   *We are undone!*   Arbuthnot, p. 269.

xiii   *Charles Edward Stuart had good reason*   For detailed modern explorations of the history of the Glorious Revolution and the subsequent Jacobite uprisings, see works by Black, Hoppit, Kishlansky and Szechi (among others). Contemporary responses can be found in Bolingbroke and Hodges.

xiv   *the desperate Highlander's trusty broadsword*   *Caledonian Mercury*, 16 October 1745, cited in Black, p. 82.

xv   *Run, you cowardly Italian!*   After Culloden, Lord Elcho termed Charles Edward Stuart 'an Italian coward and a scoundrel'. In subsequent retellings of the battle, such as Peter Watkins's 1964 film *Culloden*, this gradually became represented as 'run, you cowardly Italian' or 'there you go, you cowardly Italian'. Pittock, p. 57.

xv   *moor was covered in blood*   [Douglas], p. 198.

xv   *laid waste the country*   Smollett, 1757–8, IV, pp. 674–5.

xv   *the Butcher*   Speck. My thanks to Yolande Hodson for this information regarding the Worshipful Company of Butchers.

xvi   *the chief Author and Contriver*   *Penny London Post, or Morning Advertiser*, issue 480, 23 May 1746.

xvi   *would be so cruel, as to endeavour to extirpate*   Arbuthnot, p. 265.

xvi   *the double Game you have played*   Arbuthnot, p. 257.

xvi   *Numerous pamphlets*   For example, see [Forbes], pp. 64–5 and Arbuthnot, pp. 270–1.

xvi   *Heaps of their Men*   Arbuthnot, p. 270, [Forbes], p. 66.

xvi   *They seized livestock*   Anon [1746], p. 95.

xvii   *we had near twenty Thousand*   Anon [1746], p. 95.

xvii   *put an end . . . so effectually now*   Joseph Yorke to William Augustus, Duke of Cumberland, 8 February 1746, cited in Black, p.146.

xvii   *Loch a'Mhuillidh*   Furgol.

xviii   *Throughout the course of the rebellion*   See a range of correspondence in the National Archives, London, including SP 54/28, ff. 87–9, 103–4, 109–11, 170–2, 196–8, 208; SP 54/29, ff. 1–6, 11–13, 47–8, 53–8, 99–100, 124–5, 152–3, 185–9; SP 54/30, ff. 3–4, 10–11. For histories of the mapping and exploration of Scotland prior to the mid eighteenth century, see Adams; Grenier; Holloway and Errington; Stone; Withers, 2001. A history of the militarisation of Scotland during the Jacobite rebellions can be found in Tabraham and Grove.

xviii *this Place is not marked* Caroline Frederick Scott to Robert Napier, NA, SP 54/29, f. 100, 7 March 1746.

xviii *by the Map of the country* John Campbell to Everard Fawkener, Royal Archives, Cumberland Papers, Box 14/411, 13 May 1746; cited in Hodson, 2007, p. 7.

xviii *the Want of Roads* William Anne Keppel, 2nd Earl of Albemarle, to Thomas Pelham Holles, 1st Duke of Newcastle, 'Some Thoughts Concerning the State of the Highlands of Scotland', NA, SP 54/34, ff. 24–31 [October 1746].

xix *extensive, full of rugged, rocky mountains* Anon, 'Memorial for the Heritors & Ministers of the Church in the western Parts of the Shires of Perth, Stirling & Dumbarton', NA, SP 54/32, ff. 226–7 [June 1746].

xix *there are Hiding-places enough* George Faulkener the Dublin Journal, 2002, 20 May 1746.

xix *our Detachments* Albemarle to Newcastle, NA, SP 54/33, p. 34, 12 August 1746.

xix *Intelligence [was] very difficult to obtain* Albemarle to Newcastle, NA, SP 54/33, p. 70, 5 September 1746.

xix *He was eventually caught* Anon (from Tobermory), NA, SP 54/32, f. 57, 10 June 1746.

xix *that wily old Villain* Fawkener to Newcastle, NA, SP 54/30, ff. 212–15, 19 April 1746.

xix *a glimpse of tartan* Arbuthnot, p. 274.

xix *The taking Lord Lovat* William Augustus, Duke of Cumberland to Newcastle, NA, SP 54/32, ff. 83–4, 28 June 1746.

xix *On 18 March 1747* See Furgol.

xx *burning scent* David Watson to Robert Dundas, 3rd Lord President of the Court of Session, Private Collection (Papers of Dundas Family of Arniston, Viscounts Melville, NRAS 3246), Volume 35: Letter Book 1747–55, f. 33, 14 July 1746.

xx *I should with infinite Pleasure* Albemarle to Newcastle, NA, SP 54/33, p. 70, 5 September 1746.

xx *Christopher Saxton* See Saxton.

xx *the glory of the British name* Camden, 1600, sig. ae. 4.

xx *Great Britaine and Ireland* See Speed.

xxii *surveying instruments until 1670* Bedini, pp. 687–94; Chapman, pp. 134–5; Kuhn, 1961, p. 186; Olmsted; Taylor, E.G.R., 1966, pp. 6–17. I am very grateful indeed to Dr Nicky Reeves for this information, and recommend his PhD thesis for further reading on the eighteenth-century history of astronomical and measuring instruments and the Astronomer Royal, Nevil Maskelyne.

xxii *chromatic aberration* Meadows, pp. 305–17.

xxii *a map-maker called John Cowley* Cowley.

xxii *The Scale's but small* [Hollar]. For a description of Hollar's works and attributed works (including this one), see Vertue.

xxv *paid £373 14s* Ordnance Office: Expense Ledgers, 1791–4, NA, WO 48/266, 1791.

xxv *A General Military Map of England* William Roy to George III, 24 May 1766, in Fortescue, I, pp. 328–34.

xxvi *a Betjemanesque image* Betjeman, p. 85.

xxvii *the drawing of charts or maps* 'Cartography', *The Oxford English Dictionary*, 2nd edn, 1989, *OED Online*, Oxford University Press, 4 April 2000 <http://dictionary.oed.com/cgi/entry/50033920>

CHAPTER 1: 'A MAGNIFICENT MILITARY SKETCH'

1 *On 29 June 1704* The *Oxford Dictionary of National Biography*'s entry for David Watson suggests that he was born around 1713, to a man named 'Thomas Watson' (see Baigent, 'Watson, David'). But a document in the papers of the Dundas Family of Arniston, Viscounts Melville, NRAS 3246, gives a date of 29 June 1704, and also confirms that Watson's father was called Robert (see NRAS 3246, Bundle 205: Miscellaneous Papers, 1615–1807). The *DNB* entry for Robert Dundas also confirms that he married 'in 1712, Elizabeth, eldest daughter of Robert Watson of Muirhouse' (see Scott, Richard). Another document in the Dundas Papers ('Bond of Relief, Colonel Robert Watson to the Lord President 1777', in Bundle 75: Papers mainly relating to the executry of Major-General David Watson of Muirhouse), written by Watson's nephew Robert Watson, the son of Watson's brother Robert Watson, describes how 'the Lands & Barrony of Muirhouse formerly belonging to Mr James Hunter Advocate and Alexander Hunter his Son were purchased . . . by Robert Watson Merchant in Edinburgh my Grandfather conform to a decree of Sale in his favour dated the Twenty Eight day of February One thousand Seven hundred years.' Another piece of evidence in support of Watson's birth in 1704 is that his portrait was painted by Andrea Soldi in 1761, and it seems more likely to be an image of a 57-year-old man, than a 48-year-old. (However, he had been ill for a long while, and it is possible that his aged appearance was due to his poor health.)

1 *David was the baby of a large family* See Dundas Papers, NRAS 3246, Bundle 205.

1 *But at the age of eight* See NAS, RH9/17/305; Dundas Papers, NRAS 3246, Bundle 205; NAS, GD150/3152 and GD150/3141; and NAS, CS228/W/1/39.

1 *And in that same year* Private Collection, Uncatalogued MSS, *Arniston Journals and Letters,* II, p. 17, and Elizabeth Dundas to Robert Dundas (3rd Lord President), NRAS 3246, Volume 32: Letter Book 1723–43, f. 64 20 November 1733.

1 *Dundas despotism* Fry, p. ix. See also Cockburn, pp. 48–9.

1 *to describe, in full detail* Omond, p. 58.

2 *Solicitor General of Scotland* Scott, Richard.

2 *one of the ablest lawyers* Burton, pp. 249–50, cited in Fry, p. 7.

2 *ill-looking* Burton, pp. 249–50, cited in Fry, p. 7.

2 *heat and impetuosity* Ramsay, p. 67, cited in Fry, p. 7.

2 *abundance of tact* Omond, p. 60.

2 *a bill for wine* Dundas Papers, NRAS 3246, Bundle 106: Receipts and Accounts 1686–1750. See accounts for June 1730 to March 1731.

2 *honouring Bacchus for so many hours* Scott, Walter, 1842, p. 572 (note).

2 *a torrent of good sense* Carlyle, pp. 249–50.

2 *something of surrogate parents* Robert Watson (David Watson's brother) to Robert Dundas (2nd Lord President), NRAS 3246, Volume 31: Letter Book 1655–1722, ff. 93, 101 [January 1721] and 16 March 1721.

2 *It was an exciting time* See Herman.

2 *In 1707, an Act of Union* For modern explorations of the events and effects of the 1707 Anglo-Scottish Union, see Colley, 2003, pp. 10–54; Davidson; Hayton; Levack; and MacInnes.

3 *the right to choose our own governors* Price, p. 49.

3 *vibrant public sphere* Habermas. For discussions of the social context of Enlightenment map-making, see Edney, 1994b; and Withers, 2002.

3 *Sapere aude!* Kant, 1991, p. 54.

3 *Enlightenment thinkers invested maps* See Edney, 1997, pp. 39–120; Frängsmyr, Heilbron and Rider; Livingstone and Withers; Mayhew, pp. 168–92, 207–28, 246–57; Withers, 2001, pp. 112–57; and Withers, 2002, pp. 46–66.

3 *a kind of world map* Diderot and d'Alembert, I p. xv.

3 *esprit géométrique* Fontenelle, p. 151.

4 *Exactitude in Science* Borges, p. 181.

4 *the quantifying spirit* Frängsmyr, Heilbron and Rider.

4 *dramatic advances in instruments and methods* See Taylor, E.G.R., 1937 and 1966; Howse, 1989; and McConnell (among many others).

4 *'Tis truely strange* Mark [pp. 1–2], cited in Withers, 2002, p. 50.

4 *known to read a book* Omond, p. 111.

4 *Arniston's library was impressively stocked* NRAS 3246, Bundle 106: Receipts and Accounts 1686–1750; Bundle 108: Accounts, Discharges, and Receipts, 18th Century; Bundle 117: Miscellaneous Papers 17th and 18th Centuries; Bundle 120: Miscellaneous Papers 17th and 18th Centuries.

5 *an inspiring collection of surveying instruments*   NRAS 3246, Bundle 117 contains a catalogue for James Cargill, who sells 'Gunters Scales & Chains for measuring ground[,] Mathematical Instruments in Cases with Brass Ivory and Boxwood Scales[,] Woodscales Quadrants Sectors & small dividers[,] Boxwood Rules for measuring &c.' Bundle 120 contains an 'Instruments Makers Bill for Theodilite, Tap box, & Cleaning a Case of Instruments'. Volume 34: Letter Book 1745, ff. 34, 35, contain advice to Robert Dundas (2nd Lord President) regarding 'the best Globe maker', presumably in Edinburgh. And Volume 49: Personal and Household Account Book, 1733–44, f. 19, contains a receipt for 'mending the Globe'. Around 1785, David Allan painted an image of the children of Henry Dundas, 1st Viscount Melville, clustered around a globe. (Henry was the son of Ann Gordon and Robert Dundas [2nd Lord President].) I am grateful to Charles W.J. Withers (via Stephen Daniels) for this information.

5 *a series of cartographic depictions*   Many of these surveys have been omitted from the most recent NRAS catalogue of the Dundas Papers (NRAS 3246), but can still be found on an older catalogue (NRAS 0077). They include around fifteen estate maps of Arniston conducted over the course of the long eighteenth century.

5 *William Adam*   William Adam, 'Elevations and ground plans of Arniston House and grounds', NRAS 0077, Plans. (This is a version of the NRAS catalogue of the Dundas Papers that precedes the current one, NRAS 3246.)

5 *Colin Maclaurin*   NRAS 3246, Volume 51: Robert Dundas's Personal Accounts, 1738–55.

5 *A child prodigy*   Sageng.

5 *begins with demonstrating the grounds*   Scots Magazine, August 1741, cited in Gibson, pp. 71–93.

5 *Alexander Bryce, Murdoch Mackenzie, Robert Adam and James Short*   Withers, 2002, p. 49; Carr; Fleming, pp. 79–80; Clarke.

5 *long banishment at Gib[raltar]*   David Watson to Robert Dundas (3rd Lord President), NRAS 3246, Volume 32: Letter Book 1723–43, f. 86, 24 December 1734.

5 *a martial Scuffle*   David Watson to Robert Dundas (3rd Lord President), NRAS 3246, Volume 32: Letter Book 1723–43, f. 86, 24 December 1734.

5 *In November 1733*   Elizabeth Dundas (née Watson) to Robert Dundas (2nd Lord President), NRAS 3246, Volume 32: Letter Book 1723–43, f. 64, 20 November 1733.

6 *confind to her Chamber*   Robert Dundas (2nd Lord President) to Robert Dundas (3rd Lord President), NRAS 3246, Volume 32: Letter Book 1723–43, f. 65, 13 December 1733.

6 *When her two small daughters*   Robert Dundas (2nd Lord President) to Robert

Dundas (3rd Lord President), NRAS 3246, Volume 32: Letter Book 1723–43, f. 69, 5 January 1734.

6 *By 6 February 1734*   Robert Dundas (2nd Lord President) to Robert Dundas (3rd Lord President), NRAS 3246, Volume 32: Letter Book 1723–43, f. 70, 12 February 1734.

6 *the best of mothers and an incomparable wife*   Robert Dundas (2nd Lord President) to Robert Dundas (3rd Lord President), NRAS 3246, Volume 32: Letter Book 1723–43, f. 70, 12 February 1734.

6 *Sir William Gordon*   For a history of the Gordons of Invergordon, especially the branch who resided at Hallcraig and Milton, see Bulloch; Gordon, C.A. William; Gordon. Sir William was married to Isobel Hamilton, whose family owned Hallcraig, and the couple's children became entangled in a complex debate regarding its inheritance. After much wrangling, their second surviving son, Charles Hamilton Gordon, became proprietor after Sir William's death. For information about the Hamiltons of Hallcraig, see Hamilton, George. Further details about Sir William's business dealings, and those of his son Charles Hamilton Gordon (who was a prominent freemason), can be found in Slade, 1981, pp. 458–9; Slade, 1982, p. 499. Ann Gordon's sister Isobel was married to George Mackenzie, Lord Cromarty, who became embroiled on the Jacobite side during the 1745 rebellion, to the horror of her family.

6 *have the happness*   Elizabeth Dundas (née Watson) to Ann Gordon, NAS, GD235/9/2: Dundas Letters 1735–43, f. 77, 20 October 1733.

6 *His carriage overturned*   Robert Dundas (2nd Lord President) to [Ann Gordon], NAS, GD235/9/2: Dundas Letters 1735–43, f. 78, 21 October 1733.

6 *truble in his bake*   Elizabeth Dundas (née Watson) to Robert Dundas (3rd Lord President), NRAS 3246, Volume 32: Letter Book 1723–43, f. 64, 20 November 1733.

7 *I won't omitt ane opportunity*   Robert Dundas (2nd Lord President) to [Ann Gordon], NAS, GD235/9/2: Dundas Letters 1735–43, f. 78, 21 October 1733.

7 *the Loss You have . . . Made*   Ann Gordon to Robert Dundas (2nd Lord President), NRAS 3246, Volume 32: Letter Book 1723–43, f. 75, 26 March [1734].

7 *Ann and Robert's wedding*   They were married on 3 June 1734. Private Collection, Uncatalogued MSS, *Arniston Journals and Letters*, II, p. 17.

7 *I did not incline*   Robert Dundas (2nd Lord President) to Robert Dundas (3rd Lord President), NRAS 3246, Volume 32: Letter Book 1723–43, f. 83, 6 September 1734.

7 *upon the Establishment of an Engineer*   John Hay, 4th Marquess of Tweeddale, to Robert Dundas (2nd Lord President), NRAS 3246, Volume 32: Letter Book 1723–43, f. 155, 25 November 1742.

7 *immediately come here* John Hay, 4th Marquess of Tweeddale, to Robert Dundas (2nd Lord President), NRAS 3246, Volume 32: Letter Book 1723–43, f. 155, 25 November 1742.

7 *would be a good beginning* John Hay, 4th Marquess of Tweeddale, to Robert Dundas (2nd Lord President), NRAS 3246, Volume 32: Letter Book 1723–43, f. 160, 2 December 1742.

8 *A department known as the 'Office of Ordnance'* For accounts of the history of the Board of Ordnance and the Corps of Engineers, and their involvement in map-making, see (among others) Bailey; N.P., 1985; Boyd; Connolly, T.W.J.; Marshall, 1976; Marshall, 1980; Millburn; Napier; and Porter, Whitworth.

8 *that which is ordered or ordained* For a full exploration of the meanings of 'ordnance, *n.*' and 'ordinance, *n.*', see *The Oxford English Dictionary*, 2nd edn, 1989, *OED Online*, Oxford University Press, 4 April 2000, Draft revision December 2009 <http://dictionary.oed.com/cgi/entry/00333564> and Draft revision March 2010 <http://dictionary.oed.com/cgi/entry/00333544>. See also Hewitt, 2007, pp. 3–5.

9 *ought to be well skilled* 'Instructions to our Principal Engineer', Royal Warrant of 1683, cited in Hogg, Appendix IV.

9 *the Corps often attracted those* For the social context of admission into the Corps of Engineers, see Marshall, 1976, especially pp. 54–86, 87–92, 131–50.

9 *After attaining a sought-after place* David Watson to Robert Dundas (3rd Lord President), NRAS 3246, Volume 33: Letter Book 1743–4, f. 32, 22 June 1743.

10 *a resounding victory* David Watson to [Robert Dundas (2nd Lord President)], NRAS 3246, Volume 33: Letter Book 1743–4, f. 33, 29 June 1743; David Watson to Robert Dundas (3rd Lord President), NRAS 3246, Volume 33: Letter Book 1743–4, f. 35, 5 July 1743.

10 *we do not live in an age* Andrew Mitchell to Robert Dundas (3rd Lord President), NRAS 3246, Volume 34: Letter Book 1745, f. 89, 12 November 1745.

10 *Mr Watson has gained universal character* Andrew Mitchell to Robert Dundas (3rd Lord President), NRAS 3246, Volume 34: Letter Book 1745, f. 92, 14 November 1745.

10 *The Pretender's son [has] near 3,000 rebels* Keene, p. 70, cited in Black, pp. 75–6.

10 *David Watson was among them* John Gordon to Robert Dundas (3rd Lord President), NRAS 3246, Volume 34: Letter Book 1745, f. 96, 16 November 1745.

10 *warm Jackets, adequate blankets* John Gordon to Robert Dundas (3rd Lord President), NRAS 3246, Volume 34: Letter Book 1745, f. 96, 16 November 1745. See also William Augustus, Duke of Cumberland, to Thomas Pelham Holles, 1st Duke of Newcastle, NA, SP 54/32, ff. 83–8, 28 June 1746; William Anne Keppel, 2nd Earl of Albemarle, to Thomas Pelham Holles, 1st Duke of Newcastle, NA, SP 54/33, f. 3, 1 August 1746.

10 *the happyness this poor Country enjoy'd* [David Watson] to Robert Dundas (3rd Lord President), NRAS 3246, Volume 35: Letter Book 1746–55, f. 32, 1 June 1746.

10 *our Interview with the Rebels* [David Watson] to Robert Dundas (3rd Lord President), NRAS 3246, Volume 35: Letter Book 1746–55, f. 32, 1 June 1746.

12 *temptations of pleasure or game* [David Watson] to Robert Dundas (3rd Lord President), NRAS 3246, Volume 35: Letter Book 1746–55, f. 34, 7 July 1746.

12 *the destruction of the Ancient Seat* [David Watson] to Robert Dundas (3rd Lord President), NRAS 3246, Volume 35: Letter Book 1746–55, f. 32, 1 June 1746.

12 *the rest of the good people of Lochaber* [David Watson] to Robert Dundas (3rd Lord President), NRAS 3246, Volume 35: Letter Book 1746–55, f. 32, 1 June 1746.

12 *the Highlanders [are] the most despicable enemy* [David Watson] to Robert Dundas (3rd Lord President), NRAS 3246, Volume 35: Letter Book 1746–55, f. 32, 1 June 1746.

12 *the sole motive of restoring quiet* [David Watson] to Robert Dundas (3rd Lord President), NRAS 3246, Volume 35: Letter Book 1746–55, f. 34, 7 July 1746.

12 *Together with his brother-in-law* See letters from Robert Dundas (2nd Lord President) and Robert Dundas (3rd Lord President) to Philip Yorke, 1st Earl of Hardwicke (Lord Chancellor), BL, Add. MS 53446, ff. 178–9, 207–12, 220–1, 228–9, 270–1; BL, Add. MS 53447, ff. 9, 49–50, 112–13, 128–30, 157–8, 168–9, 180–1, 206–9, 317–18. See also David Watson to Robert Napier, NA, SP 54/40/42, f. 141, October 1749.

12 *lawless* For example, see Anon, 'Memorial for the Heritors & Ministers of the Church in the western Parts of the Shires of Perth, Stirling & Dumbarton', NA, SP 54/32, ff. 226–7, [June 1746]; Albemarle to Newcastle, NA, 'Some Thoughts Concerning the State of the Highlands', SP 54/34, ff. 24–31 (f. 25), [October 1746].

12 *Suggestions included the* For example, see NA, SP 54/35, ff. 46–9, 50–1, 113–16, 123–8; BL, Add. MS 53446, ff. 180–6, 191–3.

12 *for the purposes of civilizing, and promoting the happiness* 'Instructions by the Commissioners for managing the forfeited Estates in Scotland annexed to the Crown', NAS, E726/1, [July 1755].

12 *Many considered that the Highland landscape* Fraser, p. 258.

13 *Such Noble-men as these* Anon, 1690, p. 6.

13 *the Benefit [which] must arise* [David Watson], 'Some Observations concerning the Highlands of Scotland', BL, Add. MSS 35890, ff. 158–9, [September 1747].

13 *found themselves greatly embarrassed* John Watson [David Watson's brother], 'Memorial', NRAS 3246, Bundle 89, p. 6. There is a copy of this in NA, T1/486, p. 2.

13 *the Inconvenience was perceived* John Watson [David Watson's brother], 'Memorial', NRAS 3246, Bundle 89, p. 6. There is a copy of this in NA, T1/486, p. 2.

13 *broken glades and bare hills of dark heath* Scott, Walter, 1999, p. 178.

14 *windings of the majestic Clyde* Scott, Walter, 1999, p. 321.

14 *on 4 May 1726* Documentation surrounding William Roy's childhood and early youth is scarce. The date of his birth is shown on Carluke's Official Parish Register for 12 May 1726, 629/0010 0058 <www.scotlandspeople.gov.uk>. Details about John Roy's employment are given in his Last Will and Testament, 24 November 1750, Lanark Commissary Court, CC14/5/17. The nineteenth-century local historian, Daniel Reid Rankin, described how the name of John Roy, William's father, 'occurs frequently in the sederunt of heritors, as acting for Sir William Gordon, and his son Mr Charles Hamilton Gordon of Hallcraig and Milton, from 1739 onwards' (Rankin, June 1872, pp. 562–3). Rankin's 'Notes for a history of Carluke 1870–1878', Glasgow Manuscripts Library, MS Murray 153–55, Vol. 1, ff. 8, 11, describes in detail the residents of the Upper Ward of Lanarkshire in the early 1700s. He also provides information about the Roys' relationship with the other principal families of the area, the Gordons and the Lockharts, in Rankin, March 1853, p. 148. The Lockharts were based at Lee Castle, a few miles south of the Roys and Gordons, and were related by marriage to the Dundases (see letters from John Lockhart to Robert Dundas (2nd Lord President), NRAS 3246, Volume 34: Letter Book 1745, ff. 123, 156 and 162, 1, 19 and 22 December 1745; and Volume 35: Letter Book 1746–55, f. 140, 9 April 1755).

14 *the Duty and Office of a Land Steward* Laurence.

14 *it is necessary* Ley, p. 73.

14 *it is not only necessary* Laurence, p. 78.

14n *In 1987, after being bought* Harris, Ron.

15 *John Roy would have first measured* I am indebted to Stuart Hepburn and the Royal Geographical Society North West Region, for their 'DIY Mapping Day' on 30 May 2009, in which they expertly taught me how to use these instruments and techniques.

16 *It has been suggested that he worked* For these different accounts of William Roy's early life, see respectively Chalmers, II, p. 64; Skelton, 1967a, p. 16 (note 6); Skelton, 1967b, p. 2; Gardiner, p. 439. See also Hodson, 2007, p. 7; Moir, D.G., I, p. 105.

16 *The Upper Ward of Lanarkshire erupted* The Dundases had many connections with the Upper Ward of Lanarkshire. Not only was Robert Dundas (2nd Lord President) related by marriage to the Gordons at Hallcraig, but his son, Robert 'Robin' Dundas (3rd Lord President), also married in 1741 a woman called Henrietta Baillie, the heiress to the Bonnington estate about seven miles

south of Hallcraig. Robin appears to have lived at Bonnington between 1741 and 1753, when his father died and he inherited Arniston. Robin played an active and central role in the life of the Upper Ward, and rallied the locals (from afar – he went to Berwick during the rebellion) to rise against the Jacobites during the Forty-Five. This was not hard: the Upper Ward had a history of staunch Presbyterianism. See Andrew Orr (Presbyterian minister at Carluke) to Robert Dundas (3rd Lord President), NRAS 3246, Volume 34: Letter Book 1745, f. 122, 31 December 1745; John Lockhart to Robert Dundas (3rd Lord President), f. 123, 1 December 1745; Andrew Orr to Robert Dundas (3rd Lord President), f. 145, 16 December 1745; John Smith to Robert Dundas (3rd Lord President), f. 146, 16 December 1745; John Smith to Robert Dundas (3rd Lord President), f. 153, 18 December 1745.

16 *the only communication that is open* John Lockhart to Robert Dundas (3rd Lord President), NRAS 3246, Volume 34: Letter Book 1745, f. 123, 1 December 1745.

16 *to act for the Defence of His Majesty* Andrew Orr to Robert Dundas (3rd Lord President), NRAS 3246, Volume 34: Letter Book 1745, f. 145, 16 December 1745.

16 *heavy rains* John Lockhart to Robert Dundas (3rd Lord President), NRAS 3246, Volume 34: Letter Book 1745, f. 162, 22 December 1745.

17 *He seems to have avoided direct confrontation* The possibility has been suggested by Yolande Hodson that Roy may have been on the Jacobite side during the Forty-Five (see Gilchrist). There were certainly one or two Lanarkshire men called 'William Roy' (a common name) who were arrested after the rebellion. NA, TS 20/79, f. 1, provides a list of 'Rebel prisoners' and includes 'William Roy, from Lanerk, a Private man', who had served for five months, and was twenty-three years of age. TS 20/78, f. 15, is a 'Humble Petition of the Distressed Prisoners now Confin'd in Tilbury Fort', signed by 'Willaim [*sic*] Roy', but this man has made a mark besides his initials, indicating he was illiterate. It seems unlikely to me that he could have gone from a state of illiteracy to making the Military Survey of Scotland within a year. However, it is possible that the William Roys of TS 20/78 and TS 20/79 were different men, and that one of them was the William Roy of the Military Survey of Scotland.

17 *it may have been that this extended family* Earlier notes describe the intimacy between the Dundases and various families who lived within the Upper Ward of Lanarkshire. This closeness is increased by the fact that the uncle of Robert Dundas (2nd Lord President), John Sinclair, married Martha Lockhart, who inherited from her father the lands of Castlehill around Cambusnethan House, a few miles to the west of Wishaw, only three miles north of Hallcraig. Dundas was close to his cousin, Martha's son John Lockhart, who took his mother's

surname and inherited the lands of Castlehill. Furthermore, in 1721 or 1722 David Watson's older brother, Robert, had married Henrietta Baillie, the daughter of Sir William Baillie of Lamington and Henrietta Lindsay. The Baillie family were resident at the Lamington estate in South Lanarkshire, and Henrietta Baillie was the aunt of the Henrietta Baillie of Bonnington who married Robert Dundas (3rd Lord President). That these families socialised together in the Upper Ward of Lanarkshire is indicated by a letter sent just after Christmas in 1742, from Robert Dundas (2nd Lord President) to Ann Dundas (née Gordon). Robert was staying with his son at Bonnington, and he wrote that 'we are going just now to dine at Milnton [*sic*] with Charles [Hamilton Gordon]', and he planned to stay the night at that estate. (Robert Dundas [2nd Lord President] to Ann Dundas (née Gordon), 27 December 1742, NAS, GD235/9/2, f. 87.

17 *proper Survey of the Country*   John Watson, 'Memorial', NRAS 3246, Bundle 89, p. 6.

17 *He was at Fort Augustus*   Christian, p. 19 and Arrowsmith, p. 7.

18 *the Highlands are but little known*   Burt, 1754, I p. 5.

18 *that nervous expressive tongue*   Buchanan, p. 112, cited in Youngson, p. 13.

18 *those regions in which*   Polybius, Book 34, p. 307.

18 *Britain's roads were notoriously bad*   For general histories of Britain's roads and their construction in this period, see Moir, Esther; Kendrick, p. 3; Reader; Taylor, Christopher; and West.

18 *mere beds of torrents and systems of ruts*   Thomas De Quincey, cited in Burke, Thomas, p. 93 and Esther, 1964, p. 8.

19 *Turnpike Act*   For a history of Britain's turnpike roads, see Albert.

19 *to inspect the present situation of the Highlanders*   Wade, II, p. 268. For general histories of George Wade, see Baker; McCall, Colin; Pollard; and Salmond.

19 *the number of men able to bear arms*   Wade, II, p. 270.

19 *the little regard they ever paid*   Wade, II, p. 272.

19 *caused an exact Survey*   George Wade, 'Reports of Maj.-Gen. George Wade to George I and George II on the Highlands of Scotland, the condition of the people, the measures taken for disarming, etc.', BL, Kings MS 101, f. 17. My thanks to Carolyn Anderson for drawing my attention to this document.

20 *Had you seen these roads*   Pollard, p. 28; Le Fanu, p. 242.

20 *It is probable that*   Christian, p. 18; O'Dell, p. 58.

20 *Thanks to a break-in*   I am grateful to Jean Barr of the Carluke Parish Historical Society for showing me the documentation pertaining to the theft of a miniature of Roy, by Maria Cosway, in 1929.

20 *Measure the roads from Inverness*   William Roy, 'Measurements of His Majesties Military Road from Dunkeld to Blair in Atholl, 1747', Atholl Estates, NRAS 234, Bundle 683: Miscellaneous Legal Estate, Political, and other Papers with no apparent common factor, *c.*1696–1760.

20 *Orders and Instructions*   David Watson, 'Orders and Instructions to be Observed by Col. Watson's Assistants in Reconoitring [*sic*], Examining, Describing, Representing, and Reporting, any Country, District, or Particular Spot of Ground, they may at any time be ordered to Reconnoitre and Report', NA, OS 3/5.

20 *William Roy began the daunting task*   Roy, 'Measurements of His Majesties Military Road', NRAS 234, Bundle 683; Arrowsmith, p. 7.

21 *He measured the distance*   My account of Roy's surveying technique during the Military Survey is extrapolated from information in Arrowsmith, pp. 7–8; Christian, p. 19; David Dundas, 'Memorandums respecting the map of Scotland Jan 12 1806 Given to Arrowsmith', NAS, cited in Hodson, 2007, p. 10; Skelton, 1967b, p. 4; and Watson, 'Orders and Instructions', NA, OS 3/5.

21 *theodolite*   'Theodolite', *The Oxford English Dictionary*, 2nd edn, 1989, <http://dictionary.oed.com/cgi/entry/50250603>. See also Digges, 1571, Book 1: Longimetra; Digges, 1927; Turner, Gerard, 1991, p. 315; and Turner, Gerard, 1998, p. 39.

22 *a circle divided into 360 degrees*   Digges, 1571.

23 *John Ogilby*   Ogilby, 1675.

25 *Ideally, traverse surveys*   Arrowsmith, p. 7.

25 *to make this way*   Johnson, 1984, p. 56.

25 *In the early 1740s, the Perthshire hydrographer*   Chalmers, II, p. 63; Sher.

26 *as I was sitting in my hutt*   John Russell to William Skinner, BL, Add. MS 17499, ff. 41–2, 2 July 1747.

26 *scandalous Scrawl and form*   William Skinner to Mr Fern, BL, Add. MS 174500, f. 22, 14 April 1750.

26 *the remote situation*   Wade, cited in Baker and Baker, pp. 22–3.

26 *had marched at least six miles*   Johnson, 1984, p. 58.

26 *to be particularly attentive*   Watson, 'Orders and Instructions', NA, OS 3/5.

26 *on Christmas Day 1748 his father died*   John Roy, Testament, Lanark Commissary Court, CC14/5/17, 24 November 1750.

27 *a scene the most wild and romantic*   Roy, 1793, p. 59.

27 *love much talking*   William Roy to James Hope, NRAS, Hopetoun MSS, Bundle 590, 9 December 1765, cited in Hodson, 2007, p. 7.

27 *a much truer notion may be formed*   Roy, 1793, p. 59.

27 *In 1748*   Charles Bush to William Skinner, BL, Add. MS 17499, f. 72, 7 June 1748.

27 *The Surveying Scheme has given me*   David Watson to William Skinner, BL, Add. MS 17499, ff. 71–2, 7 June 1748.

27 *having three more Assistants*   Surveyor-General's Notes, NA, WO 47/34, f. 179, 27 March 1750, and NA, WO 47/35, f. 233, 27 March 1750.

27 *Woolwich had been designed*   For accounts of the history of the Royal Military Academy at Woolwich, see Edney, 1994a, and Smyth, John.

27 *the people of its Military Branch* First Charter of the Royal Military Academy, Woolwich, cited in 'The Royal Military Academy 1741–1939', <http://webarchive. nationalarchives.gov.uk/20060130194436/http://atra.mod.uk/atra/rmas/history/ history3.htm> [accessed 03 April 2010].

28 *When Watson applied* Surveyor-General's Notes, NA, WO 47/34, f. 315, 19 June 1750.

28 *Deputy Head Master* Surveyor-General's Notes, NA, WO 47/34, f. 315, 19 June 1750.

28 *Idle, Lunactic, and lost to Debauchery* Surveyor-General's Notes, NA, WO 47/34, ff. 315, 317, 315, 19 June 1750.

28 *Watson duly made his choice* Surveyor-General's Notes, NA, WO 47/34, f. 179, 27 March 1750.

28 *Thomas Howse, and . . . John Manson* Arrowsmith, p. 7; Chalmers, II, p. 62; Hodson, 2007, p. 10; O'Donoghue, Yolande, p. 1; Skelton, 1967a, p. 7; Skelton, 1967b, p. 3; Whittington and Gibson, p. 63. There is some debate about the spelling of Howse's surname, and he often appears in Board of Ordnance documentation as 'Howes'. For example, on 27 September 1749, William Roy and John Manson witnessed a legal document signed by him (Ordnance Quarter Books, NA, WO 54/498, p. 9, 27 September 1749).

28 *additional Servants, Guides, Interpreters & otherwise* David Watson, 'Accompt of money expended by Major-General David Watson upon account of a Survey made of Scotland', NRAS 3246, Bundle 89: Account of Money Expended by Major-General David Watson, f. 1.

28 *also sent a draughtsman* For accounts of Paul Sandby's role in the Military Survey, see Bonehill and Daniels, pp. 13–5, 74–5, 82–103; Christian, pp. 18–22. For general accounts of Sandby's career, see Ball; Bonehill and Daniels; Herrmann; Irwin; Sandby, 1892.

29 *In 1749 Watson sent Sandby* Sandby was sent to work on the Military Survey alongside another draughtsman, called Charles Tarrant. See Charles Bush to William Skinner, BL, Add. MS 17500, f. 10, 8 March 1750; William Skinner to Board of Ordnance, BL, Add. MS 17500, f. 24, 3 May 1750; Board of Ordnance to William Skinner, BL, Add. MS 17500, f. 26, 5 May 1750. See also Christian, p. 19. Sandby's map of Tiorim and Muydart is held in the National Map Library of Scotland, 'Plan of Castle Tyrim in Muydart. Plan of Castle Duirt in the Island of Mull', MS 1648 Z. 03/28e.

30 *Displaying such an acute awareness* Sandby's political cartoons can be found, among other places, in '13 Prints by P. Sandby satirising W. Hogarth', Bodleian Library, Johnson b.221/1–13. Daniels offers an interpretation of the nationalist potential of Sandby's art in Daniels, 2006, pp. 23–60.

30 *He also helped found the Royal Academy of Art* See Hoock, 2003.

30 *the father of English watercolour*   Ball, p. xv.

31 *in the field*   Tabraham, p. 28.

31 *David Watson paid regular visits*   David Watson, 'Accompt of money expended by Major-General David Watson upon account of a Survey made of Scotland', NRAS 3246, Bundle 89: Account of Money Expended by Major-General David Watson, f. 1.

31 *in the Governor's House of Edinburgh Castle*   Tabraham, pp. 27–9.

31 *remote Corners of the Highlands*   David Watson, 'Accompt of money expended by Major-General David Watson upon account of a Survey made of Scotland', NRAS 3246, Bundle 89: Account of Money Expended by Major-General David Watson, f. 1.

31 *the really laborious part of the business*   Robert Kearsley Dawson, cited in Portlock, p. 133.

31 *it was our practice in walking*   Robert Kearsley Dawson, cited in Portlock, pp. 146–7.

33 *large and conical stones*   Robert Kearsley Dawson, cited in Portlock, p. 136.

33 *around sixty personnel*   Hodson, 2007, p. 8.

33 *Some of this money came from*   David Watson appears to have spent a great deal of his own and his family's money on the Military Survey. See Estate Account Book of Robert Dundas 1737–1782, NRAS 3246, Bundle 172, pp. 64–8, 110, 138, and NRAS 3246, Bundle 89: Account of Money Expended by Major-General David Watson. After Watson's death, his brother John and Robert Dundas (3rd Lord President) successfully attempted to seek remuneration from the Board of Ordnance. See NRAS 3246, Bundle 89: Account of Money Expended by Major-General David Watson; NRAS 3246, Bundle 75: Papers relating mainly to the executry of Major-General David Watson of Muirhouse; and NA, T1/486, items 1–6.

34 *a fifteen-year-old nephew called David Dundas*   Omond, p. xxxii. William Roy's close relations with the Dundas family of Arniston continued long after the Military Survey of Scotland. Letters in NRAS 3246, Bundle 120, describe how he helped Francis Dundas, the son of Robert Dundas (3rd Lord President) and his second wife Jean Grant, to prepare for a commission in the 1770s (William Roy to Robert Dundas [3rd Lord President], NRAS 3246, Bundle 120, 14 March 1777; and Francis Dundas to Robert Dundas, NRAS 3246, Bundle 120, 11 March 1777).

34 *David Dundas would become Commander-in-Chief*   Houlding.

34 *mapping their fertile, elegant estate*   David Dundas's 1752 survey of Arniston is listed in NRAS 0077, under Plans. There is a photostat in NAS, RHP5246/2.

34 *many others . . . whose superior taste*   Fleming, pp. 5–6.

34 *it has often been in my power*   John Crawford to Robert Dundas (2nd Lord Arniston), NRAS 3246, Volume 35: Letter Book 1746–1755, f. 96A, 19 December 1752.

35 *the social status of the Board of Ordnance's employees* John Elphinstone was disinherited from the aristocratic Elphinstone family (the Lords Balmerino) in the early 1740s, and was forced to gain military employment for financial sustenance. I am very grateful to Roderick Barron for fascinating discussions about Elphinstone.

35 *the extraordinary Adam family of architects* See Gifford, especially pp. 76–107, 173–4; Fleming; and Mylne.

35 *a man of distinguished genius* John Clerk of Eldin, cited in Fleming, p. 7.

35 *a tumulus, where several urns* Sinclair, X, pp. 286–7.

35 *pointed it out in his maps* Sinclair, X, pp. 286–7.

35 *ordinary employments* Roy, 1793, p. i.

36 *While the ranges of mountains* Roy, 1793, p. i.

36 *to compare present things with past* Roy, 1793, p. i.

36 *While surveying this territory* For information about Roy's antiquarian interests and pursuits, see Adamson; Gardiner; O'Donoghue, Yolande, pp. 19–29; Rankin, March 1853; and Thomson, Thomas, p. 301.

36 *begin[s] at the Clyde* Roy, 1793, p. 157.

36 *When the reading public got wind* For reactions to Roy's *Mappa Septentrionalis* among the map-making community, see Walters, 1976, p. 123.

39 *eighty-four brown linen rolls* O'Dell, p. 58; Skelton, 1967a, pp. 10–11; Whittington, p. 12.

39 *These were later cut and reconstituted* Fleet and Kowal, p. 197.

39 *it was immense* Chalmers, II, p. 62. For further descriptions of the Military Survey's appearance, see Carlucci and Barber, p. 14; Kinniburgh, pp. 16–19; and Mallett. The map itself is held in the BL, catalogued under William Roy, 'A Very Large and Highly Finished Colored Military Survey of the Kingdom of Scotland, Exclusive of the Islands, Undertaken by Order of William Augustus, Duke of Cumberland . . .', 1747–55, in King George III's Topographical Collection. The rough copy of the northern part carries the shelfmark Maps K.Top.48.25.1a., while the fair copy has the shelfmark Maps CC.5.a.441. A reduced copy of the map bears the shelfmark Maps K.Top.48.25–1e.

39 *the greatest work of this sort* Hugh Debbieg to Henry Seymour Conway, 'Memorial', NA, CO 325/1, pp. 197–200 (p. 198), 10 March 1766.

40 *the Mountains and Ground appear shaded* Arrowsmith, p. 7.

40 *vocabulary of symbols* Skelton, 1967a, p. 9.

40 *a picture of Scotland on the eve of great changes* O'Dell, p. 63.

40 *a uniformly accurate image of the landscape* For critiques of the Military Survey's accuracy, see Fleet and Kowal; O'Dell; and Whittington.

40 *carried on with instruments of the common* Roy, 1785, p. 386.

41 *a magnificent military sketch* Roy, 1785, pp. 386–7.

41 *the maps were only temporarily excavated*   Arrowsmith, p. 6.

41 *drawn by Roy, and given by him to Lord President Dundas*   This map was included in the catalogue of the Dundas archives made in the 1960s (NRAS 0077), labelled under 'Plans' as 'Excerpt from survey of Scotland presented to Lord President Dundas by General Roy, 1755'. However, unfortunately it has been left off the most up-to-date catalogue, NRAS 3246. A photostat of the map is available from NAS, RHP5246/1. The inscription is only visible on the original map, which is held in Arniston House.

42 *The exiled Stuart court*   For the presence of maps on Jacobite memorabilia, see Sharp, pp. 100, 224–5. A useful discussion of the symbols and rhetoric of Jacobitism can be found in Erskine-Hill. The Marischal Collection at the National Map Library of Scotland consists of 137 maps published between 1573 and 1873, many of which were collected by the exiled Stuarts at the court of St-Germain-en-Laye.

42 *unsufferable tyranny*   Robert Dundas (2nd Lord President) to Hardwicke, BL, Add. MS 35446, ff. 228–9, 31 December 1747.

42 *Scotland has no unity except on the map*   Stevenson, p. 39.

42 *The history of cartography's progress*   For more on the idea that failure is a necessary component of progress, see Kuhn, 1962.

42 *public-private partnership*   Hoock, 2010, pp. 11–19.

CHAPTER 2: 'THE PROPRIETY OF MAKING A GENERAL MILITARY MAP OF ENGLAND'

43 *By December, he had wangled*   National Archives, OS 3/408. Commission as Practitioner Engineer given to William Roy on 26 December 1755.

43 *Engineer William Roy*   London Gazette, 9548, 24 January 1756.

44 *Watson was asked to conduct a reconnaissance*   O'Donoghue, Yolande, p. 30.

44 *the Position in front of Dorking*   [William Roy and David Dundas], 'Military Reconnaissances of Parts of Southern And Eastern England with Proposals for Defence', NA, OS 3/1, f. 1, [1756].

44 *And he was later asked to do the same for Ireland*   William Roy, 'Defence of England: Observations by Colonel Roy made on a tour of Ireland', NA, WO 30/115, 1766.

44 *deep valleys and intervening ridges*   William Roy, 'Military Description of the South-East Part of England', NAS, GD364/2/208, p. 22, July 1765.

44 *the fingers of a hand*   William Roy, 'Military Description of the South-East Part of England', NAS, GD364/2/208, p. 22, July 1765.

45 *he helped to found a Commission*   For an account of the Annexed Estates

Commission, see Wills. For David Watson's role, see 'Reports on Annexed Estates, Struan', in Wills. See also Minutes of Commissioners for Annexed Estates, NAS, E721/1, pp. 12–17, 37, 21 June–14 July 1755, 5 November 1755. See also David Watson to Robert Dundas (3rd Lord President), NRAS 3246, Volume 35: Letter Book 1746–55, f. 170, 13 August 1755; Hardwicke to Robert Dundas (3rd Lord President), NRAS 3246, Volume 35, f. 173, 25 September 1755; John Forbes to Robert Dundas (3rd Lord President), NRAS 3246, Volume 36: Letter Book 1756–1779, f. 7, 19 February 1756.

45 *very indifferent* David Watson to Hardwicke, BL, Add. MS 35606, f. 353, 20 July 1761.

45 *Watson's housekeeper* David Watson, 'Last Will of Major-General David Watson', NRAS 3246, Bundle 75: Papers mainly relating to the executry of Major-General David Watson of Muirhouse, 1761.

45 *Shortly before 1 November 1761* 'List of Deaths for the Year 1761', *Gentleman's Magazine*, 31, p. 539, 1761.

45 *has made [Scotland] more his study* John Forbes to Robert Dundas (3rd Lord President), NRAS 3246, Volume 36: Letter Book 1756–1779, f. 7, 19 February 1756.

45 *David Watson's last will and testament* David Watson, 'Last Will of Major-General David Watson', NRAS 3246, Bundle 75: Papers mainly relating to the executry of Major-General David Watson of Muirhouse, 1761.

46 *When military strategists* For accounts of the botched mission to Rochefort under Sir John Mordaunt, and William Roy's role, see Chalmers, II, pp. vi–vii; Gardiner, pp. 443–4; MacDonald, p. 176; Mordaunt; Towse.

46 *By March 1759* MacDonald, p. 164.

46 *Roy industriously surveyed* For Roy's role in Minden, see Hodson, 2007, pp. 17–18.

46 *zeal, assiduity and talents* 'Letter to William Roy as a Captain in Germany in 1759 and 1760', NA, OS 3/2, f. 1, 8 June 1760.

46 *Roy produced a map of Minden* Roy, 1760. See also 'Plan de la Bataille de Thonhausen pres de Minden, Gagnée le 1er Aout 1759, par W. Roy', Centre for Kentish Studies, U1350/P16.

46 *a controversial court martial* Gardiner, pp. 443–4; Harley and Walters, 1977, pp. 12–13; Mackesy; O'Donoghue, Yolande p. 30. For a later description of Roy's role, see also *Morning Herald and Daily Advertiser*, 1326, 25 January 1786.

46 *He was promoted to* Gardiner, p. 443; O'Donoghue, Yolande p. 30.

46 *Commissary General* St James's Chronicle or the British Evening Post, 2675, 23 May 1778; *Gazetteer and New Daily Advertiser*, 15484, 20 September 1778.

47 *An anecdote told almost twenty years after Minden* Gazetteer and New Daily Advertiser, 15484, 20 September 1778.

47 *this extraordinary promotion* NA, OS 3/2, f. 9, 28 September 1759.

47 *in July 1765* Royal Warrant to Master of the Ordnance, NA, WO 55/365, p. 14, 31 July 1765. The appointment is also confirmed in 'Copy Book of military reports by Roy, as surveyor-general of coasts', Centre for Kentish Studies, Pratt Manuscripts, U840/0171, 1756–79; Gardiner, p. 445; Hodson, 2007, p. 19; MacDonald, p. 177.

47 *The instruction to survey the coasts* Hodson, 2007, p. 19.

47 *it is His Majesty's pleasure* Henry Seymour Conway to William Roy, NA, WO 124/1, no. 140 (f. 198), 26 October 1765.

48 *the town and port of Dunkirk* Entry Book of Official and Some Private Correspondence [about] the demolition of fortifications, NA, WO 124/1, f. 31, Article 13th.

48 *wholesomeness of the air* Entry Book of Official and Some Private Correspondence [about] the demolition of fortifications, NA, WO 124/1, f. 31, Article 13th. Documentation about the Dunkirk expedition can be found in, among other places, NA, WO 124/1; WO 124/5; WO 124/6; SP 28/290.

48 *In November 1765* Desmaretz, Roy and Frazer to Conway, NA, WO 124/1, no. 144 (f. 200), 5 November 1765.

48 *Roy may have already met* I am grateful to Yolande Hodson for this suggestion: indeed, Roy and Lennox were both present at the Battle of Minden (see Lowe).

48 *solicited the engineer's opinion* Charles Lennox to William Roy, NA, WO 124/1, no. 152 (f. 211), 30 November 1765.

48 *Roy voiced his suspicion* William Roy to Charles Lennox, NA, WO 124/1, no. 157 (ff. 218–19), 3 December 1765.

48 *the Demolition . . . should . . . go on* William Roy to Henry Seymour Conway, NA, WO 124/1, no. 171 (ff. 235–8), 20 December 1765.

48 *had not been bettered* William Roy to Henry Seymour Conway, NA, WO 124/1, no. 177 (ff. 240–1), 4 February 1766.

48 *you will be so good* William Roy to Henry Seymour Conway, NA, WO 124/1, no. 171 (ff. 235–8), 20 December 1765.

48 *in mid February 1766* Charles Lennox to William Roy, NA, WO 124/1, no. 179 (f. 242), 14 February 1766; Henry Seymour Conway to William Roy, NA, WO 124/1, no. 180 (f. 242), 14 February 1766; Desmaretz and Fraser to George Lennox, NA, WO124/6, no. 203, 22 February 1766.

48 *a [more] thorough knowledge of the country* Roy, 1785, pp. 387–8.

48 *Ireland (which he had surveyed in 1765)* See William Roy, 'Defence of England: Observations by Colonel Roy made on a tour of Ireland', NA, WO 30/115, 1766.

48 *join the whole together* William Roy to King George III, 24 May 1766, in Fortescue, I, p. 330.

49 *Considerations on the Propriety of making a General Military Map*   William Roy to King George III, 24 May 1766, in Fortescue, I, p. 330.

49 *the honour of the nation*   Roy, 1790, p. 262.

49 *On 22 March 1754*   Allan and Abbott, p. 3.

50 *Joseph Banks*   Allan and Abbott, p. 3.

50 *Virtuoso Tribe of Arts and Sciences!*   Marquess of Rockingham, cited in Allan and Abbott, p. xxii.

50 *the Encouragement of Boys and Girls*   Minutes of first meeting of Society of Arts, cited in Allan and Abbott, p. 92.

50 *the sculpture gallery at his mansion*   Seidmann, p. 124.

50 *a Seat of Arts*   John Gwynn, cited in Allan, p. 93.

50 *whether the state of British Geography be not very low*   William Borlase, September 1755, cited in Harley, 'Society and Surveys', p. 142.

50 *'Tis to be wished*   William Borlase, September 1755, cited in Harley, 'Society and Surveys', p. 142.

51 *give proper surveyors such Encouragement*   Cited in Harley, 'Society and Surveys', p. 143.

51 *of great use in plannning*   Cited in Harley, 'Society and Surveys', p. 143.

51 *The Society of Arts advertised its first prize*   For more detailed information about the Society of Arts' competition, see Harley, 'Society and Surveys', pp. 141–57.

52 *without much doubt be traced to the offer*   Wood, p. 298, cited in Harley, 'Society and Surveys', p. 154.

53 *In 1763 William Roy*   Gardiner, p. 444. Roy conducts some barometric observations from this address. See Roy, 1777, p. 716.

53 *both General Wade and Paul Sandby*   I am indebted to Yolande Hodson for pointing this out to me.

53 *Sixteen years later, in 1779*   Gardiner, p. 444.

53 *Today, a blue plaque*   I am very grateful to the house's current owner, Jim Jordan, for showing me around William Roy's former residence.

53 *In 1767, his brother James*   Addison, p. 29 (no. 991), cited in MacDonald, p. 163.

54 *the Society of Antiquaries*   For a history of the Society of Antiquaries, see Pearce.

54 *William Roy was admitted as a Fellow of the Royal Society*   Certificates of Election and Candidatures, Royal Society, EC/1767/02.

54 *the Royal Philosophers' Club*   See Allibone, pp. 1–40.

54 *Bute was Scottish*   For contemporary accounts of the popular outrage at Bute's election, see Churchill, p. 27, and Wilkes. For a modern interpretation, see Brewer.

54–5 *Scotophobic cartoons flooded*   For example, see *The S– Puppit Shew, or the Whole Play of King Solomon the Wise*, 1762, and *William Augustus, Duke of Cumberland. The Tomb-Stone*, London: Smith, 1765, in IV, nos 4049 and 4124, in Stephens, Hawkins and George.

55 *the expressions of a beggarly Scot*   See Smollett, 1971, p. 145; Beattie. See also Grenier, p. 24, for an account of this cultural attitude.

55 *truth became our central criterion*   Stoddart, p. 32.

55 *paradise*   Holmes, pp. 1–59. For further information about Banks's life, see O'Brian.

55 *than either of the Indies*   Burt, I 1754, p. 5.

55 *Banks soon tired of*   Information in this paragraph is taken from Allibone, pp. xiii, 86–105, 106–15.

56 *that curious and useful branch of philosophy*   Roy, 1777, p. 715.

56 *I have sometimes found*   Roy, 1777, p. 728 (note).

56 *the dining room of the Spaniard's Inn*   Roy, 1777, p. 717.

56 *amateur astronomer James Lind*   Roy, 1777, p. 720.

56 *a mere lath who was married to a fat handsome wife*   Charles Burney, cited in Cooper.

56 *the summit of Arthur's Seat*   Roy, 1777, p. 720.

56 *the observatory of Hawk-hill westward*   Roy, 1777, p. 720.

56 *spirit of the kindest tolerance*   Percy Bysshe Shelley, cited in Cooper.

57 *The laws of motion*   See Smith, James Raymond, 1996, p. 23; Torge, p. 8.

57 *In 1735 and 1736*   For further discussions of the attraction of mountains, see Anon., 1816, pp. 36–51; Maskelyne, 1775a, pp. 495–9; Maskelyne, 1775b; Pringle; Smith, James Raymond, 1986, p. 95.

58 *Nevil Maskelyne*   An account of the life and achievements of Maskelyne can be found in Howse, 1989. For his role in investigating the attraction of mountains, see Maskelyne, 1775a and 1775b; Danson, 2006, pp. 141–54; Howse, 1989, pp. 129–41; Reeves, 2009b; Taylor, E.G.R., 1966, pp. 59–61.

58 *Charles Mason [. . .] Jeremiah Dixon*   For an account of Mason and Dixon's measurement of the 'Mason–Dixon Line', see Danson, 2001. Pynchon offers a post-modern fictional retelling of some of the principal concerns of Enlightenment geodesy, such as the nature of truth.

58n *Who claims Truth*   Pynchon, p. 350.

59 *a broad-back'd massy hill*   Nevil Maskelyne to James Lind, RS, MS/244, no. 12, 3 August 1773.

59 *a remarkable hill*   Maskelyne, 1775b, pp. 502–3.

60 *Mason pulled out of the experiment*   Howse, 2004.

60 *the badness of the weather*   Cited in Sillito.

60 *it is not but a week ago*   Nevil Maskelyne to James Lind, RS, MS/244, no. 13, 18 July 1774.

60 *to ascertain [Schiehallion's] dimensions*   Maskelyne, 1775b, p. 508.

60 *to determine the position*   Nevil Maskelyne to William Roy, NA, OS 3/2, 16 August 1774.

60 *was honoured . . . by visits* Maskelyne, 1775b, pp. 524–5. See also Howse, 1989, p. 138.

61 *In August Maskelyne's trial* Nevil Maskelyne to William Roy, NA, OS 3/2, 16 August 1774; Roy, 1777, pp. 721–2.

61 *the difference between Roy and Maskelyne's calculations* Roy, 1777, pp. 721–2.

61 *they often paid me visits on the hill* Cited in Sillito.

62 *a local boy called Duncan Robertson* Howse, 1989, pp. 137–8.

62 *seek you out a new fiddle* Cited in Howse, 1989, p. 137.

62 *On the trip I took to Schiehallion* Cited in Howse, 1989, pp. 137–8.

62 *the mountain Schehallien* Maskelyne, 1775b, p. 532.

63 *the mean density of the earth* Maskelyne, 1775b, p. 533.

63 *totally contrary to the hypothesis* Maskelyne, 1775b, p. 533.

63 *a residence of four months* Pringle, 1775, p. 28, cited in Reeves, 2009b, p. 323.

63 *Hutton adopted a map-making technique* See Hutton.

64 *concealed pattern and . . . obscure plan* Pope, Epistle 1, l. 6.

64 *an upsurge of nationalistic feeling in Britain* For a discussion of nationalism in the second half of the eighteenth century, see Colley, 2003; and for a discussion of nationalist feeling in relation to King George III, see Colley, 1984.

CHAPTER 3: THE FRENCH CONNECTION

66 *Le Ministre des affaires philosophiques* Lord Auckland to Lord Grenville, 6 November 1791, cited in Gascoigne, p. 3; and McConnell, p. 9 (note 13).

66 *an ashy fog* Stothers.

66 *The politician Charles James Fox* Joseph Banks to Charles Blagden, Royal Society, CB1/1/92, 31 October 1783.

66 *il est interessant* César François Cassini de Thury, Memoir sur la jonction de Douvres à Londres, RS, DM4, f. 3, [1783].

67 *a dynasty of astronomers* For accounts of the history of the Cassini dynasty, the Paris Meridian and the *Carte de France*, see Godlewska; Konvitz, pp. 1–31; Murdin; Pelletier.

67 *Libellus de locorum describendorum* See Frisius.

67 *The French meridian arc's triangulation* Murdin, pp. 17–20.

69 *By the end of 1681* For descriptions of this early triangulation, see Taylor, E.G.R., 1937; Taylor, E.G.R., 1941 and Ravenhill, 1994, p. 162.

69 *Triangulation was the surveying technique* Edney, 1997, pp. 19–21, 236–92.

70 *Since 1767* See Howse, 1989, pp. 85–96.

71 *John Bull* See Rogers, pp. 31–40.

71 *In the face of this deep-rooted* See Crosland; and de Beer.

71 *I believe that it is not strictly the case* Charles Blagden to Joseph Banks, Fitzwilliam Museum, Cambridge, Perceval Collection H-191, 10 October 1783; cited in Martin and McConnell, p. 357.

72 *without sufficient reason* 'Domestic Literature', *The New Annual Register, or General Repository of History, Politics, and Literature for the Year 1787*, 1788, pp. 240–1.

72 *that you might consider it fully* William Roy to Nevil Maskelyne, RS, DM4, f. 14, 11 December 1786.

72 *A small, marbled notebook* Martin and McConnell, pp. 151–3.

72 *as doing honor to our scientific character* Joseph Banks to Charles Blagden, RS, CB1/1/92, 31 October 1783.

72 *He had spent much of that summer measuring* Roy, 1785, p. 388.

72 *that an operation of the same nature* Roy, 1785, p. 388.

72 *the first of the kind* Roy, 1785, p. 390.

72 *accurate Survey of the British Dominions* [William Roy], 'On the Advantages that are Likely to Arise from the Operations on Hounslow Heath', RS, DM4, f. 6.

72 *redound to the credit of the Nation* Roy, 1785, p. 390.

73 *a sum of £2000* Cotton to Joseph Banks, RS, DM4, f. 39, 21 February 1793.

73 *one of the open level Counties* William Roy to King George III, 24 May 1766, in Fortescue, I, p. 331.

73 *in all the Tours I have made* William Roy to Joseph Banks, RS, DM4, f. 12, 28 June 1784.

73 *Because . . . of its vicinity to the Capital* Roy, 1785, p. 390.

74 *natural obstacles at every turn* *World and Fashionable Advertiser*, 174, 21 July 1787.

74 *On 16 April 1784* Roy, 1785, p. 391.

74 *the site of Hampton Poor House* Roy, 1785, p. 391.

74 *upwards of five miles* Roy, 1785, p. 391.

74 *Roy marked the baseline's ends* Roy, 1785, pp. 414–15.

74 *country labourers* Roy, 1785, pp. 391–2.

74 *a narrow tract along the heath* Roy, 1785, pp. 391–3.

74 *through brushwood* Roy, 1785, p. 393.

74 *certain ponds, or gravel-pits* Roy, 1785, p. 392.

74 *extreme wetness* Roy, 1785, p. 393.

76 *guarding such parts of the apparatus* Roy, 1785, p. 392.

76 *On 16 June the base* Roy, 1785, pp. 416–17.

76 *100-foot steel chain* McConnell, p. 194.

76 *15 July 1784* Roy, 1785, p. 425.

76 *met with the most hospitable supply* Roy, 1785, pp. 425–6.

76 *rather too unwieldy* Roy, 1785, p. 398.

76 *Roy inched along* Roy, 1785, pp. 402–3.

76 *a short distance of 300 feet*   Roy, 1785, p. 427.

76 *they expanded and warped*   Roy, 1785, pp. 428–9.

77 *William Calderwood*   Calderwood was the son of Margaret Steuart Calderwood, a characterful woman who took on the management of her husband's estate at Polton, and took her sons travelling around Europe to see her exiled Jacobite brother. The Polton estate was not too far from the Dundases at Arniston and Sir John Clerk at Penicuik, and it is possible that William Roy met the Calderwoods while conducting the Military Survey of Scotland. Later the Calderwoods and the Dundases would intermarry, when William Calderwood's grand-niece Lilias Calderwood-Durham would marry Robert Dundas, the great-grandson of Robert Dundas (2nd Lord President).

77 *8.30 a.m.*   Roy, 1785, p. 429–32.

77 *so straight that, when laid on a table*   Roy, 1785, p. 441.

77 *On 30 August*   Roy, 1785, pp. 458–9.

77 *Journalists for national newspapers*   St James's Chronicle, or the British Evening Post, 3644, 13 July 1784; *Felix Farley's Bristol Journal*, 1864, 17 July 1784; *Whitehall Evening Post*, 5782, 13 July 1784.

77 *The frenzy came to a head*   Roy, 1785, p. 456; McConnell, p. 195.

77 *the space of two hours*   Roy, 1785, p. 456.

77 *his Majesty (who is ever ready*   St James's Chronicle, or the British Evening Post, 3644, 13 July 1784.

78 *his attendance from morning*   Roy, 1785, p. 425.

78 *27,404.7 feet*   Roy, 1785, p. 477–8.

78 *there has never been so great a proportion*   [William Roy], 'On the Advantages that are Likely to Arise from the Operations on Hounslow Heath', RS, DM4, f. 6.

78 *Ramsden was an innkeeper's son*   For an account of Jesse Ramsden's history, see Anita McConnell's wonderful biography.

79 *John Dollond*   For an account of Dollond's life and achievements, see Clifton.

79 *In 1766*   McConnell, p. 17.

79 *at the sign of the Golden Spectacles*   McConnell, p. 18.

79 *dividing engine*   McConnell, pp. 18, 29, 36, 39–51.

79 *the natural philosopher Robert Hooke*   Hooke, p. 7, cited in Chapman, p. 135.

79 *micrometer*   See Chapman, pp. 134–5; McConnell, pp. 173–90; Roy, 1790, pp. 145–54.

80 *full of intelligence and sweetness*   Louis Dutens, cited in McConnell, p. 20.

80 *an expression of cheerfulness*   Louis Dutens, cited in McConnell, p. 20.

80 *Ramsden was late for almost everything*   McConnell, pp. 77, 87–8, 114.

80 *there are few lively persons*   Johan Lexell, May 1781, cited in Nevskaya, p. 201. Cited also in McConnell, p. 113.

80 *punctual as to the day and the hour*   Cited in McConnell, pp. 275–6.

80 *the best Instrument [he] could*   Jesse Ramsden to Royal Society, RS, MM.3.30, p. 2, 13 May 1790.

80 *might be rendered superior*   Jesse Ramsden to Royal Society, RS, MM.3.30, p. 3, 13 May 1790.

80 *The instrument that Ramsden began*   For descriptions of Jesse Ramsden's three-foot theodolite, see McConnell, pp. 191–221; Roy, 1790, pp. 135–60.

80 *brass circle, three feet*   Roy, 1790, p. 136.

80 *about 1/24,000 part of an inch*   Jesse Ramsden to Royal Society, RS, MM.3.30, p. 3, 13 May 1790.

82 *Dollond's famous achromatic lenses*   Roy, 1790, pp. 140–1.

82 *turn round very smoothly*   Roy, 1790, p. 137.

82 *large brass ruler*   Joseph Banks to John Lloyd, National Library of Wales, MS 12415.28, August 1786; cited in McConnell, p. 201.

82 *obliged . . . to take out*   Jesse Ramsden to Royal Society, RS, MM.3.30, p. 4, 13 May 1790.

82 *Ramsden is hard at work*   Charles Blagden to Joseph Banks, BL, Add. MS 33272, f. 27, 1 June 1787.

83 *He read his account*   Roy, 1785, p. 385.

83 *incomparable Engineer*   World and Fashionable Advertiser, 5, 5 January 1787.

83 *kissed the King's hand*   Morning Herald, 1899, 25 November 1786.

83 *our trusty and well-beloved*   Whitehall Evening Post, 5766, 21 September 1784.

83 *ought not to hate one another*   Cited in de Beer, p. 111.

83 *owing I suppose to the difficulty*   Joseph Banks to Cassini de Thury, Paris Observatory, D5-7, 4 June 1784.

83 *fellow Labourers*   William Roy to Jean-Dominique Cassini, PO, D5-7, 3 December 1788.

83 *some pains to investigate*   [Gen Roy's] Remarks on a Paper [put into his majesties hands by the Duke of Marlborough], RS, DM4, f. 45, 1786/7.

84 *The younger Cassini*   Charles Blagden to Joseph Banks, Fitzwilliam Museum, Cambridge, Perceval Collection H-191, 10 October 1783; cited in Martin and McConnell, p. 357.

84 *charmed at the prospect*   Jean-Dominique Cassini to Charles Blagden, RS, DM4, f. 18, 29 May 1787.

84 *this man is an electrical machine*   Jean-Dominique Cassini, cited in McConnell, p. 143.

84 *In July 1787*   Roy, 1790, p. 112.

85 *But soon Roy had tamed the theodolite*   For a description of Roy's methods, see Roy, 1787, 1790.

86 *General Roy est à quelque distance*   Charles Blagden to Jean-Dominique Cassini, RS, DM4, f. 22, 20 August 1787.

86 *Blagden suggested to Roy* Charles Blagden to Jean-Dominique Cassini, RS, DM4, f. 22, 20 August 1787.

86 *Roy even tried to pass it off as his own* William Roy to Jean-Dominique Cassini, PO, D5-7, 7 September 1787.

86 *antique and half-ruined* Cited in McConnell, p. 363.

86 *reverbaratory [sic] lamps* For descriptions of the lamps, see Roy, 1790, pp. 113, 143, 163.

86 *white lights* Roy, 1790, p. 253.

87 *placed [one lamp] over the other* Roy, 1793, p. 163.

87 *somewhat earlier than the times* William Roy to Jean-Dominique Cassini, PO, D5-7, 30 September 1787.

87 *not sufficiently clear* William Roy to Jean-Dominique Cassini, PO, D5-7, 30 September 1787.

87 *he hardly expects to be believed* Similar results are presented in Roy, 1790, pp. 143, 262.

87 *Stormy weather* William Roy to Jean-Dominique Cassini, PO, D5-7, 8 October 1787.

87 *In August 1787* Roy, 1790, pp. 121–34.

89 *from its levelness* Roy, 1787, p. 190.

89 *so much intersected* Roy, 1790, p. 122.

89 *a distance of 28,532.92 feet* Roy, 1790, p. 133.

89 *the line of verification* St James's Chronicle or the British Evening Post, 4508, 11 March 1790.

89 *as I passed through Tenterden* 'Curiousus', 'To the Editor of the Gazetteer', *Gazetteer and New Daily Advertiser*, 435, 11 January 1788.

89 *The French wrote up and published* See Cassini.

90 *a time difference of nine minutes* Roy, 1790, p. 231.

90 *in a rather devious twist* See Howse, 1989, pp. 152–3; Martin and McConnell, pp. 357, 369 (note 7); Roy, 1790, p. 231.

90 *Roy's results largely agreed* Martin and McConnell, pp. 265–7.

90 *Truth* Roy, 1790, pp. 128, 129, 185, 186, 190, 203, 224, 228, 231, 247, 248, 263, 268, 593, 594.

90 *the last exactness* Cited in Widmalm, p. 197.

90 *mathematical exactness* Cited in Widmalm, p. 197.

90 *extremely perfect* Roy, 1790, p. 136.

90 *the accuracy of this operation* New Annual Register, 1786, p. 289.

90 *conducted in a manner* 'Domestic Literature', *New Annual Register*, 1788, p. 240.

90 *among the improvements* St James's Chronicle or the British Evening Post, 4508, 11 March 1790.

90 *no measurement of a similar kind* 'Domestic Literature', *New Annual Register*, 1788, p. 240.

91 *uniform agree[ment]* Anderson, p. 27.

91 *invented an instrument* *General Evening Post*, 7900, 14 October 1784.

91 *a national work of great importance* *New Annual Register*, 1786, p. 289.

91 *may be considered in some sort* Francis Wollaston to Joseph Banks, RS, MS/820, 13 September 1790.

91 *including Roy himself* Roy, 1790, pp. 262–3.

91 *No consideration upon earth* William Roy to Joseph Banks, Natural History Museum Archives, MS DTC 7.74–8, f. 74, 6 March 1790; cited in McConnell, p. 191.

91 *too remiss and dilatory* Roy, 1787, p. 189.

91 *Nothing could equal my surprize* Jesse Ramsden to the Royal Society, RS, MM.3.30, p. 1, 13 May 1790.

92 *unskilfulness* Jesse Ramsden to the Royal Society, RS, MM.3.30, p. 7, 13 May 1790.

92 *that every part of the Instrument* Jesse Ramsden to the Royal Society, RS, MM.3.30, p. 8, 13 May 1790.

92 *an extended visit to Lisbon* Baigent, 'Roy, William'; Gardiner, p. 449; Hodson, 2007, p. 20.

92 *In the early hours of 1 July* For William Roy's Last Will and Testament, see NA, PROB 11/1194. An exploration of the contents of his library and store of mathematical instruments can be found in Harley and Walters, 1977.

92 *the Republic of Letters* *The Times*, 1726, p. 2, 6 July 1790.

CHAPTER 4: THE ARISTOCRAT AND THE REVOLUTION

93 *the young wanton* John Evelyn, 9 October 1671, cited in Curtis, p. 1.

93 *excellent and universally esteemed* Debbieg, pp. 17–18.

93 *the honour of the nation* Roy, 1790, p. 262.

95 *Jigsaws as we know them* See Hannas.

95 *put the map of Europe together* Austen, 1998, p. 15. For a wider discussion of Austen's engagement with Enlightenment concerns, see Knox-Shaw. See Batey for an exploration of her approach to landscape. See Roberts for an examination of Austen's reaction to the French Revolution.

95 *contrived by little pebbles* Tillyard, p. 414.

95 *a thorough education* Lowe. For accounts of Charles Lennox's life, see Baird, pp. 84–93; Olson.

95 *Lennox began to buy up* See Bonney, pp. 1–11; Kent, pp. 20–6;

96 *high wood* Cited in Kent, p. 36.

96 *pheasantry* Cited in Kent, pp. 20, 31.

96 *on 1 November 1758* Charles Lennox's Household Accounts, West Sussex Record Office, Goodwood Papers, Ae/1, Af/1; cited in Crone, Campbell and Skelton, p. 416.

96 *gave proofs of an extraordinary genius* See Ó Danachair, IV, pp. 62–3.

97 *maintained an illicit intercourse* Ó Danachair, IV, p. 62.

97 *on business* Ó Danachair, IV, p. 63.

97 *on different subjects* Ó Danachair, IV, p. 63.

97 *confessed all the particulars of his guilt* Ó Danachair, IV, p. 63.

98 *The alliance would be recognised* See Crone, Campbell and Skelton.

98 *not only contain an accurate plan* Gough, II, pp. 297–8, cited in Crone, Campbell and Skelton, p. 417.

99 *In his early twenties* For an account of Charles Lennox's military endeavours, see Reese, pp. 189–98.

99 *The peaceful plains of England* Lennox, 1804, p. 4.

99 *the Duke of Richmond's blackness* George III to Frederick North, Lord North, 12 June 1773, in Fortescue, II, p. 504.

99 *a serious flirtation* For a discussion of Sarah Lennox's relationship with George III, see Tillyard, pp. 119–34.

100 *different from and prettyer than* Henry Fox, cited in Tillyard, p. 122.

100 *plain* Tillyard, p. 132.

100 *diffuse throughout the Kingdom* Society for Constitutional Information, p. i.

100 *it is the Right of every Commoner* Lennox, 1783, pp. 6–7.

100 *manifold Abuses* Lennox, 1783, p. 11.

100 *his Unremitted personal ill conduct* George III to Frederick North, Lord North, 3 July 1780, in Fortescue, V, pp. 96–7.

101 *In March 1782* For an account of Lennox's role in the Ordnance, see Olson, pp. 64–75.

101 *Defences had been* See Olson, pp. 74–5, 81–4 for an account of Lennox's fortifications project.

101 *change our system* Lennox, 1804, p. 20.

101 *the first Martello tower* Clements offers a history of Britain's Martello towers.

102 *the Trigonometrical Operation begun* Out-letters from Master-General, Board of Ordnance, and Commander-in-Chief, NA, WO 46/22, 1791–2.

102 *£373 14s* Expenses Ledger of Board of Ordnance, NA, WO 48/266, 21 June 1791.

103 *small stakes* Mudge, Williams and Dalby, 1795b, p. 62.

103 *more correct maps* Mudge, Williams and Dalby, 1795b, p. 62.

103 *the Trigonometrical Survey* Surveyor-General's Minutes, NA, WO 47/118, 12 July 1791.

103 *General Survey* 'Lights on the Coast', *Sussex Weekly Advertiser*, 8 and 15 April 1793, cited in Harley and O'Donoghue, I, p.xxv; 'The General Survey of England and Wales', *The Times*, 9982, p. 1, 2 November 1816.

103 *British Survey* Portlock, p. 192.

103 *Duke of Richmond's Survey* *Lloyds Evening Post*, 5473, 25 July 1792.

103 *until 1801* [Thomas Budgen], 'Exeter', Ordnance Surveyors' Drawings, BL, OSD 40 pt. 3, 1801. I am indebted to Yolande Hodson for the information that this map contains an annotation in William Mudge's hand which refers to the 'Ordnance Survey' – but she also made the point that this could have been added at a subsequent date.

103 *in print in 1809* Arrowsmith, p. 5.

103 *engaged in constructing* Arrowsmith, p. 5.

103 *in 1810* Mudge, William, 1810.

103 *On 22 June 1791* Surveyor-General's Minutes, NA, WO 47/117, 22 June 1791.

104 *wit, rake, and dope-fiend* Tillyard, p. 158.

104 *Major Williams and Lieut. Mudge* Surveyor-General's Minutes, NA, WO 47/118, 12 July 1791.

104 *I should rejoice could I say* Joseph Banks, also cited in Markham; also cited in Close, 1969, p. 37.

105 *in the summer of 1789* For accounts of the French Revolution and its effect on British politics and culture, see Andrews, Stuart; Claeys; Dickinson; Hibbert, 1980; and Philp, among others.

105 *I have lived to see* Price, p. 54.

106 *The French had begun* For accounts of the progress of French map-makers before and during the Revolutionary years, see Ravenhill, 1994, pp. 162–3; and Konvitz.

106 *the French who had made the greatest strides* Ravenhill, 1994, p. 162.

106 *The provinces of ancien régime France were dissolved* See Alder, pp. 263–4; Konvitz, pp. 43–6; Robb, pp. 68–9.

106 *a geometrical and arithmetical constitution* Burke, 1968, p. 144.

107 *Nothing more than an accurate land surveyor* Burke, 1968, p. 286.

107 *ancien régime France contained* Alder, p. 3.

107 *for all men, for all time* Cited in Alder, p. 227.

107 *To extrapolate this length* For an account of the project to measure the metre unit, see Alder.

108 *Our well-worn terminology* Hibbert, 1980, p. 109.

108 *emblems of these distracted times* Taylor, John, 1642.

108 *A horse erect* Taylor, John, cited in Malcolm, p. 33.

108 *this topsy-turvy world* Horace Walpole to Mary Berry, 19 May 1791, in Walpole, XI, p. 270.

108 *an adept in turning*   Burney, p. 372.

108 *the servants [had] turned masters*   Taylor, Jane, p. 56.

108 *simple moral*   Taylor, Jane, p. 63.

110 *the right to choose our own governors*   Price, p. 49.

110 *settling the succession of the [Hanoverian] crown*   Burke, 1968, p. 100.

110 *look up with awe to kings*   Burke, 1968, p. 182.

110 *the manifesto of a Counter-Revolution*   Mackintosh, p. 91.

111 *carte du Paÿs*   Lennox, 1785, p. 54.

111 *Robert Edward Clifford*   See Ravenhill, 1994.

112 *the anatomy of the Veins*   John Graves Simcoe Papers, Archives of Ontario, Series A-4-1, 1801; cited in Ravenhill, 1994, p. 164.

112 *skeleton maps*   Ravenhill, 1994, p. 171.

112 *incline the Administration*   William Borlase, September 1755, cited in Harley, 'Society and Surveys', p. 142.

CHAPTER 5: THEODOLITES AND TRIANGLES

114 *the best mathematicians*   Cited in Harley and O'Donoghue, I, p. xxiv. Histories of the Ordnance Survey can be found in Close, 1969; Harley, 1986–7; Harley and O'Donoghue; Harley and Oliver; Owen and Pilbeam; Seymour.

114 *Williams staged a mock battle*   'More Particular Account of the Grand Review of the Royal Regiment of Artillery', *Morning Chronicle and London Advertiser*, 5982, 11 July 1788.

114 *of all the sham battles*   'More Particular Account of the Grand Review of the Royal Regiment of Artillery', *Morning Chronicle and London Advertiser*, 5982, 11 July 1788.

114 *Ladies Night*   *Bath Journal*, 9 January 1792.

114 *regular afternoon assemblies*   *Star*, 1776, 15 February 1794; *Gazetteer and New Daily Advertiser*, 15656, 17 April 1779; *General Evening Post*, 7976, 4 February 1796; *St James's Chronicle*, 6272, 1 March 1798.

115 *never made an observation or calculation*   Isaac Dalby to Thomas Colby, 5 February 1821, cited in Close, 1969, p. 42.

115 *too much brains in it*   William Mudge to Richard Rosdew, 20 November 1804, cited in Flint, p. 130.

115 *impressed with just ideas*   Mudge, William, 1800, p. 564.

115 *He had grown up in Plymouth*   For an account of the various extraordinary members and friends of Mudge's family, see Flint.

115 *the necessity of Government*   Mudge, Zachariah, 1739, p. 62.

115 *Obedience to Authority*   Mudge, Zachariah, 1790, p. 2.

115 *there has been ever acknowledged*   Mudge, Zachariah, 1790, p. 34.

116 *strong indications*   Flint, p. 33.

116 *very fond of that method*   Reynolds, I, p. xxxiv.

116 *the Discovery of Longitude at Sea*   Histories of the eighteenth-century 'longitude' debate are given by Sobel; and Howse, 1989, pp. 40–52, 74–84, 124–6. For information about Thomas Mudge's part, see Howse, 1989, pp. 170–7 and Seccombe.

117 *In the mid 1770s*   Accounts of Thomas Mudge's involvement in the quest to discover longitude at sea can be found in Mudge, Thomas; RS, MM/7/91, MM/7/94, MM/7/100, MM/7/114, MM/7/117–19; *The Times*, 2598, p. 2, 30 April 1793; *The Times*, 4166, p. 3, 10 April 1798.

117 *discourage the advancement*   Joseph Banks, Observations on Mr Mudge's Application to Parliament for a Reward for his Time-keepers, RS, MM/7/100.

118 *he inserted a brief aside*   Mudge, William, 1800, p. 666.

118 *William enjoyed clock-making*   Flint, p. 122.

118 *King of Denmark*   Flint, p. 151.

118 *Samuel Reynolds*   For a life of Joshua Reynolds and his family, see Leslie; Reynolds.

119 *very great danger*   Leslie, I, p. 6.

119 *every precaution*   Leslie, I, p. 6.

119 *at the very time predicted*   Leslie, I, p. 6.

119 *By the age of eight*   Edgcumbe, pp. 724–6.

119 *the schoolhouse according to rule*   Leslie, I, p. 8.

119 *his first disposition to generalize*   Reynolds, I, p. xxxiii.

119 *an ordinary painter*   Leslie, I, p. 16.

119 *painters of our own nation*   *Universal Magazine*, November 1748, cited in Postle.

119 *coarse features, slovely dress*   Knight, I, p. 9, cited in Postle.

119 *gracious assistance, patronage, and protection*   Chambers, West, Cotes and Moser, Memorial, 28 November 1768, in Sandby, 1862, I, p. 46.

120 *Mudge's Inhalers*   Flint, p. 81.

120 *Never mind!*   Flint, p. 117.

121 *excesses in new honey*   Leslie, I, pp. 216–17.

121 *What! Another, Dr Johnson?!*   Boswell, I, p. 347, cited in Flint, p. 15.

121 *esteemed an idol*   Cited in Flint, p. 20.

121 *learned and venerable old man*   Edmund Burke to Malone, 1797, cited in Reynolds, I, pp. xxxiii–iv.

121 *the wisest man*   Joshua Reynolds, cited in Flint, p. 20.

122 *I have lived in intimacy*   Edmund Burke, cited in Flint, p. 20.

122 *a sharp boy*   Flint, p. 121.

122 *not very attentive*   Flint, p. 121.

122 *comfortable and pleasant*   Risdon, p. 396.

122 *freeman of Plymouth*   For documentation regarding the Rosdews, see Plymouth and West Devon Record Office, 81/A/11–42. 1/117 refers to Richard Rosdew as coroner in 1791. He is also listed with 'An Alphabetical List of Freemen of the Borough of Plymouth', August 1817, which is held in Plymouth Local Studies Library.

122 *Margaret Jane Williamson*   Flint, p. 154.

122 *made considerable progress in mathematics*   Flint, p. 123.

124 *a first rate mathematician*   Flint, p. 123.

124 *a man of the nicest feelings*   Flint, p. 127.

124 *devoted to his own*   Flint, p. 154.

124 *this operation might not rest*   Mudge, Williams and Dalby, 1795a, pp. 446–7.

124 *the uncertain nature of accuracy*   Fascinating discussions of the idea of accuracy in the eighteenth and early-nineteenth centuries can be found in Chapman; Gooday; Hacking; Stigler; and Wise.

125 *a second scientific revolution*   Kuhn, cited in Wise, p. 3.

125 *the astronomer Francis Wollaston*   Francis Wollaston to Joseph Banks, RS, MS/820, 13 September 1790.

125 *on 15 August*   Mudge, Williams and Dalby, 1795a, p. 429. Further information about the remeasurement of the Hounslow Heath baseline can be found in NA, MR1/382.

125 *While the sun shone out*   Mudge, Williams and Dalby, 1795a, p. 429.

125 *several other members of the Royal Society*   Mudge, Williams and Dalby, 1795a, p. 429.

126 *Keg of small Beer*   Edward Williams to Joseph Banks, RS, DM4, f. 29, 8 August 1791.

126 *Thanks, for the kind Attentions*   Edward Williams to Joseph Banks, RS, DM4, f. 32, 14 December 1791.

126 *an official 'handing over' ceremony*   Harley and O'Donoghue, I, p. xxv. The origins of the Ordnance Survey are discussed in Skelton, 1962.

126 *seldom if ever had there been*   Brown, Lloyd, p. 257.

126 *laid down by the General*   [Charles Lennox], 'The Result Given by General Roy', RS, DM4, f. 34.

126 *in a very decayed state*   Mudge, Williams and Dalby, 1795a, p. 437.

126 *in a more permanent manner*   Mudge, Williams and Dalby, 1795a, p. 437.

126 *heavy iron cannon*   Mudge, Williams and Dalby, 1795a, p. 438.

126 *an operation of a delicate nature*   Mudge, Williams and Dalby, 1795a, p. 438.

126 *fixed at the extremities of the base*   Mudge, Williams and Dalby, 1795a, p. 440.

126 27,404.3155   Mudge, Williams and Dalby, 1795a, p. 433.

127 *a small oversight* Mudge, Williams and Dalby, 1795a, p. 434.

127 *of similar construction to that* Mudge, Williams and Dalby, 1795a, p. 415.

127 *Great Theodolite* Mudge, Williams and Dalby, 1797b, pp. 4, 12–13; cited in Harley and O'Donoghue, II, pp. xii, xiii, xiv.

127 *to avoid towers and high buildings* Mudge, Williams and Dalby, 1795b, p. 31.

128 *The same object* Berkeley, p. 127.

128 *soon Mudge had got the hang* For an account of the Ordnance Survey's progress and techniques between 1791 and 1794, see Mudge, Williams and Dalby, 1795a and 1795b. There are also details in NA, MR1/382.

129 *process called levelling* A description of the Ordnance Survey's method of levelling is given in Mudge, Williams and Dalby, 1795b, pp. 167–74.

129 *the water's edge* Mudge, Williams and Dalby, 1795a, p. 579.

130 *boundless, yet distinct* Charlotte Smith, 'The Emigrants', 1793, in Smith, Charlotte, 1993, p. 162.

130 *the top of a great hill* John Neal, 'Ode to Peace', 1829, in Kettell, III, p. 109.

130 *like a map* James Rhoades, 'Earth's Message to the Old', in Rhoades, p. 62.

130 *spring waggon* Portlock, p. 25.

131 *respectable Sussex women* Moir, Esther, p. 6.

131 *if you love good roads* Horace Walpole to George Montague, cited in Moir, Esther, p. 6.

131 *when the air was free* Mudge, Williams and Dalby, 1795a, pp. 456–7.

131 *consisted primarily* Portlock, pp. 24–5.

132 *Mr Howard of Old-Street* Mudge, Williams and Dalby, 1795a, p. 443.

132 *strong tin cases* Mudge, Williams and Dalby, 1795a, p. 444.

132 *found to equal everything* Mudge, Williams and Dalby, 1795a, pp. 443–4.

132 *was lighted on Shooter's Hill* Mudge, Williams and Dalby, 1795a, p. 444.

132 *a General Survey of the Kingdom* 'Lights on the Coast', *Sussex Weekly Advertiser*, 8 and 15 April 1793; cited in Harley and O'Donoghue, I, p. xxv.

132 *the inhabitants of Ditchling* *Gazetteer and New Daily Advertiser*, 19935, 31 October 1792.

132 *22 feet north-west* Mudge, Williams and Dalby, 1795a, p. 473.

133 *furnish Mr Gardner* Mudge and Dalby, pp. xi–xii.

133 *seemed to dance up and down* Mudge, Williams and Dalby, 1795a, p. 587.

134 *how wide the view!* Charlotte Smith, 'Beachy Head', 1807, in Smith, Charlotte, 1993, p. 238.

136 *had the mortification to hear* *Morning Post*, 6332, 7 August 1793.

136 *north-west chimney* Mudge, Williams and Dalby, 1795a, p. 549.

137 *As the first joint production* Hodson, 1999, makes the case for the Gardner and Gream map of 1795 as 'the First Ordnance Survey Map'.

137   *the only maps*   Mudge, Williams and Dalby, 1797a, p. 540.

137   *Longham Common*   Mudge, Williams and Dalby, 1795a, p. 458.

137   *King's Sedgemoor*   Mudge, Williams and Dalby, 1795a, p. 488.

137   *Over a period of about*   For histories of Enclosure, see Hoskins; Mingay, 1968; Mingay, 1997; Mingay, 1999; Rackham; Turner; Yelling, Michael. I am also grateful to Helena Kelly for conversations about enclosure (see Kelly).

138   *how the hedges and fences*   Moir, Esther, p. 111.

139   *All I know is, I had a cow*   Young, 1801, p. 43.

140   *what a disgrace*   Young, 1771, iii, p. 193.

140   *such a vast tract*   Young, 1768, pp. 193–7.

140   *what an amazing improvement*   Young, 1768, pp. 193–7.

140   *a commanding view*   Mudge, Williams and Dalby, 1795a, p. 476.

140   *on foot over Salisbury Plain*   Wordsworth, 'Guilt and Sorrow, or Incidents Upon Salisbury Plain', in Wordsworth, 1969, p. 19.

140   *melancholy forebodings*   Wordsworth, 'Guilt and Sorrow, or Incidents Upon Salisbury Plain', in Wordsworth, 1969, p. 18.

140   *long continuance*   Wordsworth, 'Guilt and Sorrow, or Incidents Upon Salisbury Plain', in Wordsworth, 1969, p. 19.

141   *gathering clouds grew red*   Wordsworth, 'Guilt and Sorrow, or Incidents Upon Salisbury Plain', in Wordsworth, 1969, p. 19, l. 19.

141   *the blank sky*   Wordsworth, 'Guilt and Sorrow, or Incidents Upon Salisbury Plain', in Wordsworth, 1969, p. 19, l. 23.

141   *a naked guide-post's double head*   Wordsworth, 'Guilt and Sorrow, or Incidents Upon Salisbury Plain', in Wordsworth, 1969, p. 21, l. 134.

141   *gleam of pleasure*   Wordsworth, 'Guilt and Sorrow, or Incidents Upon Salisbury Plain', in Wordsworth, 1969, p. 21, l. 135.

141   *when the cultivated ground*   Mudge, Williams and Dalby, 1795a, p. 467.

141   *an antique castle*   Wordsworth, 'Guilt and Sorrow, or Incidents Upon Salisbury Plain', in Wordsworth, 1969, p. 20, l. 114.

141   *Pile of Stonehenge!*   Wordsworth, 'Guilt and Sorrow, or Incidents Upon Salisbury Plain', in Wordsworth, 1969, p. 20, l. 118.

141   *by which the Druids*   Wordsworth, 1979, p. 454, Book 12: ll. 345–7.

141   *strange lines*   Wordsworth, 'Guilt and Sorrow, or Incidents Upon Salisbury Plain', in Wordsworth, 1969, p. 20, l. 112.

142   *to hint yet keep*   Wordsworth, 'Guilt and Sorrow, or Incidents Upon Salisbury Plain', in Wordsworth, 1969, p. 20, ll. 18–19.

142   *a sound of chains*   Wordsworth, 'Guilt and Sorrow, or Incidents Upon Salisbury Plain', in Wordsworth, 1969, p. 20, ll. 76–9.

142   *the largest known expanse*   'Salisbury Plain', p. 1.

142   *laid down*   Mudge, Williams and Dalby, 1795a, p. 589.

CHAPTER 6: THE FIRST MAP

144 *Geometrical and Trigonometrical Operations* *The Times*, 4817, 10 June 1800.

145 *a Survey of the Island* *Lloyd's Evening Post*, 5473, 25 July 1792.

145 *the novelty of half-a-dozen tents* *London Evening Post*, 11560, 15 August 1797; also in *London Packet or New Lloyd's Evening Post*, 4367, 14 August 1797.

145 *five hundred Copies* Edward Williams to Joseph Banks, RS, MM3, f. 50, 15 June 1795.

145 *On 25 June 1795* Mudge, Williams and Dalby, 1795a, p. 414.

146 *rate of expansion* Mudge, Williams and Dalby, 1795a, p. 426.

146 *the Commencement of the Trigonometrical Operation* Mudge, Williams and Dalby, 1795a, p. 441.

146 *the Improvements in the great Theodolite* Mudge, Williams and Dalby, 1795a, p. 441.

146 *Base of Verification on Salisbury Plain* Mudge, Williams and Dalby, 1795a, p. 474.

146 *Two years later* Mudge, Williams and Dalby, 1797a, p. 432.

147 *Joseph Lindley and William Crosley* Harley, 1966a, pp. 372–8.

147 *Thomas Vincent Reynolds* For William Roy's Last Will and Testament, see NA, PROB 11/1194.

147 *the Board of Ordnance had appointed Reynolds* See Thomas Vincent Reynolds, 'Queries Humbly Submitted to general the Duke of Richmond Relative to the Compilation of a Military Map of the Southern District', NA, OS 3/5.

147 *Military Map of Kent, Sussex, Surrey* Extracts of Minutes, NA, WO 47/2365, 19 December 1793.

147 *the road-book* Moir, Esther, pp. 9–10.

147 *the very ingenious Major Mudge* Paterson, p. xviii.

147 *The Measurements of the Heights of Mountains* Paterson, p. xviii.

147 *may depend upon every information* Paterson, p. xviii.

148 *In 1797 a physician* Maton, pp. 18–19.

148 *for the first time* Close, 1969, p. 55.

148 *very excellent trigonometrical survey* Beeke, pp. 8–9.

149 *to give the crown* Moon, p. 229, cited in Bayly and Prior. See Bayly and Prior for a brief account of Cornwallis's life and achievements.

149 *union of the parties* Mudge and Dalby, p. xii.

149 *for public use* Mudge and Dalby, p. xiii.

149 *Ordnance Survey Maps* See Mudge's annotation to [Thomas Budgen], 'Exeter', Ordnance Surveyors' Drawings, BL, OSD 40 pt 3, 1801.

150 *in June 1795* For an account of the Ordnance Survey's progress in 1795 and 1796, see Mudge, Williams and Dalby, 1797a, pp. 432–541.

150 *an eighteen-inch theodolite* Close, 1969, p. 34; Mudge, Williams and Dalby, 1797a, p. 432.

150 *General Instructions for the Officers of Engineers* Charles Lennox, 3rd Duke of Richmond, 'General Instructions for the Officers of Engineers Employed in Surveying', NA, WO 30/115, ff. 175–82, 17 July 1785. Harley and O'Donoghue, I, p. xxx, point out that another version of these Instructions is contained in NA, WO 30/54, art. 22, catalogued under the name of William Roy.

151 *small Theodolets* Charles Lennox, 'General Instructions for the Officers of Engineers Employed in Surveying', NA, WO 30/115, f. 179.

151 *proceed around* Charles Lennox, 'General Instructions for the Officers of Engineers Employed in Surveying', NA, WO 30/115, f. 179.

151 *it is likely* See Close, 1969, p. 38, for the assertion that the Ordnance Survey did not use the plane-table surveying method. See Harley and O'Donoghue, I, p. xxx, for further discussion of the Interior Surveyors' methods.

151 *works of art* Hodson, 1989, p. 15.

152 *Britain's new Hydrographic Office* See Day, Archibald; and Cook.

152 *in a masterly manner* Mudge and Dalby, p. xiii.

152 *Died, Last week in London* Sun, 1669, 29 January 1798.

152 *He had preferred to talk instead* General Evening Post, 7976, 4 February 1796.

152 *you will accordingly take on yourself* Flint, pp. 123–4.

153 *had the honour to kiss* Oracle and Public Advertiser, 19872, 3 March 1798.

153 *By now he had* Flint, pp. 154–5.

153 *no longer able to endure the fatigues* Mudge, William, 1800, p. 4.

153 *the extent of his service* Mudge, William, 1800, p. 4.

153 *he applied to the Royal Society* William Mudge to Joseph Banks, RS, MM3, f. 61, 15 November 1798.

154 *to possess some general Map* Mudge and Dalby, p. xiii.

154 *it has been very justly expected* Mudge and Dalby, p. xii–xiii.

154 *The climate of that decade* See Barrell, 2006; Hilton, pp. 39–109.

155 *Alien Office* For histories of espionage in this period, and the 'spirit of despotism', see Barrell, 2004; Barrell, 2006; Dandeker; Guest; Porter, Bernard.

155 *System of Preventitive [sic] Police* William Wickham to Portland, BL, Add. MS 33107, f. 3, 3 January 1801.

155 *pernicious seditious & evil disposed Person* 'Assizes for trial of Mr. Edward Swift, Berks', NA, TS 11/944/3433.

155 *panopticism* See Bentham; and Foucault. For a discussion of these ideas of panopticism applied to cartography, see Harley, 1988; Harley, 'Deconstructing the Map'; and Mitchell.

155 *astonishment* William Wordsworth, 'To the Editor of the Morning Post', in Wordsworth, 1977, pp. 157–66 (pp. 160–1).

156 *It was the same in France* Robb, p. 5.

156 *wandering on the hills* Coleridge, Samuel Taylor, 1983, I, pp. 196–7.

156 *surely a French Jacobin* Cited in Eagleston, p. 77.

156 *to some principal at Bristol* Cited in Eagleston, p. 80.

156 *a Sett of violent Democrats* Walsh to George III, 16 August 1797, cited in Eagleston, p. 82.

156 *Spy Nosey* Coleridge, Samuel Taylor, 1983, I, pp. 196–7.

156 *studies, as the artists call them* Coleridge, Samuel Taylor, 1983, I, pp. 196–7.

157 *it was my purpose* Coleridge, Samuel Taylor, 1983, I, pp. 196–7.

157 *a Secret Service to Survey* Documentation regarding Hugh Debbieg's 'Secret Service' can be found in NA, Chatham Papers, PRO 30/8/129, ff. 2–31, 1788–1801; and NA, TS 11/944/3436. This particular citation is found in 'Legal papers regarding case of Hugh Debbieg vs Lord Howe', NA, TS 11/944/3436, f. 1, 5 February 1782.

157 *immense invasion* Ravenhill, 1994, p. 165. See Ravenhill, 1994, pp. 159–72 for information regarding Clifford's activities.

157 *a box 5 feet long* Robert Edward Clifford, Ugbrooke Park Archives, Box 15, 30 May 1803; cited in Ravenhill, 1994, p. 168.

157 *the Plans of the Vendée* General Simcoe to Lord Clifford, John Graves Simcoe Letters, Archives of Ontario, Series A-4-1, 23 May 1803; cited in Ravenhill, 1994, p. 168.

157 *a grand military expedition* Letter, 23 August 1801, John Graves Simcoe Letters, Archives of Ontario, Series A-4-1; cited in Ravenhill, 1994, p. 168.

158 *a system of* TERROR *The Cabinet*, cited in Hilton, p. 65.

158 *the employment of spies* Knox, p. 107.

158 *If this Military Projector* Anon, 1786, p. 48.

158 *It has also been suggested* Dandeker, pp. 49–51.

158 *a Royal Warrant in 1804* Simon Woolcot to William Mudge, 4 May 1804, cited in Close, 1969, p. 50.

158 *a considerable degree of uneasiness* Simon Woolcot to William Mudge, 4 May 1804, cited in Close, 1969, p. 50.

159 *these maps the Board of Ordnance Courier and Evening Gazette*, 2180, 15 August 1799.

159 *An accurate Topographical Survey The Times*, 4525, 3 July 1799.

159 *William Faden* For an account of Faden's life, see Worms. For the history of his partnership with Jefferys, see Harley, 1966b; and Pedley.

159 *the place and quality of Geographer in Ordinary* Worms.

160 *the maps needed to be portable* William Mudge to Lieutenant-General of the Board of Ordnance, NA, WO 47/2372, 10 April 1799; cited in Harley and O'Donoghue, I, pp. xxviii.

160 *the publication process*   For descriptions of the methods and meanings of map-printing techniques in the eighteenth and nineteenth centuries, see Harley, 1968; Johns; Mumford, 1968; Mumford, 1972; Raisz, p. 22; Robinson, Arthur, 1975; Stone; Woodward.

161 *the engravers had to traipse*   Harley and O'Donoghue, I, p. xxxii.

162 *regional map-makers had been thinking*   For accounts of the history of toponymy on British maps, see Andrews, J.H., 1992; Andrews, J.H., 2006, pp. 86, 92, 120–8, 156, 167–8, 281–98; Doherty, pp. 17–21; 55–77; 140–56; Harley, 1971; Harley and Walters, 1982; Ó Cadhla, pp. 5, 26, 28, 81, 104, 171, 218–44; Withers, 2000.

162 *the most carefull observer*   Norden, p. 23, cited in Harley, 1971, p. 92.

162 *knowing Gentlemen*   Camden, 1971, Preface, cited in Harley, 1971, p. 92.

162 *to make the work as perfect*   William Mudge, Ordnance Office and War Office: Correspondence, NA, WO 44/299; cited in Harley, 1971, p. 93.

164 *like a gauzy and radiant fabric*   Conrad, 1988, p. 8.

165 *the map of Kent was by much*   Mulgrave, cited in Flint, p. 144.

165 *presented the Map to his Majesty*   William Mudge to Lieut-Col Hadden, 14 January 1802, cited in Close, 1969, p. 56.

165 *I think [he] still remains*   William Mudge to Lieut-Col Hadden, 14 January 1802, cited in Close, 1969, p. 56.

166 *Arch Dukes wanted to see*   William Mudge, 29 February 1816, cited in Close, 1969, p. 65.

166 *to procure a map*   *The Times*, 10355, p. 1, 16 January 1818.

166 *the Board of Ordnance's headquarters*   See Harley and O'Donoghue, I, p. xxxiii.

166 *£3 3s*   Harley and O'Donoghue, I, p. xxxiii. For details regarding the sale of maps during the early Enlightenment, before the emergence of the maps discussed in this book, see Tyacke.

166 *£6 6s*   *The Times*, 9982, p. 1, 2 November 1816.

166 *the Sheffield warehouse apprentice Joseph Hunter*   Joseph Hunter, Journal, BL, Add. MS 24880, p. 6, 5 April 1797.

167 *Do you know what a map is?*   John O'Donovan to Thomas Aiskew Larcom, 16 April 1834, in O'Donovan, 2001a, pp. 44–5.

167 *an estate agent's advertisement*   *The Times*, 11377, p. 4, 15 October 1821.

167 *the 8th Earl of Wemyss*   Annotated volume in the Dunimarle Library of the Erskines of Torrie in Fife: Armstrong, Mostyn John, *A Scotch Atlas*, London: 1777, DH LIB 433.

167 *On 21 November 1783*   For accounts of the history of hot-air ballooning, see Holmes, pp. 125–63; and Rolt.

168 *first English female aerial traveller*   Sage.

168 *in crossing over Westminster*   Sage, p. 19.

168 *turned on its axis*   Sage, pp. 22–3.

169 *looking down on an enormous map* Poole, p. 65.

169 *a narrative of a balloon excursion* Baldwin.

Chapter 7: 'A Wild and Most Arduous Service'

170 *making observations to determine* Mudge, Williams and Dalby, 1797b, pp. 12–13.

170 *that, desirous of proving* Mudge, William, 1800, p. 182.

171 *Colonel Williams and Captain Mudge* *Observer*, 298, 20 August 1797.

171 *Mudge dutifully spent the spring* For an account of the Ordnance Survey's progress and methods in this period, see Mudge, William, 1800; and Mudge, William, 1801a.

171 *between Hampstead Heath* Mudge, William, 1801a, pp. 28–9.

172 *the dome of St Paul's Cathedral* Mudge, William, 1801a, p. 29. For a description of the Ordnance Survey's activities in London, and a facsimile of their productions, see Hodson, 1991c.

172 *What a shock! For eyes and ears!* Wordsworth, 1979, p. 263, ll. 685–6.

172 *the lighted shops of the Strand* Charles Lamb to William Wordsworth, in Lamb, pp. 265–7, 30 January 1801.

172 *in the season of 1799* See Close, 1969, p. 43.

172 *of the drawing room in the Tower* *Courier and Evening Gazette*, 2180, 15 August 1799.

173 *It does not appear that any advantage* William Mudge to Lieutenant-General of the Board of Ordnance, NA, WO 47/2372, 10 April 1799; also cited in Harley and O'Donoghue, I, p. xxviii.

174 *relinquish[ing] the prosecution* William Mudge to Lieutenant-General of the Board of Ordnance, NA, WO 47/2372, 10 April 1799; also cited in Harley and O'Donoghue, I, p. xxviii.

174 *about one third* William Mudge to Lieutenant-General of the Board of Ordnance, NA, WO 47/2372, 10 April 1799; also cited in Harley and O'Donoghue, I, p. xxviii.

174 *an estimate of 33s* Extracts of Board of Ordnance Minutes January–June 1799, NA, WO 47/2372, 28 April 1799; also cited in Harley and O'Donoghue, I, p. xxix.

174 *fifty more years* Harley and O'Donoghue, I, p. xxix.

174 *much applaud[ed] the zeal* Extracts of Board of Ordnance Minutes January–June 1799, NA, WO 47/2372, 28 May 1799; also cited in Harley and O'Donoghue, I, p. xxix.

174 *Corps of Royal Military Surveyors and Draftsmen* Ordnance Office: Warrants, NA, WO 55/421; also cited in Harley and O'Donoghue, I, p. xxix.

174 *rules and disciplines of war* Ordnance Office: Warrants, NA, WO 55/421; also cited in Harley and O'Donoghue, I, p. xxix.

174 *32s 6d per mile* Close, 1969, p. 64.

175 *represent the Towns, Villages, Woods, Rivers* William Gardner, Extracts of Board of Ordnance Minutes January–June 1799, NA, WO 47/2372, 9 April 1799; also cited in Harley and O'Donoghue, I, p. xxix.

175 *He hired Thomas Foot* Extracts of Board of Ordnance Minutes, January–April 1801, NA, WO 47/2379, 3 March 1801.

175 *Agent for the sale of Ordnance maps* *The Times*, 9982, 2 November 1816.

175 *There are also numerous contenders* See Hodson, 1999.

176 *the First Series or Old Series* Harley and Oliver, VI, p. xiii. There are numerous invaluable guides to the maps of the First Series: see Harley, 1975; Harley and Phillips; Hellyer; Hellyer and Oliver; Hodson, 1991a; Messenger, 1991a and 1991b; Oliver, 2005; Oliver and Hellyer. Wheeler, 1990, offers a discussion of the accuracy of the Ordnance Survey's Old Series maps.

176 *He did not even have a permanent base* William Mudge to Richard Rosdew, 8 January 1805, cited in Flint, pp. 133–4.

176 *he now had five children* Flint, pp. 153–6.

176 *a more meritorious, sensible and affectionate* Flint, p. 141.

177 *he often complained of depression of spirits* Isaac Dalby to Thomas Colby, 5 February 1821, cited in Close, 1969, p. 42.

177 *well grounded in the rudiments of mathematics* William Mudge, 12 January 1802, cited in Portlock, p. 13.

177 *Thomas Colby came from a deeply* For accounts of Thomas Colby's life, see Portlock; Baigent, 'Colby, Thomas'.

177 *overheated enthusiasm* Portlock, p. 7.

177 *a surveyor called Robert Dawson* For accounts of Dawson's achievements, see Hodson, 1991b; and Baigent, 'Dawson, Robert'.

179 *an old pair of pistols* Portlock, p. 15.

179 *producing a Fracture* William Mudge to Robert Morse, 16 December 1803, cited in Close, 1969, p. 85.

179 *almost dying* Portlock, p. 15.

180 *violently injured* William Mudge to Robert Morse, 16 December 1803, cited in Close, 1969, p. 85.

180 *apprehensive . . . to report his death* William Mudge to Robert Morse, December 1803, cited in Portlock, p. 38.

180 *a constitution of unusual strength* Portlock, p. 15.

180 *triumph[ed] over the effects* Portlock, p. 15.

180 *slight peculiarities of dress* Portlock, p. 8.

180 *come back, my boy* Cited in Portlock, p. 5.

180 *a fearful indent* Portlock, p. 15.

180 *the Brain it seems remains free* William Mudge to Robert Morse, in Ordnance Survey Correspondence, NA, WO 55/960, 16 December 1803; cited in Close, 1969, pp. 85–6.

180 *Spirits [were] raised* Hall, in Ordnance Survey Correspondence, NA, WO 55/960, 20 January 1804.

180 *quick manner of speaking* Hall, in Ordnance Survey Correspondence, NA, WO 55/960, 20 January 1804.

180 *over-strained prejudice* Portlock, p. 7.

181 *shall not want the necessary sittings* William Mudge to Richard Rosdew, 20 November 1804, cited in Flint, p. 130.

181 *I should be very glad* William Mudge to Richard Rosdew, 20 December 1804, cited in Flint, p. 132.

181 *I am not yet settled* William Mudge to Richard Rosdew, 8 January 1805, cited in Flint, p. 134.

181 *Holles Street* Flint, p. 139.

181 *How retrograde* William Mudge, January 1806, cited in Flint, p. 138.

182 *Trellech Beacon* Mudge and Colby, pp. 10–11, 74.

182 *Contemporary visitors reported* Gilpin, 1789a, pp. 22–3.

183 *the number of turnpike roads had exploded* Jarvis, pp. 19–20.

183 *John Loudon McAdam* For an account of McAdam's life and achievements, see Reader.

183 *all the roads in England* Thomas De Quincey, cited in Burke, Thomas, p. 93, and in Moir, Esther, p. 8.

183 *between 1790 and 1836* See Bagwell, especially p. 48; Copeland, especially p. 85; and Jarvis, pp. 19–21.

183 *a once-boasted, though now unfortunate* Joseph Budworth, cited in Jarvis, p. 9.

183 *In 1757* See Burke, Edmund, 1990.

184 *like a living thing* William Wordsworth, 1979, p. 51, Book 1: l. 384.

184 *mountainous and other similarly dramatic sites* Further explorations of this rise in popularity of mountains can be found in Macfarlane; and Stafford.

184 *In the summer of 1770* Gilpin, 1789a, pp. 1–2.

184 *Berkshire, Gloucestershire and the Vale of Severn* Gilpin, 1789a, pp. 3–14.

184 *a new object* Gilpin, 1789a, p. 1.

184 *that of examining the face* Gilpin, 1789a, p. 1.

184 *By 'picturesque'* See Gilpin, 1768; Gilpin, 1789a and 1789b; Gilpin, 1792. For the history of the picturesque and picturesque tourism, see also Andrews, Malcolm; Bermingham; and Copley and Garside.

184 *that peculiar kind of beauty* Gilpin, 1768, p. 2.

184 *of the most beautiful kind* Gilpin, 1789a, p. 18.

184 *two side-screens* Gilpin, 1789a, p. 18.

184 *front-screen* Gilpin, 1789a, p. 18.

184 *the most beautiful and picturesque view* Gilpin, 1789a, p. 45.

185 *Thousands flocked* For an account of the effect of Gilpin's ideas on British tourism, see Andrews, Malcolm.

185 *If the hues are well sorted* Gilpin, 1789b, I, p. 124. For a lengthy discussion of the Claude Glass and Mirror, and its significance, see Maillet.

185 *magnifiers for botany* James Plumptre, Narrative of a Pedestrian Journey . . . in the Summer of 1799, Cambridge University Library, Add. MS 5814, f. 35; cited in Moir, Esther, pp. 4–5. For further evidence of amusement at tourists' excessive accoutrements, see [Plumptre] in Works Cited, below.

185 *Mudge superimposed in his imagination* For analyses of the relationship between cartographic and painterly representations of the British landscape in this period, see Alfrey; and Cosgrove.

186 *a simple red square* See Thomas Budgen, 'Plan of Chepstow', BL, OSD 175, 1812.

186 *Gilpin instead advised travellers* See Andrews, Malcolm, p. 62.

186 *the Picturesque Point* Gray, p. 360 (note).

186 *latent and overarching geometry* A discussion of geometry, and the issues most pertinent to the discipline in this period, can be found in Mlodinow.

186 *some grand scene* Gilpin, 1792, pp. 49–50.

186 *future almost inaccessible* William Mudge to Thomas Colby, 15 July 1818, in Close, 1969, p. 51.

187 *triangles of sight lines* Information regarding the Ordnance Survey's progress through Wales can be found in Mudge and Colby.

187 *the discovery of an appropriate baseline* Mudge and Colby, pp. 82–3.

188 *This lay in the shadow of* I am very grateful to Alistair Pegg for information regarding the topography and history of the area surrounding Rhuddlan.

188 *at Misterton Carr near Doncaster* Mudge and Colby, pp. v–vi, 82.

188 *a wild and most arduous service* Robert Dawson, cited in Portlock, p. 7.

188 *the mountain tapered* Kohl, pp. 68–9, cited in Harley and Oliver, VI, pp. vii–viii.

188 *several tents* Kohl, pp. 68–9, cited in Harley and Oliver, VI, pp. vii–viii.

189 *Snowdon . . . lies right in the centre* Kohl, pp. 68–9, cited in Harley and Oliver, VI, pp. vii–viii.

189 *Survey of Norfolk* Close, 1969, p. 50. See also Thomas Colby, 'Précis of the Progress of the Ordnance Survey of England and Wales 1783–1834', NA, WO 44/714, f. 1.

189 *the range of mountains* Robert Dawson to Thomas Colby, 4 July 1816, cited in Close, 1969, p. 51.

189 *to select those who draw well* Hobbs to Thomas Colby, 1 March 1821, Ordnance Survey Letter Book, NA, OS 3/260, f. 235; cited in Hodson, 1989, p. 16.

190 *aim[ing] at a large and striking example* Robert Dawson to Thomas Colby, 8 December 1815, cited in Close, 1969, p. 81. This is also commented on by Alfrey, p. 25.

190 *a degree of perfection* Flint, pp. 140–1.

190 *a plan of Snowdonia* Robert Dawson, 'Plan of Cader Idris', BL, OSD 319, part 1.

190 *2914 feet* See Compton, p. 5.

191 *towards the expression of Ground* Robert Kearsley Dawson, Essays towards the expression of Ground in Topographical Plans, BL, Maps C.21.e.7.

191 *mathematical forms* Robert Kearsley Dawson, 'Mathematical Forms Applicable to Hills', Essays towards the expression of Ground in Topographical Plans, BL, Maps C.21.e.7.

191 *conical and hemispherical hill[s]* Robert Kearsley Dawson, 'Natural Form Corresponding to Fig. 3 of the Mathematical Forms. Conical Hill' and 'Hemispherical Hill', Essays towards the expression of Ground in Topographical Plans, BL, Maps C.21.e.7.

191 *I feel it my duty to state* Gosset to Richard Zachariah Mudge, 11 June 1832, Letters copied from a book in the possession of the Ordnance Survey relating to the origin of the Geological Survey and the Geological Museum, ranging from March 1830 to March 1845, National Museum of Wales, Cardiff, Department of Geology, De la Beche Papers. Cited in Harley and Oliver, VI, p. xii.

191 *I should be very sorry to see* Richard Zachariah Mudge to Robert Dawson, 16 June 1832, National Museum of Wales, Cardiff; cited in Harley and Oliver, VI, p. xii.

191 *The requirements of art and cartography* For discussions of the relationship between art and cartography, and the use of symbols by both, see Harvey; Keates; MacEachren; Robinson, Arthur, 1952; and Wright.

192 *They duly misunderstood Llandovery* Harley, 1971, p. 92.

192 *Ruddy pene* Harley and Walters, 1982, p. 100.

192 *murdered by English map-makers* Lewis Morris, 'Proposals for taking, by Subscription, a General Survey of Mona', National Library of Wales, MSS 1731–3; cited in Walters, 1970; and Harley, 1971, p. 93.

192 *much indebted* Robert Dawson, 'Plan of Arran Mowddy', BL, OSD 303, 1819; cited in Harley, 1971, p. 98.

192 *Mr Colby in the Northern part* Ordnance Survey Letter Book, 1817–1822, Ordnance Survey Headquarters Southampton, ff. 219–21; cited in Harley, 1971, p. 95.

192 *Crosswoodig* Cited in Harley and Walters, 1982, p. 109.

192 *on one side of it* Ordnance Survey Letter Book, 1817–1822, Ordnance Survey Headquarters Southampton, ff. 219–21; cited in Harley and Walters, 1982, pp. 108–9.

193 *You have called it Caerdiff* Lewis Weston Dillwyn, letter accompanies BL, two-

inch Hill Sketches serial 459 (one-inch sheet no. 36); cited in Harley and Walters, 1982, p. 108.

193 *that in the Ordnance Map* Ordnance Survey Letter Book, 1817–1822, Ordnance Survey Headquarters Southampton, f. 217.

193 *we have long been aware of the difficulty* Letters copied from a book in the possession of the Ordnance Survey relating to the origin of the Geological Survey and the Geological Museum, ranging from March 1830 to March 1845, National Museum of Wales, Cardiff, Department of Geology, De la Beche Papers, pp. 36–7; cited in Harley, 1971, pp. 98–9.

193 *who is employed in the County* Letters copied from a book in the possession of the Ordnance Survey relating to the origin of the Geological Survey and the Geological Museum, ranging from March 1830 to March 1845, National Museum of Wales, Cardiff, Department of Geology, De la Beche Papers, pp. 36–7; cited in Harley, 1971, p. 99.

194 *Cities, Market-Towns, Parishes* Ravenhill, 1978, p. 424.

194 *the words Lan and Llan* Alfred Thomas to Richard Zachariah Mudge, 29 October 1831, Letters copied from a book in the possession of the Ordnance Survey relating to the origin of the Geological Survey and the Geological Museum, ranging from March 1830 to March 1845, National Museum of Wales, Cardiff, Department of Geology, De la Beche Papers, pp. 35–6; cited in Harley and Walters, 1982, p. 111.

194 *I cannot help feeling* J.R. Haslam to Richard Zachariah Mudge, 9 November 1831, Letters copied from a book in the possession of the Ordnance Survey relating to the origin of the Geological Survey and the Geological Museum, ranging from March 1830 to March 1845, National Museum of Wales, Cardiff, Department of Geology, De la Beche Papers, pp. 38–9; cited in Harley and Walters, 1982, p. 112.

195 *the Public are hereby informed* *The Times*, 10355, p. 1, 16 January 1818.

CHAPTER 8: MAPPING THE IMAGINATION

196 *Early in the mapping season* Although the exact date that the Ordnance Surveyors were in Kettlewell has not been established, Mudge and Colby, pp. 41–3, indicate it was towards the end of the triangulation conducted between 1800 and 1809.

196 *At first, they followed a track* My fictionalised description of the surveyors' route is extrapolated from the current Wharfedale and Nidderdale landscape and the Ordnance Survey's own First Series map of Wharfedale, Sheet 97, published in 1860–1. Admittedly, this was published half a century after the

Trigonometrical Survey's presence, but this passage is intended to be read in the spirit of the chapter's title: as material for the 'imagination' and an exploration of its cartographic energy, rather than as a meticulously authentic representation. Whereas the rest of the book is based on archival and historical research, this chapter makes greater use of fictional and poetic descriptions of the act of mapping, and some of my account of the surveyors' activities is based on that provided by authors. Nevertheless, Hag Dyke cottage was built before 1730, and still survives as a Scout Hostel, and it provides a constant in the landscape between the Ordnance Surveyors' experience and the present-day walker's.

196 *a great number of huge rocks* Mudge and Colby, p. 79.

197 *These reservoirs were built* See Chris Hawkesworth's video, *A Century of Reservoirs*, and 'History of Scar House Reservoir', <http://www.nidderdaleaonb.org.uk/Scar%20House%20%20History.pdf> [accessed 12 April 2010].

198 *Mudge and Colby's team arrived* Again, the exact date that the Ordnance Surveyors were in Kettlewell has not been established, but Broglio, p. 70; Wyatt, 2001, pp. 1.5–1.6; and Mudge and Colby, pp. 41–2, indicate between them that it was in 1808.

198 *the men first made their way to Bootle* William Wordsworth, 1993, p. 29, indicates that the Reverend Dr James Satterthwaite, who had the living at Bootle, was acquainted with the surveyors' presence in Bootle.

198 *guide-posts and milestones* Moir, Esther, pp. 8–9.

199 *In the summer of 1811* Wyatt, 2001, pp. 1.1–1.3.

200 *Eight years previously* See Barker, Juliet, pp. 418–19; cited in Wyatt, 2001, p. 1.2. For biographical details about Wordsworth and his trip to Bootle, see Gill.

200 *Ocean's ceaseless roar* William Wordsworth, 'To Sir George Howland Beaumont', in Wordsworth, 1969, p. 408.

200 *bleakest point of Cumbria's shore* William Wordsworth, 'To Sir George Howland Beaumont', in Wordsworth, 1969, p. 408.

200 *grim neighbour! huge Black Combe* William Wordsworth, 'To Sir George Howland Beaumont', in Wordsworth, 1969, p. 408.

200 *the loveliest spot* William Wordsworth, 'A Farewell', in Wordsworth, 1969, p. 84.

200 *rough is the time* William Wordsworth, 'To Sir George Howland Beaumont', in Wordsworth, 1969, p. 408.

200 *Reverend Dr James Satterthwaite* Wordsworth, 1993, p. 29; cited in Wyatt, 2001, p. 1.5.

201 *the best authority* William Wordsworth, 'An Unpublished Tour', 1974, II, p. 302. For a discussion of Wordsworth's acquaintance with, and representations of, Mudge, see Wiley, especially pp. 30, 158.

201 *a two-volume collection of poetry*   See Wordsworth, 1815.

201 *from the summit of Black Combe*   Wordsworth, 'View from the Top of Black Combe', 1815, I, p. 305.

201 *low dusky tracts*   Wordsworth, 'View from the Top of Black Combe', 1815, I, p. 305.

202 *that experienced observer*   Wordsworth, 1977, p. 8.

202 *the solitary Mountain Black Combe*   Wordsworth, 1977, p. 8.

202 *a revelation infinite*   Wordsworth, 'View from the Top of Black Combe', 1815, I, p. 306.

202 *grand terraqueous spectacle*   Wordsworth, 'Inscription: Written With a Slate Pencil on a Stone, on the Side of the Mountain of Black Combe', 1815, II, p. 285.

202 *display august of man's inheritance*   Wordsworth, 'View from the Top of Black Combe', 1815, I, p. 306. For an extended discussion of the nationalist resonance of Wordsworth's Black Combe poems, see the chapter called 'Abandoning Utopia' in Wiley.

203 *Mentions of maps . . . poems and plays*   See Hewitt, 2007, pp. 139–40, for a rudimentary discussion of the apparent upsurge in mentions of maps in literature in the late eighteenth century.

203 *map-minded*   Edney, 1994a.

203 *maps were popular subjects for embroidery*   See, for example, Sampler, 1780, V&A Museum No. 497–1905.

203 *A New Geographical Pastime*   Board Game – A New Royal Geographical Pastime for England & Wales, published by Robert Sayer, London, 1 June 1787. Hand-coloured engraved paper on linen. V&A Museum No. E. 5307–1960. On a different but related note, it is also interesting to think about how surveyors' viewing practices may be related to what Peter de Bolla calls 'the birth of visual culture', which he pinpoints in the 1760s: the decade in which William Roy articulated his first coherent proposal for a complete national survey (see de Bolla).

203 *reading, with a Map*   Fennell, p. 14 (Act I, scene ii).

203 *at some leisure hour*   Wordsworth, *The Excursion*, in Wordsworth, 1949–54, V, p. 256.

204 *shatter'd map*   Thomas Dermody, 'The Invalid', in Dermody, p. 89.

204 *lost lovers*   See Aslam.

204 *a mere geographer*   Thomas Tickell, 'An Epistle', in Johnson, 1810, XXI, p. 109.

204 *prospect poetry or painting*   For discussions of the aesthetic and political significance of the prospect in poetry and art, see Barrell, 1972; Barrell, 1980; Barrell, 1983.

204 *Hope's deluding glass*   John Dyer, 'Grongar Hill', in Dyer and Akenside, p. 6.

204 *commanding height*   See Barrell, 1972, p. 24. It is noticeable how regularly the adjective 'commanding' crops up in the works of Roy, Mudge and Colby in

this context. See Roy, 1790, p. 264; Mudge, William, 1800, pp. 579, 580; Mudge, Williams and Dalby, 1795a, p. 476; Mudge, Williams and Dalby, 1975b, pp. 64, 153; Mudge, Williams and Dalby, 1797a, p. 436; Mudge, Williams and Dalby, 1797b, pp. 6, 77.

204 *Hertford's ancient town* John Hughes, 'A Monumental Ode', in Hughes, II, p. 101.

204 *all the coast of Galloway* Robert Dundas (son of 3rd Lord President), Journal of Lake District Travels, NRAS 3246, Bundle 174, 'chapter 29'.

205 *true philanthropy* Smith, Charlotte, 1798, pp. 351–2.

205 *genuine liberty* Anon, 1802, pp. 28, 29.

205 *Europe's continent* Anon, 1802, p. 56.

205 *poor dwindled map* Ireland, pp. 21–2.

205 *perspective* For discussions of the geometric correspondences between art and maps, see Alfrey; and Cosgrove.

205 *the art of delineating* Chambers, II, 'Perspective'.

205 *a childhood enthusiast of perspective* Edgcumbe.

205 *reduce the idea of beauty to general principles* Reynolds, I, pp. xcvii–xcviii (note 54).

206 *telescopes, and crucibles, and maps* Wordsworth, 1979, p. 168.

206 *marked by a microscopic acuteness* Coleridge, Samuel Taylor, 1956–71, I, pp. 354–5.

206 *unthreatened, unproclaimed* Wordsworth, 'Inscription: Written With a Slate Pencil on a Stone, on the Side of the Mountain of Black Combe', 1815, II, p. 286.

206 *as if the golden day* Wordsworth, 'Inscription: Written With a Slate Pencil on a Stone, on the Side of the Mountain of Black Combe', 1815, II, p. 286.

206 *esprit géométrique* Fontenelle, p. 151.

207 *God forbid that truth* Blake, p. 659. A discussion of Blake's response to Newtonian physics can be found in Ault. I am very grateful to Pete Newbon for conversations about Blake and cartography, which has greatly helped my thinking (see Newbon in Works Cited). A discussion of Blake and Reynolds' differing approaches to aesthetics can be found in Barrell, 1986a and 1986b.

207 *disposition to abstractions* Reynolds, I, pp. xcvii–xcviii (note 54).

207 *To Generalize is to be an Idiot* Blake, p. 641.

207 *Villainy* Blake, p. 639.

207 *old [Zachariah] Mudge* Reynolds, I, pp. xxxiii–xxxiv.

207 *visionary country* Malkin, p. 93.

207 *giving names of his own invention* Malkin, p. 94.

209 *a very remarkable production* Malkin, p. 94.

209 *an exercise of the mind* Malkin, p. 95.

209 *Quitted [his] house* Coleridge, Samuel Taylor, 1957, I: *Text*, entry 1207.

209 *every man [is] his own path-maker* Coleridge, Samuel Taylor, 1957, I: *Text*, entry 1207.

209 *Motivated variously* For an extended discussion of the rise of pedestrianism in the Romantic period, see Jarvis.

210 *Coleridge made himself a map* Coleridge, Samuel Taylor, 1957, I: *Text*, entry 1206.

210 *O What a Lake* Coleridge, Samuel Taylor, 1957, I: *Text*, entry 1213.

210 *the top of the Lake* Coleridge, Samuel Taylor, 1957, I: *Text*, entry 1213.

210 *it is impossible to conceive it* Coleridge, Samuel Taylor, 1957, I: *Text*, entry 1213.

210 *He duly sketched a map* Coleridge, Samuel Taylor, 1957, I: *Text*, entry 1213.

211 *steep as the meal* Coleridge, Samuel Taylor, 1957, I: *Text*, entry 1213.

211 *When I first came* Coleridge, Samuel Taylor, 1957, I: *Text*, entry 1213.

211 *Coleridge's use of maps* For further discussion of travellers' use of guides, see Stafford; and Vaughan.

212 *The love of maps* The emotional and psychological resonance of geography and cartography is fundamental to a literary movement known as 'psycho-geography': see Debord for one of the founding texts. See also Lynch.

212 *children's fantasy worlds* See Harty; and Coleridge, Hartley, 1986, p. 153.

212 *dreaming o'er the map of things* Wordsworth, 1979, p. 168.

CHAPTER 9: THE FRENCH DISCONNECTION

213 *In September 1811* William Mudge to Thomas Colby, 2 September 1811, in Close, 1969, pp. 57–8.

213 *I cannot tell whether they* William Mudge to Thomas Colby, 2 September 1811, in Close, 1969, pp. 57–8.

213 *appear before them* William Mudge to Thomas Colby, 2 September 1811, in Close, 1969, pp. 57–8.

213 *from his entrance into office* Flint, p. 155.

214 *on all the essentials* William Mudge to Thomas Colby, 1 October 1811, in Close, 1969, p. 58.

214 *did [his] best* William Mudge to Thomas Colby, 1 October 1811, in Close, 1969, p. 58.

214 *would not take more* William Mudge to Thomas Colby, 1 October 1811, in Close, 1969, p. 58.

214 *the Total expence* William Mudge to Thomas Colby, 1 October 1811, in Close, 1969, p. 58.

214 *instantly they had these* William Mudge to Thomas Colby, 1 October 1811, in Close, 1969, p. 58.

214 *£56,165 5s 7d* William Mudge to Thomas Colby, 1 October 1811, in Close, 1969, p. 59.

214 *Everybody I have shown it to* William Mudge to Thomas Colby, 1 October 1811, in Close, 1969, p. 59.

214 *I believe I have built* Cited in Flint, p. 144.

214 *I am conscious that* William Mudge to Thomas Colby, 2 September 1811, in Close, 1969, pp. 57–8.

214 *it is not my desire* William Mudge, 2 September 1816, in Close, 1969, p. 65.

214 *I have shortly to look* Cited in Flint, p. 143.

214 *strongly impressed* William Mudge to Thomas Colby, 1 October 1811, in Close, 1969, p. 59.

214 *approved of and sell[ing] well* William Mudge to Thomas Colby, 1 October 1811, in Close, 1969, p. 59.

214 *a depression that would stay* See Isaac Dalby to Thomas Colby, 5 February 1821, in Close, 1969, p. 42.

214 *Back in 1809* Close, 1969, pp. 40, 53; Flint, p. 139.

215 *a training academy at Addiscombe* Flint, pp. 140–1; Close, 1969, p. 41.

215 *my labours are great* William Mudge, 1811, cited in Close, 1969, p. 41.

215 *I have more business* William Mudge, 1811, cited in Close, 1969, p. 41.

215 *Phipps had been an early member* See Allibone, pp. 1–40.

215 *to withhold every map from the public* Henry Phipps to William Mudge, NA, Ordnance Survey Letter Book, OS 3/260, f. 131, 2 September 1811; cited in Close, 1969, p. 57.

215 *in the spring of 1816* Chapman to Mudge, NA, OSLB, OS 3/260, f. 125, 17 April 1816.

215 *that maps for correction* Flint, p. 143.

216 *Ponder the contents* Cited in Portlock, pp. 38–9.

216 *I believe that I have more* Cited in Close, 1932, p. 45.

216 *ten days have elapsed* William Mudge to Thomas Colby, 27 July 1811, cited in Close, 1969, p. 57.

217 *Thomas Mudge's famous chronometers* Mudge, William, 1800, p. 666.

217 *proceeded with little interruption* Mudge, William, 1804, p. 3.

217 *In November 1800* McConnell, p. 115.

217 *a very masterly and accurate* Mudge, William, 1803, p. 396.

217 *would not have been superior* Mudge, William, 1803, p. 408.

219 *advice and instruction* Mudge, William, 1803, p. 411.

219 *the most extensive arc* Mudge, William, 1803, p. 384.

219 *a baseline at Misterton Carr* Mudge and Colby, pp. v–vi, 82.

219 *he had proudly shown it off* William Mudge to Joseph Banks, BL, Add. MS 33981, f. 12, 1 April 1802; William Mudge to Joseph Banks, BL, Add. MS 33981, f. 45, 26 August 1802.

219 *any Topographical Material*   William Mudge to Joseph Banks, BL, Add. MS 33981, f. 12, 1 April 1802.

221 *shutters in the roof*   Mudge, William, 1803, p. 413.

221 *how thoroughly satisfied*   William Mudge to Joseph Banks, BL, Add. MS 33981, f. 45, 26 August 1802.

221 *the length of a degree on the meridian*   Mudge, William, 1803, pp. 488–9.

221 *of the Arc*   William Mudge to Joseph Banks, BL, Add. MS 33981, f. 45, 26 August 1802.

221 *had read them before the Royal Society*   Mudge, William, 1803, p. 383.

221 *On 4 June 1812*   Rodriguez, p. 321.

221 *Mudge was suggesting the opposite*   Rodriguez, pp. 321–6.

221 *a suspicion of some*   Rodriguez, p. 327.

221 *had not informed us*   Rodriguez, p. 327.

221 *published and explained*   Rodriguez, p. 328.

222 *extending the arc from its termination*   Portlock, pp. 39–40.

222 *it was no uncommon occurrence*   Cited in Portlock, p. 137.

222 *almost constantly covered*   *The Times*, 10726, p. 2, 16 September 1819.

222 *by the application of guy ropes*   Cited in Portlock, p. 133.

222 *resting places*   Flint, p. 146.

222 *At one camp*   Portlock, p. 138.

222 *on the mail coach*   Portlock, pp. 132–3.

222 *only a single day at Edinburgh*   Portlock, p. 132.

223 *neither rain nor snow*   Portlock, pp. 132–3.

223 *on a bundle of tent linings*   Portlock, p. 133.

223 *put up his camp bedstead*   Portlock, p. 133.

223 *a pocket compass and map*   Portlock, p. 138.

223 *several beautiful glens*   Portlock, p. 139.

223 *average thirty-nine miles*   Portlock, p. 139.

223 *a magnificent 586 miles*   Portlock, p. 148.

223 *our day of rest*   Portlock, p. 142.

223 *completely foiled*   Portlock, pp. 144–5.

224 *probably the only instance*   Portlock, p. 144.

224 *admiring for a while*   Portlock, p. 145.

224 *to our inn*   Portlock, p. 145.

224 *would break away*   Portlock, p. 137.

224 *Ramden's great three-foot theodolite*   Portlock, p. 135.

224 *Garviemore Inn*   Portlock, p. 139.

224 *I really thought it was more*   Portlock, pp. 139–40.

224 *a rough boggy tract*   Portlock, p. 140.

224 *I had no alternative*   Portlock, p. 140.

224 *I kept pace with him*   Portlock, p. 140.

225 *carte blanc[h]e*   Portlock, p. 153.

225 *the chief dish*   Portlock, pp. 153–4.

225 *those quantities were all multiplied*   Portlock, p. 154.

225 *a long table*   Portlock, p. 154.

225 *partook of the pudding*   Portlock, p. 154.

225 *In the course of this*   Caledonian Mercury, 14598, 6 July 1815.

225 *the obvious utility*   Caledonian Mercury, 14598, 6 July 1815.

226 *This operation*   Cited in Portlock, p. 69.

226 *the more precious advantage*   Cited in Portlock, p. 69.

226 *it was natural to wish*   Cited in Portlock, p. 71.

226 *all the data required*   Cited in Portlock, p. 70.

226 *I am overwhelmed*   William Mudge, cited in Flint, p. 148.

227 *really I think hereafter*   William Mudge to Thomas Colby, 22 September 1813, cited in Close, 1969, p. 60.

227 *you are very much wanted*   William Mudge to Thomas Colby, 22 September 1813, cited in Close, 1969, p. 60.

227 *I beg you will immediately*   William Mudge, 21 September 1815, cited in Close, 1969, p. 65.

227 *these joint [Anglo-French] processes*   The Times, 10157, 28 May 1817.

227 *No man Could have been*   Joseph Banks to Charles Blagden, RS, CB1/1, f. 148, 30 December 1817.

227 *What may be in the womb*   William Mudge, 6 May 1817, cited in Flint, p. 148.

227 *I have been travelling*   William Mudge, 20 May 1817, cited in Close, 1969, p. 67.

228 *I am chained up here*   William Mudge to Thomas Colby, 7 June 1817, cited in Close, 1969, p. 67.

228 *a very able man*   William Mudge to Thomas Colby, 7 June 1817, cited in Close, 1969, p. 67.

228 *If I choose to sink*   William Mudge, 6 May 1817, cited in Flint, p. 148.

228 *If my observations were bad*   Jean-Baptiste Biot, cited in Flint, p. 149.

228 *The state of his health*   Jean-Baptiste Biot to Jean Baptiste Joseph Delambre, PO, Ms 1056, p. 4, 4 July 1817. [*L'état de sa santé rendai impossible qu'il allai plus aran dans le nord.*]

228 *very tall and personable*   William Mudge to Richard Rosdew, 8 January 1805, cited in Flint, p. 133.

228 *a very good Frenchman*   William Mudge to Richard Rosdew, 8 January 1805, cited in Flint, p. 133.

228 *a young officer full of zeal*   Jean Baptiste Biot to Jean Baptiste Joseph Delambre, PO, Ms 1056, p. 4, 4 July 1817. [*le capitaine Richard Mudge, jeune homme plein d'activité, de zele . . .*].

229 *The French multiplying-circle*   William Mudge to Joseph Banks, BL, Add. MS 33981, f. 45, 26 August 1802.

229 *Any opportunity*   Portlock, p. 69.

229 *attached himself to Captain Colby*   Portlock, p. 75.

229 *the depression of English*   Olinthus Gregory, 'Vindication of the Attack on Don Joseph Rodriguez's Paper in the Philosophical Transactions', in Thomson, Thomas, III, p. 282.

229 *attempt by a foreigner*   Gregory, p. 29.

229 *style quite new*   Thomas Thomson, in Gregory, p. 57.

229 *low abuse*   Thomas Thomson, in Gregory, p. 69.

230 *Has [Gregory] made any addition*   Thomas Thomson, in Gregory, p. 70.

230 *had the ice once been melted*   Richard Mudge, cited in Portlock, p. 74.

231 *the rocks of the ancient Thule*   Colby, cited in Portlock, p. 77.

231 *where there were just tents*   *The Times*, 10278, p. 3, 16 October 1817.

231 *One great disadvantage*   William Mudge to Thomas Colby, cited in Close, 1969, p. 68.

231 *see no Reason*   Joseph Banks to Charles Blagden, RS, CB1.1, f. 149, 12 January 1818.

231 *Thomas Edmondston*   See Jean Baptiste Biot to Jean Baptiste Joseph Delambre, PO, Ms 1056, 4 July 1817; Thomas Edmondston to Jean Baptiste Joseph Delambre, PO, X5 (E1), 22 September 1820.

232 *warm hospitalities made up*   Cited in Flint, p. 149.

232 *I was very much helped*   Jean Baptiste Biot to Jean Baptiste Joseph Delambre, PO, Ms 1056, f. 7, 4 July 1817. (The hand is obscure but it seems to read: *j'ai eté comblé de facilités et de [secours] pour mes opérations. Je n'ai trouvé partout que l'empressement le plus genereux, et le plus degagé de ces prejugés antisociaux, qui peut etre tirent souvent moins de force de la realité que des preventions de ceux qui les supposent.*)

232 *the hope of facilitating*   Thomas Edmondston to Jean Baptiste Joseph Delambre, PO, X5 (E1), 22 September 1820.

232 *it is pleasing to observe*   'Literary and Scientific Intelligence', *Blackwood's Edinburgh Magazine*, 4, 1818, p. 237.

232 *Kirby or Kolby*   See Jean-Baptiste Biot to Jean Baptiste Joseph Delambre, PO, Ms 1056, 4 July 1817.

232 *one of the officers serving*   Cited in Portlock, pp. 75–6.

232 *as the queen who gave up Calais*   William Mudge, cited in Portlock, p. 82.

233 *he resolved to spend the Christmas*   Flint, p. 154.

233 *subject to the hyp*   Isaac Dalby to Thomas Colby, 5 February 1821, cited in Close, 1969, p. 42. For details regarding William Mudge's death, see Flint, p. 154.

233 *We are left in sorrow and reflection*   Robert Dawson to Thomas Colby, cited in Close, 1969, p. 70.

233 *the late celebrated and scientific*   *The Times*, 10914, p. 3, 22 April 1820; *The Times*, 11366, p. 2, 2 October 1821.

234 *such part of the Survey*   *Caledonian Mercury*, 15435, p. 2, 10 August 1820.

CHAPTER 10: 'ENSIGN OF EMPIRE'

235 *used no interest*   Thomas Colby to Arthur Wellesley, 20 June 1820, cited in Close, 1969, p. 83.

235 *His Grace is now*   Arthur Wellesley to Thomas Colby, 21 June 1820, cited in Close, 1969, pp. 83–4.

236 *the charismatic chemist Humphry Davy*   Close, 1969, p. 84.

236 *very eminent Professor of Mathematics*   Cited in Close, 1969, p. 83.

236 *No man more so*   Cited in Close, 1969, p. 84.

236 *that is all I want to know*   Cited in Close, 1969, p. 84.

236 *a Gentleman well versed in Mathematics*   Certificates of Election and Candidature, RS, EC/1819/35.

236 *unceasing attention and liberality*   See Anon, 1853.

236 *His Grace appoints you*   Board of Ordnance and Lord Raglan to Thomas Colby, 10 July 1820, cited in Close, 1969, p. 83.

236 *the appointment of an officer*   'The Morning Chronicle', *Morning Chronicle*, 15996, 3 August 1820.

237 *most slovenly*   Cited in Harley, 1971, pp. 95–6. See this article for further information about the Ordnance Survey's mapping of Lincolnshire. Descriptions of Colby's reforms can also be found in Wheeler, 2006.

237 *the direction of [Lundy] Island*   Ordnance Survey Letter Book, NA, OS 3/260, p. 112; cited in Harley and Oliver, VI, p. ix.

237 *It was with extreme regret*   Thomas Colby to Charles Budgen, Ordnance Survey Letter Book, NA, OS 3/260, p. 177; cited in Harley and Oliver, VI, p. ix.

237 *to point out clearly*   Compton, p. ii.

237 *give[s] me great pleasure*   Thomas Colby to Charles Budgen, Ordnance Survey Letter Book, NA, OS 3/260, p. 183; cited in Harley and Oliver, VI, p. ix.

238 *the idea of tolerable error*   Andrews, J.H., 2006, p. 58.

238 *in 1800 two Acts of Union*   For accounts of the Anglo-Irish Union, see Brown, Geoghegan, and Kelly; Geoghegan; Girvin; MacDonagh; Smyth, Jim; and Stewart, Bruce.

239 *a British engineer had suggested*   Greville to John Foster, Public Records Office of Northern Ireland, Foster/Massareene papers D207/48/1, October 1805.

239 *the Irish survey bill*   Cited in Close, 1969, p. 57.

239 *the mapping of Ireland*   For an exhaustively detailed study of the Irish Ordnance Survey, see Andrews, J.H., 2006. See also Ordnance Survey of Ireland; and Andrews, J.H., 1980.

239 *some mode should be devised*   Report from the select committee appointed to examine the copies of the grand jury presentments of Ireland, p. 4, HC 1814–1815 (283), vi.

239 *for all purposes of navigation*   Wakefield, I, pp. 3–7, cited in Andrews, J.H., 2006, p. 12.

240 *a map would cost £200,000*   See Andrews, J.H., 2006, p. 33.

240 *cultural nationalism*   For a discussion of Irish cultural nationalism, see Hutchinson.

240 *ensigns of empire*   Wordsworth, 1979, p. 168.

240 *William Petty*   For discussions of Petty's life and career, see Petty, 1769 and 1851; Barnard; and Fitzmaurice.

241 *colourful and eccentric*   Andrews, J.H., 2006, p. 3. An interesting account of Vallancey's surveying work can be found in O'Reilly.

241 *the most disagreeable part*   Cited in Andrews, J.H., 2006 (epigraph).

241 *Richard Lovell Edgeworth*   Uglow offers a fascinating exploration of Edgeworth's achievements. Colvin provides a briefer one. Andrews, J.H., 2006, pp. 5–6, describes his accomplishments in map-making.

241 *William Roy's Paris–Greenwich triangulation*   See Edgeworth, pp. 116–17.

242 *the first truly indigenous bid*   Andrews, J.H., 2006, p. 5.

242 *native Irish cartography*   Andrews, J.H., 2006, p. 6.

242 *as little-known as classical manuscripts*   Thomas Aiskew Larcom to L.A. Hall, Ordnance Survey Office, Dublin, Ordnance Survey Letters 566, 11 April 1849; and Thomas Aiskew Larcom, NLI, volume 7552, note dated 1862.

242 *private individuals will*   Thomas Spring Rice, *The Spring Rice Report*, H.C. 445 (1824), VIII, 79, 21 June 1824.

243 *I was obliged to hold*   Arthur Wellesley to Sir W. Knighton, 6 November 1824, in Wellesley, II, pp. 332–3.

243 *cannot be executed*   Cited in Andrews, J.H., 2006, p. 34.

243 *appear extraordinary or ridiculous*   Cited in Andrews, J.H., 2006, p. 37.

243 *Initially twenty young men*   Cited in Andrews, J.H., 2006, p. 36.

243 *for further instruction*   *Morning Chronicle*, 17141, 26 March 1824.

243 *Charles Pasley*   See Vetch, 'Pasley, Charles'.

244 *we are quite run over*   Jane Austen to Cassandra Austen, 24 January 1813, in Austen, 2003, p. 198.

244 *I am . . . much in love*   Jane Austen to Cassandra Austen, 24 January 1813, in Austen, 2003, p. 198.

244 *all the glories of* Austen, 2001, p. 152.

244 *whilst we glory in the freedom* Pasley, p. 49.

244 *military nation* Pasley, p. 456.

244 *in spirit* Pasley, p. 182.

244 *wherever we act* Pasley, p. 184 (note).

244 *as good maps* Pasley, p. 180.

245 *show[ed] him the proposed Course* Charles Pasley, Diary 1825–1826, BL, Add. MS 41984, f. 14, 23 February 1825.

245 *the name Royal Sapper* Charles Pasley, Diary 1825–1826, BL, Add. MS 41984, f. 14, 20 February 1825.

245 *seven years to map Ireland* See Andrews, J.H., 2006, p. 71.

246 *with a detachment* *Derry Journal*, cited in *Caledonian Mercury*, 16180, 9 May 1825.

246 *Major Colby, Director* *The Examiner*, 914, 7 August 1825.

246 *head-quarters of his detachments* *The Examiner*, 914, 7 August 1825.

246 *inveterate haze and fogginess* Cited in Andrews, J.H., 2006, p. 42.

247 *the weather has been extremely adverse* 'Trigonometrical Survey', *Belfast News Letter*, cited by *Caledonian Mercury*, 16399, 2 October 1826.

247 *no young man* Cited in O'Brien, p. 9; cited in Palmer.

247 *of great importance* 'The Heliotrope, A New Instrument', *Gentleman's Magazine*, 92: 2, p. 358, June – December 1822.

248 *shaped into the form* *Caledonian Mercury*, 16332, 29 April 1826.

248 *the brilliance of the light* *Caledonian Mercury*, 16332, 29 April 1826.

248 *overpowering, and as it were* McLennan, p. 73; cited in Palmer.

248 *is like the moon* *Caledonian Mercury*, 16332, 29 April 1826.

248 *so that we may consider ourselves safe* Thomas Drummond, 28 October 1825, cited in Close, 1969, p. 74.

249 *a storm of snow* Thomas Drummond, 4 November 1825, cited in Close, 1969, p. 74.

249 *my tent is blown* Thomas Drummond, 4 November 1825, cited in Close, 1969, p. 74.

249 *despair[ed] of success* Thomas Drummond, 5 November 1825, cited in Close, 1969, p. 75.

249 *a very hurried intimation* Henderson to Drummond, 12 November 1825, cited in Close, 1969, p. 75.

250 *a momentary ejaculation of anger* This anecdote is related in Portlock, p. 5.

250 *though he could not again succeed* Portlock, p. 5.

250 *buried or bottled up* Thomas Colby to Cornelia Hadden, 8 May 1840 and 28 May 1840; both cited in Close, 1969, pp. 94, 90.

250 *William Edgeworth stepped in* See Andrews, J.H., 2006, p. 27.

252 *many of the parishioners* Cited in Day, McWilliams and English, III, p. 144.

252 *every time they'd stick* Friel, 1981, pp. 11–12.

252 *perhaps from a fear*   Cited in Andrews, J.H., 2006, p. 43.

252 *Bantry Common, in County Wexford*   Cited in Andrews, J.H., 2006, p. 92.

253 *a cairn revered by pilgrims*   Cited in Day, McWilliams and English, XXXI, p. 57. See also John O'Donovan to Thomas Aiskew Larcom, 23 April 1834, pp. 54–6, in O'Donovan, 2001a.

253 *in giving their [own] names*   John O'Donovan to Thomas Aiskew Larcom, 6 August 1834, in O'Donovan, 1992, pp. 15–6.

253 *people [were] not willing*   Cited in Ó Cadhla, p. 136.

253 *some people are afraid*   John O'Donovan to Thomas Aiskew Larcom, 21 March 1834, in O'Donovan, 2001a, p. 10.

253 *you have a great deal*   John O'Donovan to Thomas Aiskew Larcom, 15 April 1834, in O'Donovan, 2001a, p. 39.

253 *drudgery and labour*   Cited in Ó Cadhla, p. 134.

253 *they brought welcome trade*   Andrews, J.H., 2006, p. 92.

253 *the residents of Glenomara*   *Dublin Evening Post*, 23 September 1828; cited in Andrews, J.H., 2006, pp. 43–4.

254 *as precise and up-to-date*   Thomas Colby, cited in Andrews, J.H., p. 44.

254 *a complex amalgamation*   For descriptions of Colby and Drummond's 'Compensation Bars', see Andrews, J.H., 2006, pp. 45–7; Keay, pp. 103–4.

255 *The Lough Foyle baseline*   See Yolland.

255 *annoyed and vexed*   'The Irish Trigonometrical Survey', *The Belfast News-Letter*, 10165, 18 November 1834.

255 *Where a difference of a few inches*   'The Irish Trigonometrical Survey', *The Belfast News-Letter*, 10165, 18 November 1834.

255 *not perceptible to the eye*   'The Irish Trigonometrical Survey', *The Belfast News-Letter*, 10165, 18 November 1834.

255 *41,640.8873 feet long*   See Andrews, J.H., 2006, pp. 44–9.

255 *You may conceive*   'The Irish Trigonometrical Survey', *The Belfast News-Letter*, 10165, 18 November 1834.

257 *William Rowan Hamilton*   For accounts of Hamilton's life and achievements, see Graves; Hamilton, William Rowan, 1931; Hankins; and O'Connell.

257 *did not look through his telescopes*   De Vere, p. 47.

257 *the geometry of light*   See Hamilton, William Rowan, 1833, especially pp. 3–5.

257 *Law of Conical Refraction*   See Babbage, p. 107; Graves, I, pp. 636–7; Hamilton, 1833, p. 5; Hamilton, William Rowan, 1931, I, pp. x–xi.

257 *an essential requisite*   Graves, I, pp. 256–7.

258 *the only operations*   William Rowan Hamilton, Icosian Game Publications, Hamilton Papers, Trinity College Dublin, MS 1492/379/1-6.

258 *none of her family*   W.E. Hamilton to William Rowan Hamilton, Hamilton Papers, TCD, MS 7762-72/1328, 26 May 1857.

258   *how much pleasure*   William Rowan Hamilton, Notebook, TCD, MS 1492/16, ff. 20–21 [n.d.].

258   *recognise[d] it as a county*   William Rowan Hamilton, Notebook, TCD, MS 1492/16, ff. 20–21 [n.d.].

258   *an 'Admirable Crichton'*   Maria Edgeworth to Honora Edgeworth, 28 August 1824, cited in Graves, I, p. 161.

258   *[she] who forms the great*   William Rowan Hamilton to Grace Edgeworth, 27 August 1824, cited in Graves, I, p. 162.

258   *carried off*   William Rowan Hamilton to Eliza Hamilton, 19 March 1828, in Graves, I, p. 291.

259   *an educational tour*   Maria Edgeworth to William Rowan Hamilton, TCD, MS 7762-72/156, 27 April 1828; Maria Edgeworth to William Rowan Hamilton, TCD, MS 7762-72/163, 20 May 1828; William Rowan Hamilton to Maria Edgeworth, TCD, MS 7762-72/162, 23 May 1828.

259   *the map-publisher Bartholomews*   I am grateful to Yolande Hodson for this information.

259   *a situation among the calculators*   William Rowan Hamilton to Maria Edgeworth, TCD, MS 7762-72/157, 30 April 1828.

259   *with the intention of reconnoitring*   William Rowan Hamilton to Revd [Thomas Romney] Robinson, 23 October 1828, cited in Graves, I, p. 303.

259   *Lieutenant Drummond at home*   William Rowan Hamilton to Revd [Thomas Romney] Robinson, 23 October 1828, cited in Graves, I, p. 303.

259   *Captain Everest*   William Rowan Hamilton to Revd [Thomas Romney] Robinson, cited in Graves, I, pp. 335–6.

259   *The upright, mutton-chopped*   For accounts of George Everest, see Edney, 1997; Keay; Smith, James Raymond, 1998.

259   *this Great Arc of India*   See Edney, 1997; and Keay.

260   *Never-rest*   See Smith, James Raymond, 1998, p. i.

260   *conversations with Colby*   Edney, 1997, p. 35; Keay, pp. 101–4.

260   *work on the Meridional Arc*   William Rowan Hamilton to George Airy, 25 July 1830, cited in Graves, I, p. 377.

260   *you are so wrapt up*   Thomas Colby to William Rowan Hamilton, TCD, MS 7762-72/385, 16 November 1833.

260   *somewhat of a universalist*   William Rowan Hamilton to Graves, 4 May 1842, cited in Graves, II, p. 376. For a history of the Royal Irish Academy, and Hamilton's role, see Ó Raifeartaigh.

260   *own personal conviction*   Thomas Colby to William Rowan Hamilton, TCD, MS 7762-72/385, 16 November 1833.

261   *Hamilton and Thomas Romney Robinson collaborated*   See Andrews, J.H., 2006, p. 97; Graves, II, pp. 281, 298–9; Portlock, pp. 89–90; Robinson, Thomas Romney.

261 *telescope correctly on the spot* Thomas Aiskew Larcom to William Rowan Hamilton, TCD, MS 7762–72/852, 13 May 1842.

261 *There can be no harm* Thomas Aiskew Larcom to William Rowan Hamilton, TCD, MS 7762–72/851, 4 May 1842.

261 *one of the friends who took* Graves, II, p. 61.

261 *Dugald Stewart* See Stewart, Dugald, II, p. 205 (note).

261 *we are not liable* See Stewart, Dugald, II, p. 205 (note).

261 *of the philosophy of mind* Graves, II, pp. 140–3.

261 *a language of pure space* See Graves, II, pp. 138–43.

261 *a new light flashed* Kant, 2003, p. 19.

261 *true method* Kant, 2003, p. 19.

262 *something simple, perfect, and one* William Rowan Hamilton, 'Appendix 2: "Waking Dream: or Fragment of a Dialogue between Pappus and Euclid, in the Meads of Asphodel"', pp. 662–71 in Graves, I, p. 664.

262 *walked back with our party* Graves, I, p. 264.

262 *very prosing* Maria Edgeworth, cited in Inglis-Jones, pp. 219–20.

262 *Mr Secretary Drummond* Thomas Drummond to William Rowan Hamilton, 11 January 1836, in Wordsworth, 1967–93, VI, p. 151.

262 *calculating celebrity* Thomas Drummond to William Rowan Hamilton, 28 September 1835, in Wordsworth, 1967–93, VI, p. 98.

262 *Thomas Spring Rice* Alan Hill (note 1), in Wordsworth, 1967–93, VI, p. 690; William Wordsworth to Thomas Spring Rice, 28 April 1839, in Wordsworth, 1967–93, VI, p. 690; William Wordsworth to Dora Wordsworth, 9 May 1839, in Wordsworth, 1967–93, VI, p. 692.

263 *short tour of Ireland* For discussions of Wordsworth's trip to Ireland and its effect, see Fackler. Further examinations of his attitude to Ireland can be found in Marjarum; and McCormack, pp. 18–42.

263 *Wordsworth made sure to buy one* See Shaver and Shaver, p. 191.

263 *the relationship between literature and science* Much work has been done in the last thirty years into the relationship between literature and science, particularly with the aim of showing that 'the simple view that Romanticism . . . was uniformly opposed to science, has been untenable for a long time' (Wyatt, 1995, p. 3). With respect to the relationship between Wordsworth and Hamilton and its context, I have found the following works helpful: Cunningham and Jardine; Eichner; Fulford; Fulford, Lee and Kitson; Haefner; Holmes; Jordanova; Richardson; Roe; Ruston; Thomas and Ober; Wyatt, 1995 and 2001. The cultural interactions of mid-nineteenth-century Irish scientists are explored in Eagleton; Foster; Green; Patten; and Ryan. A more detailed account of my research into the friendship between Wordsworth and William Rowan Hamilton can be found in Hewitt, 2006 and 2007.

263 *not entirely overlook* Wordsworth, 1979, p. 192.

263 *the relation those abstractions bear* Wordsworth, 1979, p. 116. A discussion of Wordsworth's interest in geometry can be found in Simpson.

CHAPTER II: 'ALL THE RHYMES AND RAGS OF HISTORY'

264 *peasant garb* Andrews, J.H., 2006, p. 122.

264 *He had been first educated* For an account of O'Donovan's life, see Ó Muraile.

264 *the living repertory* O'Donovan, 1856, II, p. 159 (note).

265 *Hardiman was a committed researcher* See Hardiman.

265 *After ages of neglect and decay* Hardiman, p. i.

266 *geography is a noble and practical science* Colby and Larcom, p. 7.

266 *a full face portrait* Thomas Aiskew Larcom, loose notes, Larcom Papers, NLI, MS 7790, 2 July 1861.

266 *in 1828 he married* See Baigent, 'Colby, Thomas'.

266 *A separate Boundary Commission* See Andrews, J.H., 2006, pp. 32–4, 60–2.

267 *2s a day* Andrews, J.H., 2006, p. 62.

267 *the people deny* John O'Donovan to Thomas Aiskew Larcom, 8 October 1834, in O'Donovan, 1992, p. 135.

267 *I cannot get the people to agree* John O'Donovan to Thomas Aiskew Larcom, 15 August 1836, in O'Donovan, 2001b, p. 82.

267 *Mr Griffith is an excellent and able engineer* *Morning Chronicle*, 24491, p. 4, 20 April 1848.

268 *a permanent Geological Survey* For a stimulating account of early-nineteenth-century geological mapping, see Winchester.

268 *It would have been of great importance* Portlock, pp. 42–3. For a history of the Ordnance Survey's involvement with geology, see Close, 1969, pp. 61–3.

268 *no person but a man* J. Jardine to Thomas Colby, December 1819, cited in Close, 1969, p. 61.

268 *the most minute and accurate* Thomas Colby, Annual Report, Larcom Papers, NLI, MS 7555 (1826).

268 *a grant was obtained* Cited in Close, 1969, p. 62.

269 *gentleman geologist* Secord.

269 *a regular fun-engine* Fox, I, p. 5; cited in Secord.

269 *morbid apprehension of criticism* Portlock, p. 3.

269 *It is impossible not to regret* Portlock, p. 3.

270 *paper landscape* Andrews, J.H., 2006.

270 *fully rounded national survey*   Andrews, J.H., 2006, p. 56.

270 *a great variety of materials*   Thomas Colby, Annual Report, Larcom Papers, NLI MS 7555 (1826).

270 *a great deal of information*   Thomas Colby, Annual Report, Larcom Papers, NLI MS 7555 (1826).

270 *A number of map-makers*   See Arrowsmith; Beaufort; Petty, 1851; Sampson.

270 *one of the earliest uses*   Arrowsmith, p. 5.

270 *artificial state*   Colby and Larcom, p. 17.

270 *social economy and productive economy*   Colby and Larcom, p. 8.

270 *food; fuel; dress*   Larcom, 'Heads of inquiry', NLI, MS 7550 ([1832 (?) or 18 March 1834 (?)]).

271 *would we publish any book*   John O'Donovan to Thomas Aiskew Larcom, 27 March 1834, in O'Donovan, 2001a, p. 14.

272 *men who work like machines*   Cited in Andrews, J.H., 2006, p. 137.

272 *steam engine rapidity*   Larcom to Griffith, 10 November 1836: VLB. Survey of Ireland: Return to House of Commons Order, 16 June 1837, H.C. 1837 (522), p. xxxix.

272 *to make the Maps a standard*   Thomas Aiskew Larcom to Thomas Colby, 12 May 1842, in O'Donovan, 2001a, p. xl.

272 *to trace all the mutations*   Colby and Larcom, p. 8.

272 *the name of each place*   Thomas Colby, 'Instructions for the Interior Survey of Ireland', pp. 309–21 in Andrews, J.H., 2006 (p. 311). A discussion of the toponymy of Ireland can be found in Andrews, J.H., 1992.

273 *by 1826 the number of Irishmen*   For details regarding the employment of Irish surveyors on the Ordnance Survey of Ireland, see Andrews, J.H., 2006, pp. 62–6.

273 *outnumbered the sappers*   See Andrews, J.H., 2006, pp. 91, 65.

274 *immediately, offering me a situation*   John O'Donovan to Todd, 12 May 1842, in O'Donovan, 2001a, p. xxxv.

274 *besides two or three more*   Stokes, p. 96. For an exploration of the principal personalities of the Topographical Branch, see Chuto, 1976.

274 *strange admission*   Larcom to Boteler, 4 October 1836, Ordnance Survey memorandums: Cavan, Ordnance Survey Office, p. 120.

275 *all sorts of old documents*   William F. Wakeman, cited in Stokes, p. 96.

275 *useful, laudable, and patriotic pursuit*   Cited in Andrews, J.H., 2006, p. 123.

275 *a kind of one-man local history department*   Andrews, J.H., 2006, p. 128.

275 *I was never so disgusted*   John O'Donovan to Thomas Aiskew Larcom, 15 April 1834, in O'Donovan, 2001a, p. 40.

275 *in the present artificial state*   John O'Donovan to Thomas Aiskew Larcom, 3 September 1834, in O'Donovan, 1992, p. 59.

275 *O'Donovan really writes in a way* Thomas Aiskew Larcom to George Petrie, NLI, MS 7566, 17 June 1839.

275 *defaced (or disgraced)* Thomas Aiskew Larcom to George Petrie, NLI, MS 7566, 17 June 1839.

275 *very serious, cold, and un-Irish* John O'Donovan to Thomas Aiskew Larcom, 8 September 1837, Ordnance Survey memorandums: Roscommon, Ordnance Survey Office, p. 154.

275 *as sceptical an enquirer* John O'Donovan to Thomas Aiskew Larcom, 28 October 1835, in O'Donovan, 2001a, p. 120.

275 *exceeding (excessive)* John O'Donovan to Thomas Aiskew Larcom, 28 October 1835, in O'Donovan, 2001a, p. 120.

275 *evident approach* Petrie, p. 18.

276 *a region of fancy and fable* Petrie, p. 18.

276 *Ossian, the son of Fingal* MacPherson, 1762.

276 *convincing proof* MacPherson, 1996, p. 208.

276 *The poems of Ossian* John O'Donovan to Thomas Aiskew Larcom, 26 August 1834, in O'Donovan, 1992, p. 43.

276 *incontestible authority* O'Donovan, 1862, p. 7.

276 *that one among the modern names* Thomas Aiskew Larcom to Thomas Colby, 12 May 1842, in O'Donovan, 2001a, pp. xl–xli.

276 *all the rhymes and rags of history* John O'Donovan to Thomas Aiskew Larcom, 30 August 1835, in O'Donovan, 2001a, p. 19.

277 *the people do not agree* John O'Donovan to Thomas Aiskew Larcom, 16 August 1834, in O'Donovan, 1992, p. 24.

277 *forked tongue* James Clarence Mangan, who worked alongside John O'Donovan on the Topographical Branch, used the image of a forked tongue in his *Autobiography* (p. 14), arguably as a metaphor to depict the linguistic state of Ireland. See Chuto, 1988; Hewitt, 2007, pp. 281–339.

277 *mangled* Andrews, J.H., 2006, p. 121

277 *To comply with the general custom* John O'Donovan to Thomas Aiskew Larcom, 12 July 1836, in O'Donovan, 2001b, p. 4.

277 *Money Sterling!* John O'Donovan to Thomas Aiskew Larcom, 17 September 1836, in O'Donovan, 1992, p. 100.

277 *wild rhapsody* John O'Donovan to Thomas Aiskew Larcom, 26 August 1834, in O'Donovan, 1992, p. 47.

277 *trace[d] his course* John O'Donovan to Thomas Aiskew Larcom, 26 August 1834, in O'Donovan, 1992, pp. 45–6.

278 *skin as pale and taut* Ryan, p. 164.

278 *The Man in the Cloak* Mangan, 2002a, pp. 239–66.

278 *of fantastic shape* McCall, John, 1975, p. 27.

278 *looked like the spectre*   Duffy, pp. 109–10.

278 *the animal spirits*   Duffy, pp. 109–10.

278 *fle[e] from the admiration*   Duffy, pp. 109–10.

278 *poète maudit*   Moore, p. 117.

278 *pens and ink – gratis*   John O'Donovan, cited in Ryan, p. 165.

278 *one short poem*   John O'Donovan, cited in Ryan, p. 165.

278 *the mad poet*   Ryan, p. 164.

278 *[Mangan] says that I am his enemy*   Ryan, p. 164.

278 *severe, coldly-judging*   James Clarence Mangan, 'Sketches and Reminiscences of Irish Writers: John O'Donovan', in Mangan, 2002b, pp. 215–18.

279 *the merit of fidelity*   James Clarence Mangan, 'Anthologia Hibernia: No. III', in Mangan, 2002b, p. 167.

279 *bad Etymology and sad Orthography*   James Clarence Mangan, 'Bards of this Beautiful Isle', in Mangan, 1996, p. 38.

279 *Dates, arithmetical tack*   James Clarence Mangan, 'Bards of this Beautiful Isle', in Mangan, 1996, p. 38.

279 *it was even rumoured*   See Eugene O'Curry to Thomas Davis, in O'Donoghue, D.J., pp. 120–1.

280 *Versified Paraphrase*   O'Daly, p. 87

280 *this is incorrect*   O'Daly, pp. 87 (note 5), 92 (note 2), 95 (note 2), 97 (note 2).

280 *The few broken columns*   Mangan, 1968, p. 31.

280 *a fifth province*   Friel, 1999, p. 106.

280 *has grown out of*   Friel, 1999, p. 106.

281 *into the King's good English*   Friel, 1981, p. 30.

281 *imprisoned in a linguistic contour*   Friel, 1981, p. 52. Cronin offers a compelling discussion of the importance of acts of translation to Irish history and culture.

281 *a bloody military operation*   Friel, 1981, p. 36.

281 *quisling*   Friel, 1999, p. 118.

281 *an eviction of sorts*   Friel, 1981, p. 52.

281 *his own wilful mistranslations*   See Friel, Andrews and Barry; and Friel, 1983.

281 *imperative duty*   O'Curry, p. 3.

281 *collective folly and stupid intellect*   Eugene O'Curry to John O'Donovan, in O'Donovan, 2003, I, p. 8.

281 *the first peripatetic university*   Green, p. 244.

281 *empirical truth*   Friel, 1989, p. 66.

281 *a heroic literature*   Friel, 1989, p. 67.

281 *the tiny bruises*   Friel, 1983, p. 118.

281 *inaccurate history*   See Connolly, Sean, 1993; Friel, Andrews and Barry; and Murray.

281 *was offered pieties* Friel, 1983, p. 107.

282 *a monument for future* Cited in Andrews, J.H., 2006, p. 88.

282 *our report went to give* Cited in Andrews, J.H., 2006, p. 74.

283 *You recollect the cry of haste* Thomas Colby to Charles Pasley, BL, Add. MS 41964, f. 68, 7 January 1831.

283 *Genius is essentially sympathetic* William Rowan Hamilton, cited in Graves, II, p. 152.

283 *social feelings* William Rowan Hamilton, cited in Graves, II, p. 152.

283 *a map is in its nature* Colby and Larcom, p. 7.

284 *physical features of the ground* Colby and Larcom, p. 9.

284 *aspect, climate, and geological structure* Colby and Larcom, p. 9.

284 *a full face portrait* Thomas Aiskew Larcom, loose notes, Larcom Papers, NLI, MS 7790, 2 July 1861.

284 *some historians have done so* See Carroll, especially pp. 1–10, 83–4, 96, 100–12; Ó Cadhla; Kiberd, p. 614. Gillian Doherty provides a very different interpretation of the Ordnance Survey's work.

284 *political anatomist* Petty, 1769, p. 288.

285 *in the evening* *Dublin Evening Mail*, 19 August 1835, cited in Andrews, J.H., 2006, pp. 148–9.

285 *manifestly wrong* John O'Donovan to Thomas Aiskew Larcom, 4 April 1834, in O'Donovan, 2001a, pp. 26–7.

285 *the officers of the Royal Engineers* John O'Donovan to Thomas Aiskew Larcom, 4 April 1834, in O'Donovan, 2001a, pp. 26–7.

285 *if such descriptions and remarks* John O'Donovan to Thomas Aiskew Larcom, 4 April 1834, in O'Donovan, 2001a, pp. 26–7.

285 *somewhat of a universalist* William Rowan Hamilton to Graves, 4 May 1842, cited in Graves, II, p. 376.

286 *study of science, polite literature* Ó Raifeartaigh, p. 13.

286 *all the objects of rational inquiry* Burrowes, p. xv.

286 *two distinct cultures* See Patten.

286 *revert immediately* Board of Ordnance to Thomas Colby, NA, WO 44/703, 1 July 1840.

286 *bigotry and politics* 'A Protestant Conservative' to Henry Goulburn, NLI, MS 7553, 2 May 1842.

286 *They have actually* 'A Protestant Conservative' to Henry Goulburn, NLI, MS 7553, 2 May 1842.

287 *stimulating national sentiment* Ryan, p. 167.

287 *looked over the whole atlas* Colby to Elizabeth Colby, 9 May 1833, in Close, 1969, p. 92. Accounts of the Ordnance Survey's Irish maps can be found in Andrews, J.H., 2006; and Ordnance Survey of Ireland.

287 *he ought to have been* Cited in Close, 1969, p. 92.

287 *No one who has seen*   L[each], 1879, p. 194.

287 *common workmen*   Kohl, pp. 288–9.

288 *far more immediately popular*   For details about the sales of the Ordnance Survey's Irish maps, see Andrews, J.H., 2006, p. 138.

288 *the principal gentlemen*   Cited in Andrews, J.H., 2006, 11: p. 139.

CHAPTER 12: 'A GREAT NATIONAL SURVEY'

289 *from several windows*   See 'The Tower of London – Awful Conflagration', *Derby Mercury*, 5704, 3 November 1841; *Morning Chronicle*, 31 October 1841.

290 *the reflection on the surrounding houses*   'Great Fire in the Tower of London', *Morning Chronicle*, 22450, 1 November 1840.

290 *so intense, that it was*   'The Tower of London – Awful Conflagration', *Derby Mercury*, 5704, 3 November 1841.

290 *By October 1841*   For details about the Ordnance Survey's progress between 1835 and 1870, see Oliver, 1985. For its progress between 1841 and 1846, see Parsons, 'Changing Needs', 1980.

291 *early in the afternoon*   From an undated newspaper cutting in the possession of H.C. Baker, copy in Ordnance Survey Library, Southampton. Cited in Parsons, 'Changing Needs', p. 111.

291 *not a person to hold a long parley with*   From an undated newspaper cutting in the possession of H.C. Baker, copy in Ordnance Survey Library, Southampton. Cited in Parsons, 'Changing Needs', p. 111.

291 *everything at first was confusion*   From an undated newspaper cutting in the possession of H.C. Baker, copy in Ordnance Survey Library, Southampton. Cited in Parsons, 'Changing Needs', p. 111.

291 *a large load of furniture*   From an undated newspaper cutting in the possession of H.C. Baker, copy in Ordnance Survey Library, Southampton. Cited in Parsons, 'Changing Needs', p. 111.

291 *carpenters, bricklayers*   From an undated newspaper cutting in the possession of H.C. Baker, copy in Ordnance Survey Library, Southampton. Cited in Parsons, 'Changing Needs', p. 111.

291 *at home, the steady calm reasoner*   Cited in Portlock, p. 311.

292 *datum*   See Close, 1969, pp. 55, 140

292 *shortly to be recalculated*   Cited in Parsons, 'Changing Needs', p. 123.

294 *unguided, hard-worked*   Thomas Carlyle to Jane Welsh Carlyle, 16 August 1850, in Carlyle, XXV, pp. 156–8.

294 *In 1834 the Poor Law Commission*   For accounts of the history of the Poor Law Commission, see Chadwick.

295 *In the mid 1830s* Parsons, 'Changing Needs', p. 114.

295 *The Tithe Communication Act* Parsons, 'Changing Needs', p. 113.

295 *in April 1840* *Correspondence respecting the Scale for the Ordnance Survey*, Parliamentary Papers 1854, XLI, p. 1; cited in Parsons, 'Changing Needs', p. 109.

295 *In October 1840* *Correspondence respecting the Scale for the Ordnance Survey*, Parliamentary Papers 1854, XLI, p. 1; cited in Parsons, 'Changing Needs', 1980, p. 109.

295 *In February 1841* See Parsons, 'Changing Needs', pp. 110–11.

296 *every one seemed to have a mania* MacLehose, p. 277.

296 *the railway companies became a rival* Parsons, 'Changing Needs', p. 113.

296 *Colby soon transferred his perfectionist energies* Baigent, 'Colby, Thomas'.

296 *over 28,000 sheets* Parsons, 'Changing Needs', p. 116.

297 *By 1846* Parsons, 'Changing Needs', p. 116 (see pp. 115–17 for these cited figures).

297 *the officer William Yolland* See Vetch, 'Yolland, William'.

297 *will show most distinctly that neither* Minutes of Evidence taken before the Select Committee on Army and Ordnance Expenditure, Parliamentary Papers 1849 (499) IX, p. 456; cited in Parsons, 'Superintendency', p. 119.

297 *In November 1847* For details of the Metropolitan Sanitary Commission's contact with the Ordnance Survey, see Parsons, 'Superintendency', pp. 119–20.

298 *the new brooms* *Daily News*, 509, p. 1, 14 January 1848.

298 *a company of Royal Sappers and Miners* *Daily News*, 607, 9 May 1848.

298 *a surveyor called James Steel* Parsons, 'Superintendency', p. 120.

298 *like the booming of a piece of Ordnance* Connolly, T.W.J., II, pp. 77–82.

298 *Scales Dispute* Parsons, 'Superintendency', p. 123. For discussions of the 'Scales Dispute', see Oliver, 1985; Oliver, 1991; and Parsons, 'Superintendency'.

299 *with little or scarcely any warning* Portlock, p. 313.

299 *An obituary of Thomas Colby* See Anon, 1853.

299 *as really the great work* Anon, 1853, p. 133.

299 *into harmonious action* Anon, 1853, p. 134.

299 *would alone suffice* Anon, 1853, p. 134.

300 *was not only fully aware* Anon, 1853, p. 134.

300 *his life was a course* Anon, 1853, p. 137.

300 *I am charged with every species* Hibbert, 1961, pp. 289–90.

300 *James launched himself* For James's role in the Scales Dispute, see Parsons, 'Scales Dispute'.

301 *On 17 September 1862* 'Surveyor', 'The Old Ordnance Survey of England' (Letters to the Editor), *The Times*, 24353, p. 6, 17 September 1862.

301 *the Ordnance Survey of England* 'Surveyor', 'The Old Ordnance Survey of England' (Letters to the Editor), *The Times*, 24353, p. 6, 17 September 1862.

301 *on receiving* 'Surveyor', 'The Old Ordnance Survey of England' (Letters to the Editor), *The Times*, 24353, p. 6, 17 September 1862.

301 *no one is more conscious* Henry James, 'The Ordnance Survey of England' (Letters to the Editor), *The Times*, 24357, p. 7, 22 September 1862.

302 *half a loaf* Henry James, 'The Ordnance Survey of England' (Letters to the Editor), *The Times*, 24357, p. 7, 22 September 1862.

302 *give us the means* Henry James, 'The Ordnance Survey of England' (Letters to the Editor), *The Times*, 24357, p. 7, 22 September 1862.

302 *A Government's first and indispensable duty* Arnold, pp. 491–2.

302 *this year I shall be able* Henry James, 'The Ordnance Survey of England' (Letters to the Editor), *The Times*, 24357, p. 7, 22 September 1862.

302 *Between 1864 and 1869* For details regarding the Ordnance Survey's adventures in the Middle East, see Phillips; Wilson and James; Wilson and Palmer.

302 *to improve the sanitary state* Wilson and James, Preface, p. i.

303 *The city is at present* Wilson and James, 'Water Supply'.

303 *Of the drainage of the city* Wilson and James, 'Water Supply'.

303 *proceed with caution* Cited in Phillips, p. 155.

303 *the rugged territory of Highland Scotland* I am indebted to Yolande Hodson for this fascinating information.

303 *in some of the smaller cisterns* Wilson and James, 'III: Haram-es-Sherif'.

304 *It is no easy matter* Wilson and James, 'III: Haram-es-Sherif'.

304 *the geological structure* Wilson and James, Preface, p. iii.

304 *to enable Serj. McDonald* Wilson and James, Preface, p. iii.

304 *when we come to examine* Wilson and James, 'Ancient Supply'.

304 *the name now commonly* Wilson and James, 'III: Haram-es-Sherif'.

304 *that there is a great need* Revd C. Williams, Introduction, in Wilson and Palmer.

305 *agents for the sale of Ordnance maps* *The Times*, 9982, 2 November 1816.

305 *the trigonometrical survey of England* 'The Working Man Candidate', *Northern Echo*, 9, 11 January 1870.

305 *railway system itself* 'The Working Man Candidate', *Northern Echo*, 9, 11 January 1870.

306 *the Ordnance Survey is a work* Cited in Mumford, 1980, p. 159.

306 *within the reach of all* 'The Ordnance Survey of England and Wales', *Daily News*, 808, 28 December 1848.

306 *the maps fit together* 'The Ordnance Survey of England and Wales', *Daily News*, 808, 28 December 1848.

307 *stand an hour at a time* 'Among the Luxuries of the Last Generation', *The Times*, 24357, p. 6, 22 September 1862.

EPILOGUE: MAPS OF FREEDOM

310 *provides a new right*   CRoW Act 2000, <http://www.opsi.gov.uk/Acts/acts2000/ukpga_20000037_en_1> [accessed 15 April 2010].

310 *Some philosophers*   For example, see Deleuze and Guattari, who suggest that the map 'fosters connections between fields, the removal of blockages on bodies without organs, the maximum openings of bodies without organs onto a plane of consistency . . . The map is open and connectable in all of its dimensions; it is detachable, reversible, susceptible to constant modification . . . A map has multiple entryways, as opposed to the tracing, which always comes back "to the same"' (pp. 13–14). There are many similar postmodern and other philosophical discussions of the meaning and function of maps: see Bachelard; Jameson, 1988; Jameson, 1991, pp. 45–54; LeFebvre.

310 *persuade the government*   See Free Our Data, <http://www.freeourdata.org.uk/blog/2010/01/data-gov-uk-now-thats-what-we-call-a-result/> [accessed 15 April 2010].

310 *useful to any business*   Free Our Data, <http://www.freeourdata.org.uk/blog/2010/01/data-gov-uk-now-thats-what-we-call-a-result/> [accessed 15 April 2010].

311 *great national undertaking*   Cited in Seymour, p. 133.

311 *the simple directions*   Martin Phillips, 'Internet's Wiping History off the Map', *Sun*, 30 August 2008.

311 *the equal wide survey*   Thomson, 'Summer', in Thomson, James, p. 119.

Works Cited

Texts marked in **bold** are my recommendations for further reading on the issues, periods and personalities discussed in this book.

Manuscript and contemporary newspaper sources are listed individually in the Notes. All other printed sources are given here. The *Philosophical Transactions of the Royal Society* are abbreviated to *Philosphical Transactions*.

Adams, I.H., *The Mapping of a Scottish Estate*, Edinburgh: University of Edinburgh, 1971.

Adamson, Helen C., *William Roy 1726–1790: Pioneer of Roman Archaeology in Scotland*, Glasgow: Glasgow Art Gallery & Museum, 1984.

Addison, W. Innes, *The Matriculation Albums of the University of Glasgow from 1728 to 1858*, Glasgow: University of Glasgow, 1913.

Albert, Bill, *The Turnpike Road System in England, 1663–1840*, London: Cambridge University Press, 1972.

Alder, Ken, *The Measure of All Things: The Seven-Year Odyssey that Transformed the World*, London: Little, Brown, 2002.

Alfrey, Nicholas, 'Landscape and the Ordnance Survey, 1795–1820', pp. 23–7 in Alfrey and Daniels (eds), 1990.

Alfrey, Nicholas and Stephen Daniels (eds), *Mapping the Landscape: Essays on Art and Cartography*, Nottingham: University of Nottingham, 1990.

Allan, D.G.C., 'Artists and the Society in the Eighteenth Century', pp. 91–119 in Allan and Abbott (eds), 1992.

Allan, D.G.C. and John L. Abbott (eds), *The Virtuoso Tribe of Arts and Sciences: Studies in the Eighteenth-Century Work and Membership of the London Society of Arts*, Athens and London: University of Georgia Press, 1992.

Allibone, T.E., *The Royal Society and its Dining Clubs*, Oxford: Pergamon, 1976.

Anderson, James, *Justification of Mr Murdoch Mackenzie's Nautical Survey of the Orkney Islands and Hebrides, in answer to the accusations of Doctor Anderson*, Edinburgh: William Creech, 1785.

Andrews, J.H., 'The Survey of Ireland to 1847', pp. 79–99 in Seymour (ed.), 1980.

Andrews, J.H., '"More Suitable to the English Tongue": The Cartography of Celtic Placenames', *Ulster Local Studies*, 14, pp. 7–21, 1992.

Andrews, J.H., *A Paper Landscape: The Ordnance Survey in Nineteenth-Century Ireland*, 2nd edn, Dublin: Four Courts Press, 2006. First published Oxford: Clarendon Press, 1975

Andrews, Malcolm, *The Search for the Picturesque: Landscape Aesthetics and Tourism in Britain, 1760–1800*, Aldershot: Scolar Press, 1989.

Andrews, Stuart, *The British Periodical Press and the French Revolution, 1789–99*, Basingstoke: Palgrave, 2000.

Anon, *The Character of a Jacobite, By What Name or Title soever Dignifyed or Distinguished* London: Printed for the Author, 1690.

Anon, *A Journey through Part of England and Scotland, Along with the Army under the Command of his Royal Highness the Duke of Cumberland*, London: J. Stanton [1746].

Anon, *Authentic Account of Debates in the House of Commons, on Monday, February 27, and Tuesday, February 28, 1786, on the proposed Plan for Fortifications, by His Grace the Duke of Richmond*, London: 1786.

Anon, *Americana: or a New Tale of the Genii*, Baltimore: Pechin, 1802.

Anon, 'Attraction des Montagnes, et ses Effects', *Edinburgh Review*, 26, pp. 36–51, 1816.

Anon, 'Obituary. Major-General Thomas F. Colby', *Memoirs of the Institution of Civil Engineers*, 12, pp. 132–7, 1853.

Apianus, Petrus, *Cosmographicus liber . . . studiose correctus, ac erroribus vindicatus per Gemmam Physium*, Antwerp: Grapheus, 1533.

Arbuthnot, Archibald, *The Life, Adventures, and Many and Great Vicissitudes of Fortune of Simon, Lord Lovat, the Head of the Family of Frasers*, London: R. Walker, 1746.

Arnold, Matthew, 'Ordnance Maps', *London Review*, 5, pp. 491–2, 6 December 1862.

Arrowsmith, Aaron, *Memoir Relative to the Construction of the Map of Scotland*, London: W. Savage, 1809.

Aslam, Nadeem, *Maps for Lost Lovers*, London: Faber and Faber, 2004.

Ault, Donald D., *Visionary Physics: Blake's Response to Newton*, Chicago and London: University of Chicago Press, 1974.

Austen, Jane, *Mansfield Park*, New York and London: Norton, 1998.

Austen, Jane, *Pride and Prejudice*, New York and London: Norton, 2001.

Austen, Jane, *The Letters of Jane Austen*, ed. Deirdre Le Faye, London: Folio Society, 2003.

Babbage, Charles, *The Ninth Bridgewater Treatise*, London: John Murray, 1837.

Bachelard, Gaston, *The Poetics of Space*, trans. Maria Jolas, Boston: Beacon Press, 1964.

Bagwell, Philip Sidney, *The Transport Revolution from 1770*, London: Batsford, 1974.

Baigent, Elizabeth, 'Colby, Thomas (1784–1852)', *Oxford Dictionary of National Biography*, Oxford University Press, 2004 [http://www.oxforddnb.com/view/article/5837, accessed 9 April 2010].

Baigent, Elizabeth, 'Dawson, Robert (1771–1860)', *Oxford Dictionary of National Biography*, Oxford University Press, Sept. 2004; online edn, Jan. 2008 [http://www.oxforddnb.com/view/article/7354, accessed 9 April 2010].

Baigent, Elizabeth, 'Roy, William (1726–1790)', *Oxford Dictionary of National Biography*, Oxford University Press, Sept. 2004; online edn, Jan. 2008 [http://www.oxforddnb.com/view/article/24236, accessed 9 April 2010].

Baigent, Elizabeth, 'Watson, David (1713?–1761)', *Oxford Dictionary of National Biography*, Oxford University Press, Sept. 2004; online edn, Jan. 2008 [http://www.oxforddnb.com/view/article/28834, accessed 31 March 2010].

Bailey, De Witt, 'The Board of Ordnance and small arms supply: the Ordnance system 1714–1783' (unpublished PhD thesis, University of London, 1988).

Baird, Rosemary, *Goodwood: Art and Architecture, Sport and Family*, London: Frances Lincoln, 2007.

Baker, Joan and Arthur Baker, *A Walker's Companion to the Wade Roads*, Perth: Melven Press, 1982.

Baldwin, Thomas, *Airopaidia: containing the narrative of a balloon excursion from Chester*, Chester: 1786.

Ball, Johnson, *Paul and Thomas Sandby: Royal Academicians*, Cheddar: Charles Skilton, 1985.

Barker, Juliet, *William Wordsworth: A Life*, London: Viking, 2000.

Barker, N.P., 'The Architecture of the English Board of Ordnance' (unpublished PhD thesis, University of Reading, 1985).

Barnard, Toby, 'Petty, Sir William (1623–1687)', *Oxford Dictionary of National Biography*, Oxford University Press, Sept. 2004; online edn, May 2007 [http://www.oxforddnb.com/view/article/22069, accessed 14 April 2010].

Barnes, Trevor J. and James S. Duncan (eds), *Writing Worlds: Discourse, Text and Metaphor in the Representation of Landscape*, London and New York: Routledge, 1992.

Barrell, John, *The Idea of Landscape and the Sense of Place, 1730–1840: An Approach to the Poetry of John Clare*, Cambridge: Cambridge University Press, 1972.

Barrell, John, *The Dark Side of the Landscape: The Rural Poor in English Paintings, 1730–1840*, Cambridge: Cambridge University Press, 1980.

Barrell, John, *English Literature in History, 1730–80: An Equal, Wide Survey*, London: Hutchinson, 1983.

Barrell, John, 'Sir Joshua Reynolds and the Political Theory of Painting', *Oxford Art Journal*, 9, pp. 36–41, 1986a.

Barrell, John, *The Political Theory of Painting from Reynolds to Hazlitt: The Body of the Public*, New Haven and London: Yale University Press, 1986b.

Barrell, John, 'Coffee-House Politicians', *Journal of British Studies*, 43, pp. 206–32, 2004.

Barrell, John, *The Spirit of Despotism: Invasions of Privacy in the 1790s*, Oxford: Oxford University Press, 2006.

Batey, Mavis, *Jane Austen and the English Landscape*, London: Barn Elms, 1996.

Bayly, C.A. and Katherine Prior, 'Cornwallis, Charles, first Marquess Cornwallis (1738–1805)', *Oxford Dictionary of National Biography*, Oxford University Press, Sept. 2004; online edn, May 2008 [http://www.oxforddnb.com/view/article/6338, accessed 8 April 2010].

Beattie, James, *Scoticisms, Arranged in Alphabetical Order, Designed to Correct Improprieties of Speech and Writing*, Edinburgh: W. Creech, 1787.

Beaufort, Daniel Augustus, *Memoir of a Map of Ireland*, Dublin: Slater and Allen, 1792.

Bedini, Silvio A., 'Lens Making for Scientific Instrumentation in the Seventeenth Century', *Applied Optics*, 5, pp. 687–94, 1966.

Beeke, Henry, *Observations on the Produce of the Income Tax*, London: Wright, 1799.

Bentham, Jeremy, *Panopticon: Or, the Inspection-House*, Dublin: Thomas Byrne, 1791.

Berkeley, George, *An Essay Towards a New Theory of Vision*, Dublin: Pepyat, 1709.

Bermingham, Ann, *Landscape and Ideology: The English Rustic Tradition, 1740–1860*, Berkeley and Los Angeles: University of California Press, 1986.

Betjeman, John, *Summoned by Bells*, London: Murray, 1960.

Black, Jeremy, *Culloden and the '45*, Stroud: Alan Sutton, 1990.

Blake, William, *The Complete Poetry and Prose of William Blake*, rev. edn, ed. David V. Erdman, New York: Doubleday, 1988.

Blunden, Edmund and Earl Leslie Griggs (eds), *Coleridge: Studies by Several Hands on the Hundredth Anniversary of his Death*, London: Constable, 1934.

Board, Christopher (ed.), *New Insights into Cartographic Communication* (special edn of *Cartographica*), 21, 1984.

Bolingbroke, Henry St John Viscount, *Letters, on the Spirit of Patriotism*, Dublin: J. Smith, 1749.

Bonehill, John and Stephen Daniels (eds), *Paul Sandby: Picturing Britain*, London: Royal Academy of Arts, 2009.

Bonney, George, *Goodwood: An Illustrated Survey of the Historic Sussex Home of the Dukes of Richmond and Gordon*, Derby: English Life Publications, 1951.

Borges, Jorge Luis, 'Museum: On Exactitude in Science', p. 181 in *The Aleph*, London: Penguin, 2000.

Boswell, James, *The Life of Samuel Johnson*, 2 vols, London: Charles Dilly, 1791.

Boyd, Derek, *Royal Engineers*, London: Cooper, 1975.

Brewer, John, 'The Misfortunes of Lord Bute: A Case-Study in Eighteenth-Century Political Argument and Public Opinion', *Historical Journal*, 16, pp. 3–43, 1973.

Broglio, Ron, 'Mapping British Earth and Sky', *Wordsworth Circle*, 33, pp. 70–77, 2002.

Brown, Lloyd A., *The Story of Maps*, Boston: Little, Brown, 1949.

Brown, Michael, Patrick M. Geoghegan, and James Kelly (eds), *The Irish Act of Union, 1800: Bicentennial Essays*, Dublin: Irish Academic Press, 2003.

Buchanan, John Lanne, *Travels in the Western Hebrides*, London: J. & J. Robinson, 1793.

Bulloch, John Malcolm, *The Families of Invergordon, Newhall, also Ardoch, Ross-shire, and Carroll, Sutherland*, Dingwall: Ross-shire Printing & Publishing Co., 1906.

Burke, Edmund, *Reflections on the Revolution in France*, ed. Conor Cruise O'Brien, Harmondsworth: Penguin, 1968. First published London: J. Dodsley, 1790.

Burke, Edmund, *A Philosophical Inquiry into the Origin of our Ideas of the Sublime and Beautiful*, ed. Adam Phillips, Oxford: Oxford University Press, 1990. First published London: R. & J. Dodsley, 1757.

Burke, Thomas, *Travel in England*, London: Batsford, 1942.

Burney, Frances, *The Wanderer; or, Female Difficulties*, ed. Margaret Anne Doody, Robert L. Mack and Peter Sabor, Oxford: Oxford University Press, 2001.

Burrowes, Robert, 'Preface', *Transactions of the Royal Irish Academy*, 1, pp. ix–xvii, 1787.

Burt, Edmund, *Letters from a Gentleman in the North of Scotland to his Friend in London*, 2 vols, London: Birt, 1754.

Burt, Edmund, *Letters from a Gentleman in the North of Scotland to his Friend in London*, 5th edn, 2 vols, London: Rest Fenner, 1818.

Burton, J.H. (ed.), *The Autobiography of Dr Alexander Carlyle of Inveresk 1722–1805*, Edinburgh and London: Blackwood & Sons, 1860.

Camden, William, *Britannia, or, a Chorographicall Description of the most flourishing Kingdomes, England, Scotland, and Ireland*, trans. Philemon Holland, London: Bishop & Norton, 1600.

Camden, William, *Camden's Britannia 1695. A Facsimile of the 1695 Edition Published by Edmund Gibson*, Newton Abbott: David and Charles, 1971.

Carlucci, April and Peter Barber (eds), *Lie of the Land: The Secret Life of Maps*, London: British Library, 2001.

Carlyle, Thomas, *Carlyle Letters Online [CLO]*, 2007, [http://carlyleletters.org, accessed 15 April 2010].

Carr, Margaret, 'Charting the life of the man who mapped Orkney', *The Orcadian*, 19 September 2005.

Carroll, Patrick, *Science, Culture, and Modern State Formation*, Berkeley and London: University of California Press, 2006.

Cassini, Jean Dominique, Pierre François André Méchain and Adrien Marie Legendre, *Exposé des Opérations Faites en France en 1787 pour la Jonction des Observatoires de Paris et de Greenwich*, Paris: L'Institution des Sourds-Muets, 1790.

Chadwick, Edwin, *Report to Her Majesty's Principal Secretary of State for the Home Department, from the Poor Law Commissioners, on an Inquiry into the Sanitary Condition of the Labouring Population of Great Britain*, Edinburgh: Edinburgh University Press, 1965. First published London: Poor Law Commission, 1842.

Chalmers, George, *Caledonia*, 4 vols, London: Cadell & Davies, 1807.

Chambers, Ephraim, *Cyclopaedia: or, an universal dictionary of arts and sciences*, 4th edn, 2 vols, London: Midwinter, 1741.

Chapman, Allan, 'The Accuracy of Angular Measuring Instruments used in Astronomy between 1500 and 1850', *Journal of the History of Astronomy*, 14, pp. 133–7, 1983.

Christian, Jessica, 'Paul Sandby and the Military Survey of Scotland', pp. 18–22 in Alfrey and Daniels (eds), 1990.

Churchill, Charles, *The Prophecy of Famine. A Scots Pastoral*, 4th edn, London: Kearsly, 1763.

Chuto, Jacques, 'Mangan, Petrie, O'Donovan, and a Few Others: The Poet and the Scholars', *Irish University Review*, 6, pp. 169–87, 1976.

Chuto, Jacques, 'James Clarence Mangan and the Paternal Debt', pp. 98–110 in Eyler and Garratt (eds), 1988.

Claeys, Gregory, *The French Revolution Debate in Britain: The Origins of Modern Politics*, Basingstoke: Palgrave Macmillan, 2007.

Clarke, Tristram, 'Short, James (1710–1768)', *Oxford Dictionary of National Biography*, Oxford University Press, Sept. 2004; online edn, October 2006 [http://www.oxforddnb.com/view/article/25459, accessed 2 April 2010].

Clements, William Holliwell, *Towers of Strength: The Story of the Martello Towers*, Barnsley: Leo Cooper, 1999.

Clifton, Gloria, 'Dollond family (*per.* 1750–1871)', *Oxford Dictionary of National Biography*, Oxford University Press, Sept. 2004; online edn, Jan. 2008 [http://www.oxforddnb.com/view/article/49855, accessed 5 April 2010].

Close, Charles, *The Map of England; or, About England with an Ordnance Map*, London: Davies, 1932.

Close, Charles, *The Early Years of the Ordnance Survey*, Newton Abbot: David & Charles, 1969. First published London: Institution of Royal Engineers, 1926.

Cockburn, Henry, *Memorials of his Time*, Edinburgh: Robert Grant & Son, 1946.

Colby, Thomas and Thomas Aiskew Larcom, *Ordnance Survey of the County of Londonderry*, Dublin: Hodges and Smith, 1837.

Coleridge, Hartley, 'Adolf and Annette', ed. Judith Plotz, *Children's Literature*, 14, pp. 151–61, 1986.

Coleridge, Samuel Taylor, *Collected Letters of Samuel Taylor Coleridge*, ed. Earl Leslie Griggs, 6 vols, Oxford: Clarendon Press, 1956–71.

Coleridge, Samuel Taylor, *The Notebooks of Samuel Taylor Coleridge*, ed. Kathleen Coburn, 5 vols, London: Routledge & Kegan Paul, 1957.

Coleridge, Samuel Taylor, *Biographia Literaria: or, Biographical Sketches of my Literary Life and Opinions*, ed. James Engell and W. Jackson Bate, 2 vols, Princeton: Princeton University Press, 1983.

Colley, Linda, 'The Apotheosis of George III: Loyalty, Royalty, and the British Nation 1760–1820', *Past & Present*, 102, pp. 94–129, 1984.

Colley, Linda, *Britons: Forging the Nation*, London: Pimlico, 2003. First published New Haven: Yale University Press, 1992.

Colvin, Christina Edgeworth, 'Edgeworth, Richard Lovell (1744–1817)', *Oxford Dictionary of National Biography*, Oxford University Press, Sept. 2004; online edn, Jan. 2008 [http://www.oxforddnb.com/view/article/8478, accessed 12 April 2010].

Compton, Thomas, *The North Cambrian Mountains, or a Tour through North Wales*, London: 1817.

Connolly, Sean, 'Translating History: Brian Friel and the Irish Past', pp. 149–63 in Peacock (ed.), 1993.

Connolly, T.W.J., *The History of the Corps of Royal Sappers and Miners*, 2 vols, London: Longman, Brown, Green, 1855.

Conrad, Joseph, *Mirror of the Sea*, New York: Doubleday, 1924. First published 1904–6.

Conrad, Joseph, *Heart of Darkness*, New York and London: W.W. Norton, 1988. First published Blackwood's Magazine, 1959.

Cook, A.S., 'Alexander Dalrymple (1737–1808), hydrographer to the East India Company and to the admiralty, as publisher' (unpublished PhD thesis, University of St Andrews, 1992).

Cooper, Thomas, 'Lind, James (1736–1812)', rev. Patrick Wallis, *Oxford Dictionary of National Biography*, Oxford University Press, 2004 [http://www.oxforddnb.com/view/article/16670, accessed 3 April 2010].

Copeland, John, *Roads and their Traffic 1750–1850*, Newton Abbot: David & Charles, 1968.

Copley, Stephen and Peter Garside (eds), *The Politics of the Picturesque: Literature, Landscape, and Aesthetics since 1770*, Cambridge: Cambridge University Press, 1994.

Cosgrove, Denis, 'Prospect, Perspective, and the Evolution of the Landscape Idea', *Transactions of the Institute of British Geographers*, 10, pp. 45–62, 1985.

Cosgrove, Denis and Stephen Daniels (eds), *The Iconography of Landscape: Essays on the Symbolic Representation, Design, and Use of Past Environments*, **Cambridge: Cambridge University Press, 1988.**

Cowley, John, *A Display of the Coasting Lines of Six Several Maps of North Britain*, 1734.

Crone, G.R., E.M.J. Campbell and R.A. Skelton, 'Landmarks in British Cartography', *Geographical Journal*, 128, pp. 406–26, 1962.

Cronin, Michael, *Translating Ireland: Translation, Languages, Cultures*, Cork: Cork University Press, 1996.

Crosland, Maurice, 'Relationships between the Royal Society and the Académie des Sciences in the Late Eighteenth Century', *Notes & Records of the Royal Society of London*, 59, pp. 25–34, 2005.

Cruickshanks, Eveline (ed.), *Ideology and Conspiracy: Aspects of Jacobitism, 1689–1759*, Edinburgh: Donald, 1982.

Cunningham, Andrew and Nicholas Jardine (eds), *Romanticism and the Sciences*, **Cambridge: Cambridge University Press, 1990.**

Curtis, Edith Roelker, *Lady Sarah Lennox: An Irrepressible Stuart 1745–1826*, New York: G.P. Putnam, 1946.

Dandeker, Christopher, *Surveillance, Power and Modernity: Bureaucracy and Discipline from 1700 to the Present Day*, Cambridge: Polity Press, 1990.

Daniels, Stephen, *Fields of Vision: Landscape Imagery and National Identity in England and the United States*, Cambridge: Polity Press, 1993.

Daniels, Stephen, 'A Prospect for the Nation', pp. 23–60 in Harris and Wilcox (eds), 2006.

Danson, Edwin, *Drawing the Line: How Mason and Dixon Surveyed the Most Famous Border in America*, New York: John Wiley, 2001.

Danson, Edwin, *Weighing the World: The Quest to Measure the Earth*, **Oxford: Oxford University Press, 2006.**

Davidson, Neil, *Origins of Scottish Nationhood*, London: Pluto, 2000.

Day, Angélique, Patrick McWilliams and Lisa English, *Ordnance Survey Memoirs of Ireland*, 38 vols, Dublin: Institute of Irish Studies with the Royal Irish Academy, 1990–1997.

Day, Archibald, *The Admiralty Hydrographic Service 1795–1919*, London: HMSO, 1967.

Debbieg, Hugh, *Authentic Copy of the Proceedings of a General Court Martial, held at the Horse-Guards, on Tuesday the 9th of November, 1784*, London: J. Almon, 1784.

De Beer, Gavin, *The Sciences were Never at War*, London: Thomas Nelson, 1960.

De Bolla, Peter, *The Education of the Eye: Painting, Landscape, and Architecture in Eighteenth-Century Britain*, Stanford: Stanford University Press, 2003.

De Bolla, Peter, Nigel Leask, and David Simpson (eds), *Land, Nation and Culture, 1740–1840: Thinking the Republic of Taste*, **Basingstoke: Macmillan, 2005.**

Debord, Guy, 'Introduction to a Critique of Urban Geography', pp. 5–8 in Knabb (ed.), 1981.

Deleuze, Gilles and Félix Guattari, *A Thousand Plateaus: Capitalism and Schizophrenia*, trans. Brian Massumi. London: Continuum, 2004. First published Minneapolis: University of Minnesota Press, 1987.

Dermody, Thomas, *Poems, Moral and Descriptive*, London: Crowder, 1800.

De Vere, Aubrey, *Recollections of Aubrey De Vere*, New York and London: Edward Arnold, 1897.

Dickinson, H.T., *Britain and the French Revolution*, Basingstoke: Macmillan, 1989.

Diderot, Denis and Jean Le Rond d'Alembert, *Encyclopédie, ou dictionnaire raisonné des sciences, des arts et des métiers, par une société de gens de lettres*, 17 vols, Paris: 1751–65.

Digges, Leonard, *A Geometrical Practise named Pantometria*, London: Henrie Bynemann, 1571.

Digges, Leonard, *The Theodelitus and Topographical Instrument of Leonard Digges*, Oxford: Old Ashmolean Reprints, 1927.

Doherty, Gillian M., *The Irish Ordnance Survey: History, Culture and Memory*, Dublin: Four Courts Press, 2006.

Donne, John, *The Complete Poems of John Donne*, 2 vols, London: privately printed, 1872.

[Douglas, Francis], *The History of the Rebellion in 1745 and 1746*, Aberdeen: Douglas and Murray, 1755.

Duffy, Charles Gavan, *Young Ireland: A Fragment of Irish History, 1840–45*, Dublin: Gill, 1884.

Dyer, John and Mark Akenside, *The Poetical Works of Mark Akenside and John Dyer*, ed. Robert Willmott, London and New York: Routledge, 1855.

Eagleston, A.J., 'Wordsworth, Coleridge, and the Spy', pp. 71–81 in Blunden and Griggs (eds), 1934.

Eagleton, Terry, *Scholars and Rebels in Nineteenth-Century Ireland*, Oxford: Blackwell, 1999.

Edgcumbe, John, 'Reynolds's Earliest Drawings', *Burlington Magazine*, 129, pp. 724–6, 1987.

Edgeworth, Richard Lovell, 'An Essay on the Art of Conveying Secret and Swift Intelligence', *Transactions of the Royal Irish Academy*, 6, pp. 95–141, 1797.

Edney, Matthew, 'British Military Education, Mapmaking, and Military "Map-Mindedness" in the Later Enlightenment', *Cartographic Journal*, 31, pp. 14–20, 1994a.

Edney, Matthew, 'Mathematical Cosmography and the Social Ideology of British Cartography, 1780–1820', *Imago Mundi*, 46, pp. 101–20, 1994b.

Edney, Matthew, *Mapping an Empire: The Geographical Construction of British India 1765–1843*, Chicago: University of Chicago Press, 1997.

Edwards, Jess, *Writing, Geometry and Space in Seventeenth-Century England and America*, London: Routledge, 2006.

Eichner, Hans, 'The Rise of Modern Science and the Genesis of Romanticism', *PMLA*, 97, pp. 8–30, 1982.

Erskine-Hill, Howard, 'Literature and the Jacobite Cause: Was there a Rhetoric of Jacobitism?', pp. 49–69 in Cruickshanks (ed.), 1982.

Eyler, Audrey S. and Robert F. Garratt (eds), *The Uses of the Past: Essays on Irish Culture*, London and Toronto: Associated University Presses, 1988.

Fackler, Herbert V., 'Wordsworth in Ireland, 1829: A Survey of his Tour', *Éire-Ireland: a Journal of Irish Studies*, 6, pp. 53–64, 1971.

Fennell, James, *Lindor and Clara: or, the British Officer*, London: E. & T. Williams, 1791.

Fitzmaurice, Edmond, *The Life of Sir William Petty 1623–1687*, London: John Murray, 1895.

Fleet, Christopher and Kimberly C. Kowal, 'Roy Military Survey map of Scotland (1747–1755): mosaicing, geo-referencing, and web delivery', *e-Perimetron*, 2, pp. 194–208, 2007.

Fleming, John, *Robert Adam and his Circle*, London: John Murray, 1962.

Flint, Stamford Raffles, *Mudge Memoirs: Being a Record of Zachariah Mudge, and some Members of his Family*, Truro: Netherton and Worth, 1883.

Fontenelle, Bernard Le Bovier M. de, *Oeuvres Complètes de Fontenelle*, ed. Georges-Bernard Depping, Paris: Belin, 1818.

[Forbes, Duncan], *Memoirs of the Life of Lord Lovat*, London: William Brien, 1746.

Fortescue, John (ed.), *Correspondence of King George the Third from 1760 to December 1783*, 6 vols, London: Macmillan, 1927.

Foster, John Wilson, *Nature in Ireland: A Scientific and Cultural History*, Dublin: Lilliput Press, 1997.

Foucault, Michel, *Discipline and Punish: the Birth of the Prison*, trans. Alan Sheridan, London: Allen Lane, 1977.

Fox, Caroline, *Memories of Old Friends*, ed. H.N. Pym, 2 vols, London: Smith, Elder, 1882.

Frängsmyr, T., J.L. Heilbron and Robin E. Rider (eds), *The Quantifying Spirit in the Eighteenth Century*, Berkeley: University of California Press, 1990.

Fraser, Simon, Lord Lovat, 'Memorial addressed to his Majesty George I concerning the State of the Highlands, 1724', II, pp. 254–67 in Burt, 1818.

Friel, Brian, *Translations*, London: Faber, 1981.

Friel, Brian, *Making History*, London: Faber, 1989.

Friel, Brian, *Essays, Diaries, Interviews: 1964–99*, ed. Christopher Murray, London: Faber, 1999.

Friel, Brian, J.H. Andrews, and Kevin Barry, 'Translations and a Paper Landscape: Between Fiction and History', *The Crane Bag*, 7, pp. 118–24, 1983.

Frisius, Gemma, *Libellus de locorum describendorum ratione*, in Apianus, 1533.

Fry, Michael, *The Dundas Despotism*, Edinburgh: Edinburgh University Press, 1992.

Fulford, Tim (ed.), *Romanticism and Science, 1773–1833*, London: Routledge, 2002.

Fulford, Tim, Debbie Lee and Peter J. Kitson (eds), *Literature, Science, and Exploration in the Romantic Era: Bodies of Knowledge*, Cambridge: Cambridge University Press, 2004.

Furgol, Edward M., 'Simon Fraser, eleventh Lord Lovat (1667/8–1747)', *Oxford Dictionary of National Biography*, Oxford University Press, Sept. 2004; online edn, January 2010 [http://www.oxforddnb.com/view/article/10122, accessed 31 March 2010].

Gardiner, R.A., 'William Roy, Surveyor and Antiquary', *Geographical Journal*, 143, pp. 439–50, 1977.

Gardner, William and Thomas Gream, *A Topographical Map of the County of Sussex, divided into Rapes, Deanries and Hundreds*, London: Faden, 1795.

Gascoigne, John, *Science in the Service of Empire: Joseph Banks, the British State, and the Uses of Science in the Age of Revolution*, Cambridge: Cambridge University Press, 1998.

Geoghegan, Patrick M., *The Irish Act of Union: A Study in High Politics, 1798–1801*, Dublin: Gill & Macmillan, 1999.

Gibson, George, 'Sketch of the History of Mathematics in Scotland to the end of the 18th Century: Part 2', *Proceedings of the Edinburgh Mathematical Society*, 1, pp. 71–93, 1927.

Gifford, William, *William Adam, 1689–1748: A Life and Times of Scotland's Universal Architect*, Edinburgh: Mainstream, 1989.

Gilchrist, Jim, 'Return of the Map', *Scotsman*, 15 January 2008.

Gill, Stephen, *William Wordsworth: A Life*, Oxford: Clarendon, 1989.

Gillies, John, *Shakespeare and the Geography of Difference*, Cambridge: Cambridge University Press, 1994.

Gilpin, William, *An Essay Upon Prints*, London: Robson, 1768.

Gilpin, William, *Observations on the River Wye, and Several Parts of South Wales, &c. Relative Chiefly to Picturesque Beauty; Made in the Summer of 1770*, 2nd edn, London: Blamire, 1789a.

Gilpin, William, *Observations, Relative Chiefly to Picturesque Beauty, Made in the Year 1776, on Several Parts of Great Britain*, 2 vols, London: Blamire, 1789b.

Gilpin, William, *Three Essays: On Picturesque Beauty, On Picturesque Travel, and on Sketching Landscape*, London: Blamire, 1792.

Girvin, Brian, *From Union to Union: Nationalism, Democracy, and Religion in Ireland – Act of Union to EU*, Dublin: Gill & Macmillan, 2002.

Godlewska, Anne Marie Claire, *Geography Unbound: French Geographic Science from Cassini to Humboldt*, Chicago and London: Chicago University Press, 1999.

Gooday, Graeme J.N., *The Morals of Measurement: Accuracy, Irony, and Trust in Late Victorian Electrical Practice*, Cambridge: Cambridge University Press, 2004.

Gordon, C.A., *A Concise History of the Ancient and Illustrious House of Gordon*, Aberdeen: D. Wyllie, 1890.

Gordon, William, *The History of the Ancient, Noble, and Illustrious Family of Gordon*, 2 vols, Edinburgh: Thomas Ruddiman, 1726.

Gough, Richard, *British Topography or, An Historical Account of what has been Done for Illustrating the Topographical Antiquities of Great Britain and Ireland*, 2 vols, London: Payne and Nichols, 1780.

Graves, Robert Perceval, *The Life of Sir William Rowan Hamilton*, 3 vols, Dublin: Hodges, Figgis, 1882.

Gray, Thomas, *The Poems of Mr Gray, to which are prefixed Memoirs of his life and writings by W. Mason*, London: Hughs and Dodsley, 1775.

Green, Alice Stopford, *Irish Nationality*, London: Williams & Norgate, 1911.

Gregory, Olinthus (ed.), *Dissertations and Letters . . . Tending Either to Impugn or to Defend the Trigonometrical Survey of England and Wales*, London: Law and Gilbert, 1815.

Grenier, Katherine Haldane, *Tourism and Identity in Scotland, 1770–1914: Creating Caledonia*, Aldershot: Ashgate, 2005.

Guest, Harriet, 'Suspicious Minds: Spies and Surveillance in Charlotte Smith's Novels of the 1790s', pp. 169–87 in De Bolla, Leask and Simpson (eds), 2005.

Habermas, Jürgen, *The Structural Transformation of the Public Sphere*, trans. Thomas Burger, Cambridge, Mass.: MIT Press, 1989.

Hacking, Ian, *The Taming of Chance*, Cambridge: Cambridge University Press, 1991. First published 1990.

Haefner, Joel, 'Displacement and the Reading of Romantic Space', *Wordsworth Circle*, 23, pp. 151–6, 1992.

Hamilton, George, *A History of the House of Hamilton*, Edinburgh: J. Skinner, 1933.

Hamilton, William Rowan, *On a General Method of Expressing the Paths of Light, and of the Planets, by the Coefficients of a Characteristic Function*, Dublin: P. Dixon Hardy, 1833.

Hamilton, William Rowan, *The Mathematical Papers of Sir William Rowan Hamilton*, ed. A.W. Conway and J.L. Synge, 3 vols, Cambridge: Cambridge University Press, 1931.

Hampsher-Monk, Iain (ed.), *The Impact of the French Revolution: Texts from Britain in the 1790s*, Cambridge: Cambridge University Press, 2005.

Hankins, Thomas L., *Sir William Rowan Hamilton*, Baltimore and London: Johns Hopkins University Press, 1980.

Hannas, Linda, *The English Jigsaw Puzzle, 1760–1890*, London: Wayland, 1972.

Hardiman, James, *Irish Minstrelsy, or Bardic Remains of Ireland*, London: Robins, 1831.

Harley, J.B., 'English County Map-Making in the Early Years of the Ordnance Survey: The Map of Surrey by Joseph Lindley and William Crosley', *Geographical Journal*, 132, pp. 372–8, 1966a.

Harley, J.B., 'The Bankruptcy of Thomas Jefferys: An episode in the economic history of eighteenth century map-making', *Imago Mundi*, 20, pp. 27–48, 1966b.

Harley, J.B., 'The Evaluation of Early Maps: Towards a Methodology', *Imago Mundi*, 22, pp. 62–74, 1968.

Harley, J.B., 'Place-Names on the Early Ordnance Survey Maps of England and Wales', *The Cartographic Journal*, 8, pp. 91–104, 1971.

Harley, J.B., *Ordnance Survey Maps: A Descriptive Manual*, Southampton: Ordnance Survey, 1975.

Harley, J.B. (ed.), *The Old Series Ordnance Survey Maps of England and Wales*, Vols 4–5, Kent: Harry Margary, 1986–7.

Harley, J.B., 'Maps, Knowledge, and Power', pp. 277–312 in Cosgrove and Daniels (eds), 1988.

Harley, J.B., 'Deconstructing the Map', pp. 231–47 in Barnes and Duncan (eds), 1992.

Harley, J.B., 'The Society and the Surveys of English Counties, 1759–1809', pp. 141–57 in Allan and Abbott (eds), 1992. (First published in *Journal of the Royal Society of Arts*, 111, pp. 43–6, 1963; 112, pp. 119–24, pp. 269–75, pp. 538–43, 1964.)

Harley, J.B. and Yolande O'Donoghue (eds), *The Old Series Ordnance Survey Maps of England and Wales*, vols 1–3, Kent: Harry Margary, 1975–81.

Harley, J.B. and Richard R. Oliver (eds), *The Old Series Ordnance Survey Maps of England and Wales*, vols 6–8, Kent: Harry Margary, 1989–92.

Harley, J.B. and C.W. Phillips, *The Historian's Guide to Ordnance Survey Maps*, London: The Standing Conference for Local History, 1964.

Harley, J.B. and Gwynn Walters, 'William Roy's Maps, Mathematical Instruments, and Library: the Christie's Sale of 1790', *Imago Mundi*, 29, pp. 9–22, 1977.

Harley, J.B. and Gwynn Walters, 'Welsh Orthography and Ordnance Survey Mapping 1820–1905', *Archaeologia Cambrensis*, 131, pp. 98–135, 1982.

Harris, Ron, 'Clyde Valley Mansion to be "Santa's Castle" on Japanese Island', *Hamilton Advertiser*, 24 July 1987.

Harris, Theresa Fairbanks and Scott Wilcox (eds), *Papermaking and the Art of Watercolor in Eighteenth-Century Britain: Paul Sandby and the Whatman Paper Mill*, New Haven and London: Yale University Press, 2006.

Harty, Joetta, 'The Islanders: Mapping Paracosms in the Early Writing of Hartley Coleridge, Thomas Malkin, Thomas De Quincey and the Brontës' (unpublished PhD thesis, George Washington University, 2007).

Harvey, P.D.A., *The History of Topographical Maps: Symbols, Pictures, and Surveys*, London: Thames & Hudson, 1980.

Hayman, David and Sam Slote, *Genetic Studies in Joyce*, Amsterdam: Rodopi, 1995.

Hayton, D.W., Introduction, IV, pp. 1–33 in Owens and Furbank (eds), 2000.

Helgerson, Richard, *Forms of Nationhood: The Elizabethan Writing of England*, Chicago and London: University of Chicago Press, 1992.

Hellyer, Roger, *A Guide to the Ordnance Survey 1: 25,000 First Series*, London: Charles Close Society, 2003.

Hellyer, Roger and Richard Oliver, *Military Maps: The One-Inch Series of Great Britain and Ireland*, London: Charles Close Society, 2004.

Herman, Arthur, *The Scottish Enlightenment: The Scots' Invention of the Modern World*, London: Fourth Estate, 2001.

Herrmann, Luke, *Paul and Thomas Sandby*, London: Batsford, 1986.

Hewitt, Rachel, 'Wordsworth and the Irish Ordnance Survey: "Dreaming o'er the Map of Things"', *Wordsworth Circle*, 38, pp. 80–5, 2006.

Hewitt, Rachel, 'Dreaming o'er the Map of Things: The Ordnance Survey and Literature of the British Isles 1747–1842' (unpublished PhD thesis, University of London, 2007).

Hibbert, Christopher, *The Destruction of Lord Raglan: A Tragedy of the Crimean War, 1854–55*, London: Longmans, 1961.

Hibbert, Christopher, *The French Revolution*, London: Allen Lane, 1980.

Hilton, Boyd, *A Mad, Bad, & Dangerous People? England 1783–1846*, Oxford: Clarendon Press, 2006.

Hodges, James, *The Rights and Interests of the Two British Monarchies*, Edinburgh: Donaldson, 1703.

Hodson, Yolande, *Ordnance Surveyors' Drawings, 1789–c.1840: The Original Manuscript Maps of the First Ordnance Survey of England and Wales from the British Map Library*, Reading: Research Publications, 1989.

Hodson, Yolande, *An Inch to the Mile: The Ordnance Survey One-Inch Map 1805–1974*, London: Charles Close Society, 1991a.

Hodson, Yolande, 'Robert Dawson (1771–1860), Ordnance Surveyor and Draftsman: A Brief Note on his Early Family Life', *The Map Collector*, 54, pp. 28–30, 1991b.

Hodson, Yolande, *Facsimile of the Ordnance Surveyors' Drawings of the London Area 1799–1808*, London: London Topographical Society, 1991c.

Hodson, Yolande, 'Board of Ordnance Surveys 1683–1820', [no pages] in *Proceedings of the Symposium to Celebrate the Ordnance Survey Bicentenary*, 1999.

Hodson, Yolande, 'William Roy and the Military Survey of Scotland', pp. 7–23 in Roy, 2007.

Hogg, O.F.C., *The Royal Arsenal: Its Background, Origin, and Subsequent History*, 2 vols, London: Oxford University Press, 1963.

[Hollar, Wenceslaus], *A New Map of the Cittyess of London and Westminster with the Borough of Southwark*, London: Greene [1680(?)].

Holloway, James and Lindsay Errington, *The Discovery of Scotland: The Appreciation of Scottish Scenery through Two Centuries of Painting*, Edinburgh: HMSO, 1978.

Holmes, Richard, *The Age of Wonder: How the Romantic Generation Discovered the Beauty and Terror of Science*, London: HarperCollins, 2008.

Hoock, Holger, *The King's Artists: The Royal Academy of Arts and the Politics of British Culture, 1760–1840*, Oxford: Clarendon Press, 2003.

Hoock, Holger, *Empires of the Imagination: Politics, War, and the Arts in the British World, 1750–1850*, London: Profile, 2010.

Hooke, Robert, *Some animadversions on the first part of Hevellus, his 'Machina Coelestis'*, London: 1674.

Hoppit, Julian, *A Land of Liberty: England 1689–1727*, Oxford: Oxford University Press, 2000.

Hoskins, W.G., *The Making of the English Landscape*, London: Hodder and Stoughton, 1955.

Houlding, J.A., 'Dundas, Sir David (1735?–1820)', *Oxford Dictionary of National Biography*, Oxford University Press, Sept. 2004; online edn, Jan. 2008 [http://www.oxforddnb.com/view/article/8247, accessed 3 April 2010].

Howse, Derek, *Nevil Maskelyne: The Seaman's Astronomer*, Cambridge: Cambridge University Press, 1989.

Howse, Derek, 'Mason, Charles (1728–1786)', *Oxford Dictionary of National Biography*, Oxford University Press, 2004 [http://www.oxforddnb.com/view/article/18268, accessed 4 April 2010].

Hughes, John, *Poems on Several Occasions*, 2 vols, London: Tonson and Watts, 1735.

Hutchinson, John, *The Dynamics of Cultural Nationalism: The Gaelic Revival and the Creation of the Irish State*, London: Allen and Unwin, 1987.

Hutton, Charles, 'An Account of the Calculations Made from the Survey and Measures Taken at Schehallion', *Philosophical Transactions*, 68, pp. 689–788, 1778.

Inglis-Jones, Elisabeth, *The Great Maria*, London: Faber, 1959.

Ireland, William Henry, *Henry the Second*, London: Barker, 1799.

Irwin, Robert, *The Art of Paul Sandby*, New Haven: Yale Center for British Art, 1985.

Jameson, Frederic, 'Cognitive Mapping', pp. 347–60 in Nelson and Grossberg (eds), 1988.

Jameson, Frederic, *Postmodernism: or, the Cultural Logic of Late Capitalism*, London and New York: Verso, 1991.

Jarvis, Robin, *Romantic Writing and Pedestrian Travel*, Basingstoke: Macmillan, 1997.

Johns, Adrian, *The Nature of the Book: Print and Knowledge in the Making*, Chicago: University of Chicago Press, 1998.

Johnson, Samuel (ed.), *The Works of the English Poets, from Chaucer to Cowper*, 21 vols, London: Johnson, 1810.

Johnson, Samuel, *Journey to the Western Islands of Scotland*, London: Penguin, 1984.

Jordanova, Ludmilla (ed.), *Languages of Nature: Critical Essays on Science and Literature*, London: Free Association, 1986.

Kant, Immanuel, 'An Answer to the Question: "What is Enlightenment?"', pp. 54–60 in Kant, 1991. First published as 'Beantwortung der Frage: Was ist Aufklärung', *Berlinische Monatsschrift*, 4, pp. 481–94, 12 December 1784.

Kant, Immanuel, *Kant: Political Writings*, trans. H.B. Nisbet, ed. Hans Reiss, Cambridge: Cambridge University Press, 1991.

Kant, Immanuel, *Critique of Pure Reason*, trans. Norman Kemp Smith, Basingstoke and New York: Palgrave Macmillan, 2003.

Keates, J.S., 'The Cartographic Art', pp. 37–43 in Board (ed.), 1984.

Keay, John, *The Great Arc: The Dramatic Tale of How India was Mapped and Everest was Named*, London: HarperCollins, 2000.

Keene, Benjamin, *The Correspondence of Sir Benjamin Keene*, ed. Richard Lodge, Cambridge: Cambridge University Press, 1933.

Kelly, Helena, 'The Politics of Space: Enclosure in English Literature, 1789–1815' (unpublished PhD thesis, University of Oxford, 2009).

Kendrick, Peter S. *et al.*, *Roadwork: Theory and Practice*, 5th edn, Oxford: Elsevier Butterworth-Heinemann, 2004.

Kent, John, *Records and Reminiscences of Goodwood and the Dukes of Richmond*, London: Sampson Low, 1896.

Kettell, Samuel (ed.), *Specimens of American Poetry*, 3 vols, Boston: S.G. Goodrich, 1829.

Kiberd, Declan, *Inventing Ireland: The Literature of the Modern Nation*, London: Cape, 1995.

Kinniburgh, I.A.G., 'Map Exhibitions in Edinburgh, July–August 1964', *Cartographic Journal*, 1, pp. 16–19, December 1964.

Kishlansky, Mark, *A Monarchy Transformed: Britain 1603–1714*, London: Penguin, 1996.

Klein, Bernhard, *Maps and the Writing of Space in Early Modern England and Ireland*, Basingstoke: Macmillan, 2000.

Knabb, Ken (ed.), *Situationist International Anthology*, Berkeley: Bureau of Public Secrets, 1981.

Knight, Ellis Cornelia, *Autobiography of Miss Cornelia Knight*, ed. J.W. Kaye, 2 vols, London: W.H. Allen, 1861.

Knox, Vicesimus, *Spirit of Despotism*, Morristown, New Jersey: Mann, 1799.

Knox-Shaw, Peter, *Jane Austen and the Enlightenment*, Cambridge: Cambridge University Press, 2004.

Kohl, J.G., *England and Wales*, London: Cass, 1968. First published 1844.

Konvitz, Josef W., *Cartography in France 1660–1848: Science, Engineering, and Statecraft*, Chicago and London: University of Chicago Press, 1987.

Kuhn, Thomas, 'The Function of Measurement in Modern Physical Science', *ISIS*, 52, pp. 161–93, 1961.

Kuhn, Thomas, *The Structure of Scientific Revolutions*, Chicago: University of Chicago Press, 1962.

Lamb, Charles, *The Letters of Charles and Mary Anne Lamb*, ed. Edwin W. Marrs Jr, 3 vols, Ithaca: Cornell University Press, 1975.

Laurence, Edward, *The Duty and Office of a Land Steward*, London: J. & J. Knapton, 1731.

L[each], G.A., 'Obituary: Thomas Aiskew Larcom', *Royal Engineers Journal*, p. 194, 1879.

Le Fanu, W.R., *Seventy Years of Irish Life: Being Anecdotes and Reminiscences*, London: Macmillan, 1894.

LeFebvre, Henry, *The Production of Space*, trans. Donald Nicholson-Smith, Oxford: Blackwell, 1991.

Lennox, Charles, *An Authentic Copy of the Duke of Richmond's Bill, for a Parliamentary Reform*, London: J. Stockdale, 1783.

Lennox, Charles, *An Answer to 'A Short Essay on the Modes of Defence Best Adapted to the Situation and Circumstances of this Island'*, London: J. Almon, 1785.

Lennox, Charles, *Thoughts on the National Defence*, London: W. Bulmer, 1804.

Lernout, Geert (ed.), *The Crows Behind the Plough: History and Violence in Anglo-Irish Poetry and Drama*, Amsterdam: Rodopi, 1991.

Leslie, Charles Robert, *Life and Times of Sir Joshua Reynolds*, 2 vols, London: John Murray, 1865.

Levack, Brian P., *The Formation of the British State: England, Scotland, and the Union 1603–1707*, Oxford: Clarendon Press, 1987.

Ley, Charles, *The Nobleman, Gentleman, Land Steward, and Surveyor's Compleat Guide*, London: 1786.

Livingstone, David N. and Charles W.J. Withers (eds), *Geography and Enlightenment*, Chicago: University of Chicago Press, 1999.

Lowe, William C., 'Lennox, Charles, third duke of Richmond, third duke of Lennox, and duke of Aubigny in the French nobility (1735–1806)', *Oxford Dictionary of National Biography*, Oxford University Press, Sept. 2004; online edn, Oct. 2008 [http://www.oxforddnb.com/view/article/16451, accessed 4 April 2010].

Lynch, Kevin, *The Image of the City*, Cambridge, Mass.: MIT Press, 1960.

McCall, Colin, *Routes, Roads, Regiments and Rebellions: A brief history of the life and work of General George Wade (1673–1748), the Father of the Military Roads in Scotland*, Matlock: SOLCOL, 2003.

McCall, John, *James Clarence Mangan: His Life and Writings*, Dublin: Carraig Books, 1975.

McConnell, Anita, *Jesse Ramsden (1735–1800): London's Leading Scientific Instrument Maker*, Aldershot: Ashgate, 2007.

McCormack, W.J., *Ascendancy and Tradition in Anglo-Irish Literary History from 1789 to 1939*, Oxford: Clarendon Press, 1985.

MacDonagh, Oliver, *Ireland: The Union and its Aftermath*, Dublin: University College Dublin Press, 2003.

MacDonald, George, 'General William Roy and his "Military Antiquities of the Romans in North Britain"', *Archaeologica*, 68, pp. 161–228, 1917.

MacEachren, Alan M., *How Maps Work: Representation, Visualization, and Design*, New York: Guilford Press, 1995.

Macfarlane, Robert, *Mountains of the Mind: A History of a Fascination*, London: Granta, 2003.

Mackesy, Piers, 'Germain, George Sackville, first Viscount Sackville (1716–1785)', *Oxford Dictionary of National Biography*, Oxford University Press, Sept. 2004; online edn, May 2009 [http://www.oxforddnb.com/view/article/10566, accessed 3 April 2010].

Mackintosh, Robert James, *Memoirs of . . . Sir James Mackintosh*, London: E. Moxon, 1835.

MacInnes, Allan I., *Union and Empire: The Making of the United Kingdom in 1707*, Cambridge: Cambridge University Press, 2007.

MacLehose, James, *Memoirs and Portraits of One Hundred Glasgow Men*, Glasgow: MacLehose, 1886.

McLennan, John Ferguson, *Memoir of Thomas Drummond, Under-Secretary to the Lord Lieutenant of Ireland, 1835 to 1840*, Edinburgh: Edmondston & Douglas, 1867.

MacPherson, James, *Fingal: an Ancient Epic Poem, in Six Books: Together with Several Other Poems, Composed by Ossian, the Son of Fingal*, London: Becket, 1762.

MacPherson, James, *The Poems of Ossian and Related Works*, ed. Howard Gaskill, Edinburgh: Edinburgh University Press, 1996.

Maillet, Arnaud, *The Claude Glass: Use and Meaning of the Black Mirror in Western Art*, trans. Jeff Fort, New York: Zone Books, 2004.

Malcolm, James Peller, *An Historical Sketch of the Art of Caricaturing*, London: Longman, Hurst *et al.*, 1813.

Malkin, Benjamin Heath, *A Father's Memoirs of his Child*, London: 1806.

Mallett, Robert Joseph, 'The Military Survey of Scotland 1747–1755: an analysis utilising the dual concepts of map and content' (unpublished doctoral thesis, University of Sheffield, 1987).

Mangan, James Clarence, *The Autobiography of James Clarence Mangan*, ed. James Kilroy, Dublin: Dolmen, 1968.

Mangan, James Clarence, *The Collected Works of James Clarence Mangan: Poems 1818–1837*, ed. Jacques Chuto *et al.*, Dublin: Irish Academic Press, 1996.

Mangan, James Clarence, *The Collected Works of James Clarence Mangan: Prose 1832–1839*, ed. Jacques Chuto *et al.*, Dublin: Irish Academic Press, 2002a.

Mangan, James Clarence, *The Collected Works of James Clarence Mangan: Prose 1840–1882*, ed. Jacques Chuto *et al*, Dublin: Irish Academic Press, 2002b.

Marjarum, E. Wayne, 'Wordsworth's View of the State of Ireland', *PMLA*, 55, pp. 608–11, 1940.

Mark, George, *For Publishing by Subscription, An Accurate Map or Geometrical Survey of the Shires, of Lothian, Tweddale, and Clydsdale*, 1728.

Markham, Clements, *Major James Rennell and the Rise of Modern English Geography*, London: Cassell, 1895.

Marshall, Douglas William, 'The British Military Engineers 1741–1783: A Study of Organization, Social Origin, and Cartography' (unpublished PhD thesis, University of Michigan, 1976).

Marshall, Douglas William, 'Military Maps of the Eighteenth Century and the Tower of London Drawing Room', *Imago Mundi*, 32, pp. 21–44, 1980.

Martin, Jean-Pierre and Anita McConnell, 'Joining the Observatories of Paris and Greenwich', *Notes & Records of the Royal Society*, 62, pp. 355–72, 2008.

Maskelyne, Nevil, 'A Proposal for Measuring the Attraction of some Hill in this Kingdom by Astronomical Observations', *Philosophical Transactions*, 65, pp. 495–9, 1775a.

Maskelyne, Nevil, 'An Account of Observations Made on the Mountain Schehallion for Finding its Attraction', *Philosophical Transactions*, 65, pp. 500–42, 1775b.

Maton, William George, *Observations Relative Chiefly to the Natural History, Picturesque Scenery, and Antiquities of the Western Counties of England*, Salisbury: J. Eaton, 1797.

Mayhew, Robert J., *Enlightenment Geography: The Political Languages of British Geography 1650–1850*, Basingstoke: Macmillan, 2000.

Meadows, A.J., 'Observational Defects in Eighteenth-Century British Telescopes', *Annals of Science*, 26, pp. 305–17, 1970.

Messenger, Guy, *The Sheet Histories of the Ordnance Survey One-Inch Old Series Maps of Devon and Cornwall: A Cartobibliographic Account*, London: Charles Close Society, 1991a.

Messenger, Guy, *The Sheet Histories of the Ordnance Survey One-Inch Old Series Maps of Essex and Kent: A Cartobibliographic Account*, London: Charles Close Society, 1991b.

Millburn, John R., 'The Office of Ordnance and the Instrument-Making Trade in the Mid-Eighteenth Century', *Annals of Science*, 45, pp. 221–93, 1988.

Mingay, George Edmund, *Enclosure and the Small Farmer in the Age of the Industrial Revolution*, London: Macmillan, 1968.

Mingay, George Edmund, *Parliamentary Enclosure in England: An Introduction to its Causes, Incidence and Impact, 1750–1850*, London: Longman, 1997.

Mingay, George Edmund, *A Social History of the English Countryside*, London: Routledge, 1999.

Mitchell, W.J.T., *Landscape and Power*, 2nd edn, Chicago: Chicago University Press, 2002.

Mlodinow, Leonard, *Euclid's Window: The Story of Geometry from Parallel Lines to Hyperspace*, London: Penguin, 2002.

Moir, D.G. (ed.), *The Early Maps of Scotland*, 3rd edn, Edinburgh: Royal Scottish Geographical Society, 1973.

Moir, Esther, *The Discovery of Britain: The English Tourists 1540–1840*, London: Routledge & Kegan Paul, 1964.

Moon, Penderel, *The British Conquest and Dominion of India*, London: Duckworth, 1989.

Moore, Brian, *The Mangan Inheritance*, London: Flamingo, 1995.

Mordaunt, John, *The Proceedings of a General Court Martial held in the Council Chamber at Whitehall, on Wednesday the 14th, and continued by several Adjournments to Tuesday the 20th of December 1757, upon the Trial of Lieutenant-General Sir John Mordaunt*, Dublin: J. Hoey, 1758.

Mudge, Thomas, *A Description with Plates, of the Time-Keeper invented by the late Mr Thomas Mudge*, London: Printed for the Author, 1799.

Mudge, William, 'An Account of the Trigonometrical Survey, Carried on in the Years 1797, 1798, and 1799', *Philosophical Transactions*, 90, pp. 539–728, 1800.

Mudge, William, *An Account of the Operations Carried on for Accomplishing a Trigonometrical Survey of England and Wales; continued from the year 1797, to the end of the year 1799*, vol. 2, London: Bulmer and Faden, 1801a.

Mudge, William, *General Survey of England and Wales. An Entirely New & Accurate Survey of the County of Kent*, London: Faden, 1801b.

Mudge, William, 'An Account of the Measurement of an Arc of the Meridian, Extending from Dunnose in the Isle of Wight . . . to Clifton in Yorkshire', *Philosophical Transactions*, 93, pp. 383–508, 1803.

Mudge, William, *An Account of the Measurement of an Arc of the Meridian, Extending from Dunnose in the Isle of Wight to Clifton in Yorkshire, in the Course of the Operations carried on for the Trigonometrical Survey of England, in the Years 1800, 1801, and 1802*, London: Bulmer and Faden, 1804.

Mudge, William, *Ordnance Survey of the Isle of Wight*, London: Ordnance Survey, 1810.

Mudge, William and Thomas Colby, *An Account of the Trigonometrical Survey, Carried on by Order of the Master-General of His Majesty's Ordnance, in the Years 1800, 1801, 1803, 1804, 1805, 1806, 1807, 1808, and 1809*, vol. 3, London: Bulmer and Faden, 1811.

Mudge, William and Isaac Dalby, *An Account of the Operations Carried on for Accomplishing a Trigonometrical Survey of England and Wales; from the commencement in the year 1784 to the end of the year 1796*, vol. 1, London: Bulmer and Faden, 1799.

Mudge, William, Edward Williams and Isaac Dalby, 'An Account of the Trigonometrical Survey carried on in the Years 1791, 1792, 1793, and 1794', *Philosophical Transactions*, 85, pp. 414–591, 1795a.

Mudge, William, Edward Williams and Isaac Dalby, *An Account of the Trigonometrical Survey carried on in the Years 1791, 1792, 1793, and 1794 . . . from the Philosophical Transactions*, London: 1795b.

Mudge, William, Edward Williams and Isaac Dalby, 'An Account of the Trigonometrical Survey, carried on in the Years 1795 and 1796', *Philosophical Transactions*, 87, pp. 432–541, 1797a.

Mudge, William, Edward Williams and Isaac Dalby, *An Account of the Trigonometrical Survey, carried on in the Years 1795 and 1796*, London: 1797b.

Mudge, Zachariah, *Sermons on Different Subjects*, London: S. Birt, 1739.

Mudge, Zachariah, *A Sermon on Liberty*, London: F. Knight, 1790.

Mumford, Ian, 'Engraved Ordnance Survey One-Inch Maps: The Problem of Dating', *The Cartographic Journal*, 5, pp. 44–6, 1968.

Mumford, Ian, 'Lithography, Photography, and Photozincography in English Map Production before 1870', *The Cartographic Journal*, 9, pp. 30–6, 1972.

Mumford, Ian, 'The Last Years of James's Superintendency and the Transfer to Civil Control 1864–1880', pp. 158–67, in Seymour (ed.), 1980.

Murdin, Paul, *Full Meridian of Glory: Perilous Adventures in the Competition to Measure the Earth*, New York: Copernicus Books, 2009.

Murray, Christopher, 'Brian Friel's *Translations* and the Problem of Historical Accuracy', pp. 61–78 in Lernout (ed.), 1991.

Mylne, Robert Scott, *The Master Masons to the Crown of Scotland and their Works*, Edinburgh: Scott & Ferguson, 1893.

Napier, Gerald, *Follow the Sapper: An Illustrated History of the Corps of Royal Engineers*, Chatham: Institution of Royal Engineers, 2005.

Nelson, Cary and Lawrence Grossberg (eds), *Marxism and the Interpretation of Culture*, Urbana and Chicago: University of Illinois Press, 1988.

Nevskaya, N.I., *USSR Academy of Sciences: Scientific Relations with Great Britain*, Moscow: Nauka, 1977.

New Annual Register, or General Repository of History, Politics, and Literature for the Year 1785, London: 1786.

New Annual Register, or General Repository of History, Politics, and Literature for the Year 1787, London: 1788.

Newbon, Pete, 'William Blake's Infernal Method' (unpublished MA thesis, University of Cambridge, 2006).

Newton, Isaac, *Philosophiae Naturalis Principia Mathematica*, London: Joseph Streat, 1687.

Norden, John, *Norden's Preparative to his Speculum Britanniae*, London: 1596.

O'Brian, Patrick, *Joseph Banks*, London: Collins Harvill, 1986.

O'Brien, Richard Barry, *Thomas Drummond Under-Secretary in Ireland 1835–40: Life and Letters*, London: Kegan Paul, 1889.

Ó Cadhla, Stiofán, *Civilizing Ireland: Ordnance Survey 1824–1841. Ethnography, Cartography, Translation*, Dublin: Irish Academic Press, 2007.

O'Connell, Seán, *William Rowan Hamilton: Portrait of a Prodigy*, Dublin: Boole Press, 1983.

O'Curry, Eugene, *Lectures on the Manuscript Materials of Ancient Irish History*, Dublin: Duffy, 1861.

O'Daly, Aengus, *The Tribes of Ireland: A Satire*, Dublin: O'Daly, 1852.

Ó Danachair, Donal (ed.), *The Newgate Register*, 6 vols, Ex-classics, <http://www.exclassics.com>, 2009.

O'Dell, Andrew, 'A View of Scotland in the Middle of the Eighteenth Century', *Scottish Geographical Magazine*, 69, pp. 58–63, 1953.

O'Donoghue, D.J., *The Life and Writings of James Clarence Mangan*, Edinburgh: Geddes, 1897.

O'Donoghue, Yolande, *William Roy 1726–1790: Pioneer of the Ordnance Survey*, London: British Museum Publications, 1977.

O'Donovan, John, *Annals of the Kingdom of Ireland*, 2nd edn, 7 vols, Dublin: Hodges, Figgis, 1856.

O'Donovan, John, *On the Life and Labours of John O'Donovan*, London: Richardson, 1862.

O'Donovan, John, *John O'Donovan's Letters from County Londonderry (1834)*, ed. Graham Mawhinney, Draperstown: Ballinascreen Historical Society, 1992.

O'Donovan, John, *Ordnance Survey Letters: Down. Letters Containing Information Relative to the Antiquities of the County of Down Collected during the Progress of the Ordnance Survey in 1834*, ed. Michael Herity, Dublin: Four Masters Press, 2001a.

O'Donovan, John, *Ordnance Survey Letters: Meath*, ed. Michael Herity, Dublin: Four Masters Press, 2001b.

O'Donovan, John, *Ordnance Survey Letters: Kilkenny*, ed. Michael Herity, Dublin: Four Masters Press, 2003.

Ogilby, John, *Britannia, Volume the First*, London: Hollar, 1675.

Ogilby, John, *Britannia Depicta*, London: Bowen, 1720.

Oliver, Richard, 'The Ordnance Survey in Great Britain, 1835–1870' (unpublished PhD thesis, University of Sussex, 1985).

Oliver, Richard, 'The Battle of the Scales: Contemporary Opinions and Modern Reconsiderations', *Sheetlines*, 31, pp. 59–64, 1991.

Oliver, Richard, *Ordnance Survey Maps: A Concise Guide for Historians*, London: Charles Close Society, 2005.

Oliver, Richard and Roger Hellyer, 'The one-inch Old Series: more discoveries – yet more questions', *Sheetlines*, 80, pp. 26–39, 2007.

Olmsted, John W., 'The Application of Telescopes to Astronomical Instruments, 1667–1669: A Study in Historical Method', *ISIS*, 40, pp. 213–25, 1949.

Olson, Alison Gilbert, *The Radical Duke: Career and Correspondence of Charles Lennox, third Duke of Richmond*, Oxford: Oxford University Press, 1961.

Omond, George W.T., *The Arniston Memoirs: Three Centuries of a Scottish House 1571–1838*, Edinburgh: David Douglas, 1887.

Ó Muraile, Nollaig, 'O'Donovan, John (1806–1861)', *Oxford Dictionary of National Biography*, Oxford University Press, Sept. 2004; online edn, Jan. 2008 [http://www.oxforddnb.com/view/article/20561, accessed 15 April 2010].

Ó Raifeartaigh, T., *The Royal Irish Academy: A Bicentennial History, 1785–1985*, Dublin: Royal Irish Academy, 1985.

Ordnance Survey, *First Series of Ordnance Survey Maps*, London and Southampton: Ordnance Survey, 1805–1873.

Ordnance Survey, *First Edition Six-Inch Maps of Ireland*, Dublin: Ordnance Survey, 1833–46.

Ordnance Survey of Ireland, *An Illustrated Record of Ordnance Survey in Ireland*, Dublin: Ordnance Survey of Ireland, 1991.

O'Reilly, William, 'Charles Vallancey and the *Military Itinerary* of Ireland', *Proceedings of the Royal Irish Academy*, 106, pp. 125–217, 2006.

Owen, Tim and Elaine Pilbeam, *Ordnance Survey: Map Makers to Britain since 1791*, London and Southampton: HMSO and Ordnance Survey, 1992.

Owens, W.R. and P.N. Furbank (eds), *Political and Economic Writings of Daniel Defoe*, London: Pickering and Chatto, 2000.

Palmer, Stanley H., 'Drummond, Thomas (1797–1840)', *Oxford Dictionary of National Biography*, Oxford University Press, Sept. 2004; online edn, Jan. 2008 [http://www.oxforddnb.com/view/article/8084, accessed 14 April 2010].

Parsons, E.J.S., 'Changing Needs in Great Britain', pp. 109–18, in Seymour (ed.), 1980.

Parsons, E.J.S., 'The Superintendency of Lewis Alexander Hall', pp. 119–28 in Seymour (ed.), 1980.

Parsons, E.J.S, 'The Scales Dispute – Henry James 1854–1863', pp. 129 in Seymour (ed.), 1980.

Pasley, Charles W., *Essay on the Military Policy and Institutions of the British Empire*, 2nd edn, London: A.J. Valpy, 1811.

Paterson, Daniel, *A New and Accurate Description of the Roads in England and Wales, and Part of the Roads of Scotland*, London: Longman, Rees, Faden, 1803.

Patten, Eve, 'Ireland's "Two Cultures" Debate: Victorian Science and the Literary Revival', *Irish University Review*, 33, pp. 1–13, 2003.

Peacock, Alan (ed.), *The Achievement of Brian Friel*, Gerrards Cross: Smythe, 1993.

Pearce, Susan, *Visions of Antiquity: The Society of Antiquaries of London 1707–2007*, London: Society of Antiquaries of London, 2007.

Pedley, Mary, 'Maps, War, and Commerce: Business Correspondence with the London Map Firm of Thomas Jefferys and William Faden', *Imago Mundi*, 48, pp. 161–73, 1996.

Pelletier, Monique, *La Carte de Cassini: l'extraordinaire aventure de la carte de France*, Paris: Presses de l'école nationale des ponts et chaussées, 1990.

Petrie, George, *Letter to Sir William Rowan Hamilton*, Dublin: Graisberry, 1840.

Petty, William, *Tracts; Chiefly Relating to Ireland*, Dublin: Boulter Grierson, 1769.

Petty, William, *The History of the Survey of Ireland: Commonly Called the Down Survey, AD 1655–6*, ed. T.A. Larcom, Dublin: Irish Archaeological Society, 1851.

Phillips, C.W., 'The Surveys of Jerusalem and Sinai', pp. 154–7 in Seymour (ed.), 1980.

Philp, Mark (ed.), *The French Revolution and British Popular Politics*, Cambridge: Cambridge University Press, 1991.

Pittock, Murray, G.H., 'Charles Edward Stuart', *Études Écossaises*, 10, pp. 57–71, 2005.

[Plumptre, James], *The Lakers: A Comic Opera, in Three Acts*, London: Clarke, 1798.

Pollard, Michael, *Walking the Scottish Highlands: General Wade's Military Roads*, London: Andre Deutsch, 1984.

Polybius, *The Histories*, Loeb Classical Library, London: Heinemann, 1922–7.

Poole, John, *Crotchets in the Air; or, an (un)scientific Account of a Balloon-Trip*, London: Henry Colburn, 1828.

Pope, Alexander, *Essay on Man*, London: J. Wilford, 1733–4.

Porter, Bernard, *Plots and Paranoia: A History of Political Espionage in Britain 1790–1988*, London and New York: Routledge, 1992.

Porter, Whitworth, *History of the Corps of Royal Engineers*, 2 vols, London: Longmans, Green, 1889.

Portlock, Joseph, *Memoir of the Life of Major-General Colby*, London: Seeley, Jackson & Halliday, 1869.

Postle, Martin, 'Reynolds, Sir Joshua (1723–1792)', *Oxford Dictionary of National Biography*, Oxford University Press, Sept. 2004; online edn, Oct. 2009 [http://www.oxforddnb.com/view/article/23429, accessed 7 April 2010].

Price, Richard, 'A Discourse on the Love of our Country', pp. 39–55 in Hampsher-Monk (ed.), 2005.

Pringle, John, *A Discourse on the Attraction of Mountains, Delivered at the Anniversary Meeting of the Royal Society, November 30, 1775,* London: Royal Society, 1775.

Proceedings of the Symposium to Celebrate the Ordnance Survey Bicentenary held at the Royal Geographical Society, London, on 23 May 1799: Ordnance Survey Past, Present and Future, London: 1999.

Pynchon, Thomas, *Mason & Dixon*, London: Jonathan Cape, 1997.

Rackham, Oliver, *The History of the Countryside*, London: Dent, 1986.

Raisz, Erwin, *General Cartography*, New York: McGraw-Hill, 1948.

Ramsay, John, *Scotland and Scotsmen in the Eighteenth Century*, Edinburgh: Blackwood & Sons, 1888.

Rankin, Daniel Reid, 'Report of a recent examination of the Roman camp at Cleghorn . . . with notices of General Roy and his family', *Proceedings of the Society of Antiquaries of Scotland*, pp. 145–8, March 1853.

Rankin, Daniel Reid, 'Notices of Major-General William Roy, from the parish registers of Carluke and other sources', *Proceedings of the Society of Antiquaries of Scotland*, pp. 562–6, June 1872.

Ravenhill, William, 'John Adams, His Map of England, its Projection, and his *Index Villaris* of 1680', *Geographical Journal*, 144, pp. 424–37, 1978.

Ravenhill, William, 'The Honorable Robert Edward Clifford, 1767–1817: a Cartographer's Response to Napoleon', *Geographical Journal*, 160, pp. 159–72, 1994.

Reader, W.J., *Macadam: the McAdam Family and the Turnpike Roads 1798–1861*, London: Heinemann, 1980.

Reese, M.M., *Goodwood's Oak: The Life and Times of the Third Duke of Richmond, Lennox, and Aubigny*, London: Threshold, 1987.

Reeves, Nicky, 'Constructing an Instrument: Nevil Maskelyne and the Zenith Sector, 1760–1774' (unpublished PhD thesis, University of Cambridge, 2009a).

Reeves, Nicky, '"To demonstrate the exactness of the instrument": Mountainside Trials of Precision in Scotland, 1774', *Science in Context*, 22, pp. 323–40, 2009b.

Reynolds, Joshua, *The Works of Sir Joshua Reynolds*, 3 vols, London: Cadell & Davies, 1798.

Rhoades, James, *Words by the Wayside*, London: Chapman and Hall, 1915.

Richardson, Alan, *British Romanticism and the Science of the Mind*, Cambridge: Cambridge University Press, 2005.

Risdon, Tristram, *The Chorographical Description or Survey of the County of Devon*, London: Rees & Curtis, 1811.

Robb, Graham, *The Discovery of France*, London: Picador, 2007.

Roberts, Warren, *Jane Austen and the French Revolution*, London: Macmillan, 1979.

Robinson, Arthur H., *The Look of Maps: An Examination of Cartographic Design*, Madison: University of Wisconsin Press, 1952.

Robinson, Arthur H., 'Mapmaking and Map Printing: The Evolution of a Working Relationship', pp. 1–24 in Woodward (ed.), 1975.

Robinson, Thomas Romney, 'On the difference of Longitude between the Observatories of Armagh and Dublin', *Transactions of the Royal Irish Academy*, 19, pp. 121–46, 1843.

Rodriguez, Don Joseph, 'Observations on the Measurement of Three Degrees of the Meridian Conducted in England by Lieut. Col. William Mudge', *Philosophical Transactions*, 102, pp. 321–51, 1812.

Roe, Nicholas (ed.), *Samuel Taylor Coleridge and the Sciences of Life*, Oxford: Oxford University Press, 2001.

Rogers, Ben, *Beef and Liberty: Roast Beef, John Bull and the English Nation*, London: Chatto & Windus, 2003.

Rolt, L.T.C., *The Balloonists: The History of the First Aeronauts*, Stroud: Sutton, 2006.

Roy, William, *Plan of the Battle of Thonhausen gained August 1 1759 ... by W. Roy*, London: T. Major, 1760.

Roy, William, 'Experiments and Observations Made in Britain, in Order to Obtain a Rule for Measuring Heights with the Barometer', *Philosophical Transactions*, 67, pp. 653–787, 1777.

Roy, William, 'An Account of the Measurement of a Base on Hounslow-Heath', *Philosophical Transactions*, 75, pp. 385–480, 1785.

Roy, William, 'An Account of the Mode Proposed to be Followed in Determining the Relative Situation of the Royal Observatories of Greenwich and Paris', *Philosophical Transactions*, 77, pp. 188–469, 1787.

Roy, William, 'An Account of the Trigonometrical Operation, whereby the Distance between the Meridians of the Royal Observatories of Greenwich and Paris has been Determined', *Philosophical Transactions*, 80, pp. 111–614, 1790.

Roy, William, *The Military Antiquities of the Romans in North Britain*, London: Society of Antiquaries, 1793.

Roy, William, *The Great Map: The Military Survey of Scotland 1747–1755*, Edinburgh: Birlinn, 2007.

Ruston, Sharon, *Shelley and Vitality*, Basingstoke: Palgrave Macmillan, 2005.

Ryan, Desmond, *The Sword of Light: From the Four Masters to Douglas Hyde, 1636–1938*, London: Arthur Barker, 1939.

Sage, L.A., *A Letter addressed to a Friend*, 2nd edn, London: J. Bell, 1785.

Sageng, Erik Lars, 'MacLaurin, Colin (1698–1746)', *Oxford Dictionary of National Biography*, Oxford University Press, Sept. 2004; online edn May 2006 [http://www.oxforddnb.com/view/article/17643, accessed 1 April 2010].

'Salisbury Plain', [http://www.english-nature.org.uk/citation/citation_photo/1006531.pdf accessed 18 April 2010].

Salmond, J. B., *Wade in Scotland*, Edinburgh: Moray Press, 1934.

Sampson, George Vaughan, *A Memoir Explanatory of the Chart and Survey of the County of London-Derry*, London: 1814.

Sandby, William, *The History of the Royal Academy of Arts from its Foundation in 1768 to the Present Time*, 2 vols, London: Longman, Green, 1862.

Sandby, William, *Thomas and Paul Sandby: Royal Academicians*, London: Seeley, 1892.

Saxton, Christopher, *An Atlas of England and Wales, containing 35 coloured maps* [London]: 1579.

Scott, Richard, 'Dundas, Robert, Lord Arniston (1685–1753)', *Oxford Dictionary of National Biography*, Oxford University Press, 2004 [http://www.oxforddnb.com/view/article/8257, accessed 1 April 2010].

Scott, Walter, *Guy Mannering*, Edinburgh: Robert Cadell, 1842.

Scott, Walter, *Old Mortality*, Oxford: Oxford University Press, 1999. First published 1816.

Seccombe, Thomas, 'Mudge, Thomas (1715/16–1794)', rev. David Penney, *Oxford Dictionary of National Biography*, Oxford University Press, 2004 [http://www.oxforddnb.com/view/article/19486, accessed 7 April 2010].

Secord, J.A., 'Beche, Sir Henry Thomas De la (1796–1855)', *Oxford Dictionary of National Biography*, Oxford University Press, 2004 [http://www.oxforddnb.com/view/article/1891, accessed 15 April 2010].

Seidmann, Gertrude, '"A Very Ancient, Useful and Curious Art": The Society and the Revival of Gem-Engraving in Eighteenth-Century England', pp. 120–31 in Allan and Abbott (eds), 1992.

Seymour, W.A. (ed.), *A History of the Ordnance Survey*, Folkestone: Dawson, 1980.

Sharp, Richard, *The Engraved Record of the Jacobite Movement*, Aldershot: Scolar Press, 1996.

Shaver, Chester L. and Alice C. Shaver, *Wordsworth's Library: A Catalogue Including a List of Books Housed by Wordsworth for Coleridge from c.1820 to c.1830*, New York and London: Garland, 1979.

Shaw, Henry, *Shaw's Authenticated Report of the Irish State Trials*, Dublin: Shaw, 1844.

Sher, Richard B., 'Alexander Bryce (*bap.*1713, *d.*1786)', *Oxford Dictionary of National Biography*, Oxford University Press, 2004 [http://www.oxforddnb.com/view/article/64365, accessed 3 April 2010].

Sillito, Richard M., 'Maskelyne on Schiehallion, or one man's geophysical year', RMS Archive [http://www.sillittopages.co.uk/schie/schie57.html].

Simpson, Michael, 'Strange Fits of Parallax: Wordsworth's Geometric Excursions', *Wordsworth Circle*, 35, pp. 19–24, 2003.

Sinclair, John, *The Statistical Account of Scotland. Drawn up from the Communications of the Ministers of Different Parishes*, 21 vols, Edinburgh: William Creech, 1791–9.

Skelton, R.A, 'The Origins of the Ordnance Survey of Great Britain', *Geographical Journal*, 128, pp. 415–30, 1962.

Skelton, R.A., 'The Military Survey of Scotland 1747–1755', *Scottish Geographical Magazine*, 83, pp. 5–16, 1967a.

Skelton, R.A., *The Military Survey of Scotland 1747–1755*, Edinburgh: Royal Geographical Society, 1967b.

Slade, H. Gordon, 'Cluny Castle, Aberdeenshire', *Proceedings of the Society of the Antiquaries of Scotland*, 111, pp. 454–92, 1981.

Slade, H. Gordon, 'Tillycairn Castle, Aberdeenshire', *Proceedings of the Society of the Antiquaries of Scotland*, 112, pp. 497–516, 1982.

Smith, Charlotte, *The Young Philosopher*, London: Cadell and Davies, 1798.

Smith, Charlotte, *The Poems of Charlotte Smith*, ed. Stuart Curran, New York and Oxford: Oxford University Press, 1993.

Smith, James Raymond, *From Plane to Spheroid: Determining the Figure of the Earth from 3000 BC to the 18th Century: Lapland and Peruvian Survey Expeditions*, Rancho Cordova: Landmark, 1986.

Smith, James Raymond, *Introduction to Geodesy: the History and Concepts of Modern Geodesy*, New York: Wiley, 1996.

Smith, James Raymond, *Everest: The Man and the Mountain*, Latheronwheel, Caithness: Whittles, 1998.

Smollett, Tobias, *A Complete History of England Deduced from the Descent of Julius Caesar, to the Treaty of Aix la Chapelle, 1748*, 4 vols, London: Rivington and Fletcher, 1757–8.

Smollett, Tobias, *The Life and Adventures of Ferdinand Count Fathom*, Oxford: Oxford University Press, 1971. First published London: T. Johnson, 1753.

Smyth, Jim (ed.), *Revolution, Counter-Revolution, and Union: Ireland in the 1790s*, Cambridge: Cambridge University Press, 2000.

Smyth, John George, *Sandhurst: The History of the Royal Military Academy, Woolwich*, London: Weidenfeld & Nicolson, 1961.

Sobel, Dava, *Longitude: The True Story of a Lone Genius Who Solved the Greatest Scientific Problem of his Time*, London: Fourth Estate, 1996.

Society for Constitutional Information, *Tracts Published and Distributed Gratis by the Society for Constitutional Information*, London: W. Richardson, 1783.

Speck, W.A., *The Butcher: The Duke of Cumberland and the Suppression of the '45*, Oxford: Blackwell, 1981.

Speed, John, *The Theatre of the Empire of Great Britaine*, London: Sudbury & Humble, 1611.

Stafford, Barbara Maria, *Voyage Into Substance: Art, Science, Nature, and the Illustrated Travel Account, 1760–1840*, Cambridge, Mass.: MIT Press, 1984.

Stephens, Frederic G., Edward Hawkins, and Dorothy George (eds), *Catalogue of Political and Personal Satires in the British Museum*, 11 vols, London: British Museum, 1870–1954.

Stevenson, Robert Louis, *The Silverado Squatters*, New York: Charles Scribner, 1902.

Stewart, Bruce (ed.), *Hearts and Minds: Irish Culture and Society under the Act of Union*, Gerrards Cross: Colin Smythe, 2002.

Stewart, Dugald, *Elements of the Philosophy of the Human Mind*, 3 vols, Edinburgh: Ramsay, 1792–1827.

Stigler, Stephen M., *The History of Statistics: The Measurement of Uncertainty before 1900*, Cambridge, Mass. and London: Harvard University Press, 1986.

Stoddart, David, *On Geography and its History*, Oxford: Blackwell, 1986.

Stokes, William, *The Life and Labours in Art and Archaeology of George Petrie*, London: Longmans, Green, 1868.

Stone, J.C., 'The Influence of Copper-Plate Engraving on Map Content and Accuracy: Preparation of the Seventeenth-Century Blaeu Atlas of Scotland', *The Cartographic Journal*, 30, pp. 3–12, 1993.

Stothers, Richard B., 'The great dry fog of 1783', *Climactic Change*, 32, pp. 79–89, 1996.

Szechi, Daniel, *The Jacobites: Britain and Europe 1688–1788*, Manchester: Manchester University Press, 1994.

Tabraham, Chris, 'The Military Context of the Military Survey', pp. 24–35 in Roy, 2007.

Tabraham, Chris and Doreen Grove, *Fortress Scotland and the Jacobites*, London: Batsford, 1995.

Taylor, Christopher, *The Roads and Tracks of Britain*, London: Dent, 1979.

Taylor, E.G.R., 'Robert Hooke and the Cartographical Projects of the Late Seventeenth Century 1666–1696', *Geographical Journal*, 90, pp. 529–40, 1937.

Taylor, E.G.R., 'Notes on John Adams and Contemporary Map Makers', *Geographical Journal*, 94, pp. 182–4, 1941.

Taylor, E.G.R., *The Mathematical Practitioners of Hanoverian England 1714–1840*, Cambridge, Cambridge University Press, 1966.

Taylor, Jane, *Signor Topsy-Turvy's Wonderful Magic Lantern; or, the World Turned Upside Down*, London: Tabart, 1810.

Taylor, John, *Mad Fashions, Odd Fashions, All Out of Fashions; or, the Emblems of these Distracted Times*, London: Taylor, 1642.

Thomas, W.K. and Warren U. Ober, *A Mind For Ever Voyaging: Wordsworth at Work Portraying Newton and Science*, Alberta: University of Alberta Press, 1989.

Thomson, James, *The Poetical Works of James Thomson*, London: Pickering, 1830.

Thomson, Thomas, *Annals of Philosophy; or, Magazine of Chemistry, Mineralogy, Mechanics, Natural History, Agriculture, and the Arts*, 3 vols, London: Baldwin, 1814.

Thomson, William, *Prospects and Observations; on a Tour in England and Scotland*, London: 1791.

Tillyard, Stella, *Aristocrats: Caroline, Emma, Louisa and Sarah Lennox, 1740–1832*, London: Vintage, 1995.

Torge, Wolfgang, *Geodesy*, 2nd edn, Berlin: De Gruyter, 1991.

Towse, Clive, 'Mordaunt, Sir John (1696/7–1780)', *Oxford Dictionary of National Biography*, Oxford University Press, 2004 [http://www.oxforddnb.com/view/article/19169, accessed 3 April 2010].

Turner, Gerard L'Estrange, 'Some notes on the development of surveying and the instruments used', *Annals of Science*, 48, pp. 313–17, 1991.

Turner, Gerard L'Estrange, *Scientific Instruments, 1500–1900: An Introduction*, London: Philip Wilson, 1998.

Turner, Michael Edward, *Enclosure in Britain 1750–1830*, London: Macmillan, 1984.

Tyacke, Sarah, 'Map-Sellers and the London Map Trade *c.* 1650–1710', pp. 63–80 in Wallis and Tyacke (eds), 1973.

Uglow, Jenny, *The Lunar Men: The Friends who Made the Future, 1730–1810*, London: Faber, 2002.

Vaughan, John, *The English Guide Book, c.1780–1870: An Illustrated History*, Newton Abbot: David & Charles, 1974.

Vertue, George, *A Description of the Works of the Ingenious Delineator and Engraver Wenceslaus Hollar*, 2nd edn, London: Bathoe, 1759.

Vetch, R.H., 'Pasley, Sir Charles William (1780–1861)', rev. John Sweetman, *Oxford Dictionary of National Biography*, Oxford University Press, 2004 [http://www.oxforddnb.com/view/article/21500, accessed 14 April 2010].

Vetch, R.H., 'Yolland, William (1810–1885)', rev. H.C.G. Matthew, *Oxford Dictionary of National Biography*, Oxford University Press, 2004; online edn, May 2006 [http://www.oxforddnb.com/view/article/30221, accessed 15 April 2010].

Wade, George, 'An Authentic Narrative of Marshal Wade's Proceedings in the Highlands of Scotland', II, pp. 268–337 (Appendix 3) in Burt, 1818.

Wakefield, Edward, *An Account of Ireland, Statistical and Political*, 2 vols, London: Longman, Hurst, Rees, Orme and Brown, 1817.

Wallis, Helen and Arthur H. Robinson (eds), *Cartographical Innovations: An International Handbook of Mapping Terms to 1900*, Tring: Map Collector Publications, 1987.

Wallis, Helen and Sarah Tyacke (eds), *My Head is a Map: Essays and Memoirs in Honour of R.V. Tooley*, London: Francis Edwards and Carta Press, 1973.

Walpole, Horace, *The Yale Edition of Horace Walpole's Correspondence*, 34 vols, ed. W.S. Lewis, London: Oxford University Press, 1895–1979.

Walters, Gwynn, 'The Morrises and the Map of Anglesey', *Welsh History Review*, 5, pp. 164–71, 1970.

Walters, Gwynn, 'Thomas Pennant's Map of Scotland, 1777: A Study in Sources, and an Introduction to George Paton's Role in the History of Scottish Cartography', *Imago Mundi*, 28, pp. 121–8, 1976.

Wellesley, Arthur, *Despatches, Correspondence and Memoranda of Field Marshal Arthur, Duke of Wellington*, 4 vols, London: Murray, 1867–71.

West, Graham, *The Technical Development of Roads in Britain*, Aldershot: Ashgate, 2000.

Wheeler, R.C., 'Topographical Accuracy of the Old Series one-inch map: artistic licence in the drawing office', *Sheetlines*, 28, pp. 9–10, 1990.

Wheeler, R.C, 'The transformation of the Ordnance Survey under Colby: A view from the Fens', *Sheetlines*, 76, pp. 46–51, 2006.

Whittington, G. and A.J.S. Gibson, *The Military Survey of Scotland 1747–1755: A Critique*, London: Historical Geography Research Group of the Institute of British Geographers, 1986.

Widmalm, Sven, 'Accuracy, Rhetoric, and Technology: The Paris–Greenwich Triangulation, 1784–1788', pp. 179–206 in Frängsmyr, Heilbron and Rider (eds), 1990.

Wiley, Michael, *Romantic Geography: William Wordsworth and Anglo-European Spaces*, Basingstoke and London: Macmillan, 1998.

Wilkes, John, *The North Briton, from No. I to No. XLVI inclusive*, London: Bingley, 1769.

Wills, Virginia (ed.), *Reports on the Annexed Estates 1755–1769: From the records of the Forfeited Estates preserved in the Scottish Record Office*, Edinburgh: HMSO, 1973.

Wilson, Charles and Henry James, *Ordnance Survey of Jerusalem*, Southampton: Ordnance Survey, 1886.

Wilson, Charles and H.S. Palmer, *Ordnance Survey of the Peninsula of Sinai*, Southampton: Ordnance Survey Office, 1869.

Winchester, Simon, *The Map that Changed the World*, London: Penguin, 2001.

Wise, M. Norton (ed.), *The Values of Precision*, Princeton: Princeton University Press, 1995.

Withers, Charles W.J., 'Authorizing Landscape: "Authority", Naming and the Ordnance Survey's Mapping of the Scottish Highlands in the Nineteenth Century', *Journal of Historical Geography*, 26, pp. 532–54, 2000.

Withers, Charles W.J., *Geography, Science, and National Identity: Scotland since 1520*, Cambridge: Cambridge University Press, 2001.

Withers, Charles W.J., 'The Social Nature of Mapmaking in the Scottish Enlightenment, c.1682–c.1832', *Imago Mundi*, 54, pp. 46–66, 2002.

Wood, Henry Trueman, *History of the Royal Society of Arts*, London: John Murray, 1913.

Woodward, David (ed.), *Five Centuries of Map Printing*, Chicago: University of Chicago Press, 1975.

Wordsworth, William, *Poems by William Wordsworth*, 2 vols, London: Longman, Hurst, Rees, Orme, & Brown, 1815.

Wordsworth, William, *The Poetical Works of William Wordsworth*, ed. Ernest de Selincourt and Helen Darbishire, 5 vols, Oxford: Clarendon Press, 1949–54.

Wordsworth, William, *The Letters of William and Dorothy Wordsworth*, ed. Ernest de Selincourt, 2nd edn, 8 vols, Oxford: Clarendon Press, 1967–1993.

Wordsworth, William, *Wordsworth: Poetical Works*, ed. Thomas Hutchinson, rev. Ernest de Selincourt, London: Oxford University Press, 1969.

411

Wordsworth, William, *The Prose Works of William Wordsworth*, ed. W.J.B. Owen and J.W. Smyser, 3 vols, Oxford: Clarendon Press, 1974.

Wordsworth, William, *Guide to the Lakes*, ed. Ernest de Selincourt, Oxford: Oxford University Press, 1977.

Wordsworth, William, *The Prelude 1799, 1805, 1850*, ed. Jonathan Wordsworth, M.H. Abrams, and Stephen Gill, New York and London: W. W. Norton, 1979.

Wordsworth, William, *The Fenwick Notes of William Wordsworth*, ed. J. Curtis, London: Classical Press, 1993.

Worms, Laurence, 'Faden, William (1749–1836)', rev. *Oxford Dictionary of National Biography*, Oxford University Press, 2004; online edn, Oct. 2009 [http://www.oxforddnb.com/view/article/37406, accessed 8 April 2010].

Wright, John K., 'Map Makers are Human: Comments on the Subjective in Maps', *Geographical Review*, 32, pp. 527–44, 1942.

Wyatt, John, *Wordsworth and the Geologists*, Cambridge: Cambridge University Press, 1995.

Wyatt, John, 'Wordsworth's Black Combe Poems: The Pastoral and the Geographer's Eye', *Signatures*, 3, pp. 1.1–1.20, 2001.

Yeakell, Thomas and William Gardner, *The First (Second, Third, Fourth) Sheet of an Actual Topographical Survey of the County of Sussex*, [London]: 1778–83.

Yelling, James Alfred, *Common Field and Enclosure in England, 1450–1850*, London: Macmillan, 1977.

Yolland, William, *An Account of the Measurement of the Lough Foyle Base in Ireland*, London: Longman, 1847.

Young, Arthur, *A Six Week's Tour through the Southern Counties of England and Wales*, London: W. Nicoll, 1768.

Young, Arthur, *The Farmer's Tour through the East of England*, 4 vols, London: W. Strahan, 1771.

Young, Arthur, *An Inquiry into the Propriety of Applying Wastes to the Better Maintenance and Support of the Poor*, Bury: J. Rackham, 1801.

Youngson, A.J. (ed.), *Beyond the Highland Line: Three Journals of Travel in Eighteenth Century Scotland. Burt, Pennant, Thornton*, London: Collins, 1974.

Acknowledgements

The work that has gone into this book was undertaken over a master's course, a PhD thesis and one-and-a-bit research fellowships, so there are a very large number of debts to acknowledge. Without generous funding this book would have been impossible: I am very grateful to the Arts and Humanities Research Council for master's and doctoral funding at the Universities of Oxford (Corpus Christi College) and London (Queen Mary); the British Academy for a Small Research Grant; the University of Glamorgan's Research Centre for Literature, Arts and Science and HEFCW for a two-year Research Fellowship during which this book was officially begun; and the Leverhulme Trust and the School of English and Drama at Queen Mary, the University of London, for an Early Career Research Fellowship, during which it was finished. I was immensely fortunate to be awarded first prize in the Royal Society of Literature Jerwood Awards for Non-Fiction, which was a wonderful financial and moral boost, so many thanks are due both to the Royal Society of Literature and to the Jerwood Foundation.

Existing work on the Ordnance Survey's 'biography' and the history of eighteenth-century cartography has proved inspirational. I wish to acknowledge the work of Yolande Hodson and the late J.B. Harley, especially their masterful edition of the Ordnance Survey's First Series, published by Harry Margary. The essays that preface each volume are wonderful. I have also learned a lot from J.H. Andrews's writings on the nineteenth-century history of the Ordnance Survey in Ireland; Charles Close's accessible accounts of the Ordnance Survey's early history; Matthew Edney's work on the Enlightenment context of triangulation; the late E.G.R. Taylor's explorations of the mathematical practitioners and instruments of Hanoverian England; and all the contributors to W.A. Seymour's exhaustive *History of the Ordnance Survey*. Biographies of Jesse Ramsden by Anita McConnell and of Nevil Maskelyne by Derek Howse have been invaluable. I have also been significantly influenced by Linda Colley's insightful works on eighteenth-century British cultural history and nationalism.

For providing stimulating places in which to conduct my research, I am grateful to the staff of the British Library (especially those of the Rare Books and Maps reading rooms); Cambridge University Library (the West Room, Maps and Rare Books rooms); the Bodleian Library in Oxford; the library staff at Corpus Christi College, Oxford; the National Archives in Kew; the National Archives of Scotland (especially the staff of West Register House); the National Library of Ireland; the National Map Library of Scotland; the Manuscript Library at Trinity College, Dublin; the Ordnance Survey at Southampton; and the Paris Observatory.

For assistance with locating images, and kind permission to reproduce them in this book, I thank the British Library; the Syndics of Cambridge University Library; Derrick Chivers; Dominic Fontana at the University of Portsmouth; London's Guildhall Library; the Joint Services Command and Staff College, Shrivenham (especially Chris Hobson, Head of Library Services); Laing Art Gallery, Tyne and Wear Archives and Museums; London Metropolitan Archives; Giles de Margary at Harry Margary; the National Galleries of Scotland; National Gallery of Ireland; the Trustees of the National Library of Scotland; the National Portrait Gallery, London; the Ordnance Survey, Southampton (especially Glen Hart); MAPCO: Map and Plan Collection Online (www.archivemaps.com); Routledge; the Royal Collection and HM the Queen; the Royal Society; the Royal Society of Antiquaries; Senate House Library, University of London (especially the Special Collections Administrator, Tansy Barton); Tate Gallery, London; V&A Museum; and the West Sussex Record Office. I am also grateful to Faber and Faber for permission to reproduce extracts from Brian Friel's *Translations*.

I have been very lucky to have a wonderful, energetic and inspiring agent in Tracy Bohan (Wylie Agency), and I am exceptionally grateful to the team at Granta, especially my editor, Sara Holloway, for her close reading of early drafts of this book and clear constructive comments. Benjamin Buchan has been an attentive copyeditor, with a great eye for clarity. I am also indebted to Mandy Woods for proofreading, Brigid Macleod for sales, Pru Rowlandson for publicity, Sarah Wasley for images, Amber Dowell for assistance, and Christine Lo for managing the editorial and production process. I owe a great personal and intellectual debt to Yolande Hodson, whose own research on the Ordnance Survey's early history is awe-inspiring in its depth and detail, and whose generous comments on the entire text of this book have been invaluable. I am grateful, too, to Carolyn Anderson, John Barrell, Valentine Cunningham, Jon Elek, Markman Ellis, Marilyn Gaull, Paul Hamilton, Mike Heffernan, Nick Hewitt, Holger Hoock, Anne Janowitz, John Mullan, Richard Oliver, Nicky Reeve, Catherine Delano Smith and Katherine Sutherland for their comments on versions of the material in this book that have appeared in conference papers, journal articles, doctoral work and early drafts of these chapters. And I am very grateful indeed to William Brown, who valiantly read the whole thing while laid up on the sofa with a mangled knee.

Other forms of assistance have also been greatly appreciated. Thanks are due to the members of the Charles Close Society for the Study of Ordnance Survey Maps, for thought-provoking conversations; Rosemary Baird for a fascinating tour of Goodwood House; Jean Barr for her hospitality and introduction to Carluke local history; John Bonehill for useful discussions about Paul Sandby; Rose Dixon for help with pictures; Chris Fleet at the National Map Library in Edinburgh for stimulating talk about the Military Survey of Scotland; Rita and Ben Henderson at Bunrannoch House for hospitality in a very beautiful part of the United Kingdom; Stuart Hepburn and the North West Region of the Royal Geographical Society and Institute of British Geographers, for showing me how to make maps using eighteenth-century techniques; Jim Jordan, for giving me a tour of William Roy's one-time abode on Argyll Street; Helena Kelly for thoughts about enclosure; Ian and Sue Newbon for warmth and kindness; to Elaine Owen, for a fascinating discussion about Roy and the early OS in general; Alistair Pegg for a virtual tour of Rhuddlan; Rebecca Peters for research assistance; Aurélie Petiot for help with translations; and Robert Prisk for his loan of a handheld GPS navigator. Special thanks are due to Althea Dundas Bekker and Henrietta Dundas for their astonishing hospitality at Arniston House, and their generous permission to consult the fascinating Dundas Archives and to reproduce certain images, for which I am immensely grateful.

My warmest thanks are to those who have supported me during the seven years over which this book was formed. I am indebted to Norma Clarke and Barbara Taylor for consistent encouragement; to Evonne Cameron-Phillips for more than I can say; and to my family, Kim Brown, Sarah Hewitt, Nick Hewitt, Willy Brown and Jackie Scott. I owe so much to my friends, Andrew Spencer, Alistair Pegg, Dominic Lash, Josie Camus, Isla Mackay, Sally Thomson, Will Tosh, and to Rebecca Linden and Joel Cooper, and little Ella Cooper, who I hope will grow up to enjoy the landscapes of the British Isles as much as I do. But most of all, I am so grateful to Pete Newbon, for love and happiness.

Illustration Credits

All images have been reproduced by permission where required. Where no copyright holder is stated, the image is in the public domain.

E. Paul Sandby, *Plan of Castle Tyrim in Muydart. Plan of Castle Duirt in the Island of Mull*, 1748. Reproduced by permission of the Trustees of the National Library of Scotland. Shelfmark MS.1648 Z.03/28e.

F. Paul Sandby, *A View Near Loch Rannoch*, 1749. Reproduced by permission of the British Library. Shelfmark K.Top.50.83.2.

G. William Roy, *A Very Large and Highly Finished Colored Military Survey of the Kingdom of Scotland*, 1747–1755. Detail: area around Inverness and Culloden (Grid Reference NH7146). Reproduced by permission of the British Library. Shelfmark Maps C.9.b.26/2e.

H. George Cruickshank, *The Antiquarian Society*, 1812. Coloured engraving. Reproduced with permission of Mr Derrick Chivers. Photograph copyright of Society of Antiquaries of London.

I. Sir Joshua Reynolds, *Sir Joseph Banks, Bt*, 1771–3. Oil on Canvas. © National Portrait Gallery, London.

J. Louis van der Puyl, *Nevil Maskelyne*, 1785. Oil on canvas. © The Royal Society.

K. Robert Home, *Jesse Ramsden*, c. 1790. Oil on canvas. © The Royal Society.

L. Thomas Yeakell and William Gardner, *An Actual Topographical Survey of the County of Sussex*, 1778–1783. Detail from Sheet 1, 1778. Reproduced with permission of the West Sussex Record Office. Shelfmark WSRO PM 48. Image © Dominic Fontana.

M. J.J. Hall, *Isaac Dalby*, 1816. Oil on canvas. Reproduced with permission of the Joint Services Command and Staff College, Shrivenham.

N. William Mudge, *General Survey of England and Wales. An Entirely New & Accurate Survey of the County of Kent*, 1801. Detail from Sheet 1: Isle of Dogs. © Crown copyright Ordnance Survey. Reproduced by the kind permission of the Ordnance Survey.

O. Thomas Baldwin, *Airopaidia: Containing the Narrative of a Balloon Excursion from Chester, the Eighth of September, 1785*, 1786: 'A Balloon Prospect from Above the Clouds' and 'The Explanatory Print'. Courtesy of Senate House, University of London, Porteus Library. Shelfmark: Porteus Library 05, SR.

P. James Walker Tucker, *Hiking*, 1936. Reproduced by permission of Laing Art Gallery, Tyne and Wear Archives and Museums.

Q. Robert Dawson, 'Cader Idris', 1816. Reproduced by permission of the British Library. Shelfmark OSD 319 pt 1.

R. Benjamin Robert Haydon, *William Wordsworth*, 1842. Oil on canvas. © National Portrait Gallery, London.

S. Ann Rhodes, 'Sampler, English, 1780'. © V&A Images, Victoria and Albert Museum, and Joseph Williamson. Museum No. 497-1905.

T. Robert Sayer, 'Board Game – A New Royal Geographical Pastime for England

and Wales; English', 1 June 1787. Hand-coloured engraved paper on linen. ©
V&A Images, Victoria and Albert Museum. Museum No. E.5307-1960.

U. William Blake, *Newton*, 1795/*c*.1805. Colour print finished in ink and watercolour
on paper. © Tate, London 2010.

V. *[Townland Survey of the County of] Londonderry. Surveyed in 1830–2, Six Inches to One
Statute Mile*, 1833. Detail. © Crown copyright Ordnance Survey. Reproduced by
the kind permission of the Ordnance Survey.

W. Charles Grey, *John O'Donovan*. Oil on canvas. Photo © National Gallery of
Ireland.

ILLUSTRATIONS IN TEXT

21. James Northcote, *William Mudge*, 1804, in Stamford Raffles Flint, *Mudge Memoirs*, Truro: Netherton & Worth,1883. Reproduced by permission of the Bodleian Library, University of Oxford. Shelfmark: 2182 M. e.28.   123

22. William Roy, 'An Account of the Trigonometrical Operation, Whereby the Distance between the Meridians of the Royal Observatories of Greenwich and Paris has been Determined', *Philosophical Transactions*, 80, pp. 111–614, 1790: 'Various Articles of Machinery Used in the Course of the Operation'. Courtesy of Senate House, University of London, Special Collections.   128

23. William Mudge, Edward Williams and Isaac Dalby, 'An Account of the Trigonometrical Survey Carried on in the Years 1791, 1792, 1793, and 1794 . . .', *Philosophical Transactions*, 85, pp. 412–512, 1795: 'A Plan of the Principal Triangles in the Trigonometrical Survey, 1791–1794'. Courtesy of Senate House, University of London, Special Collections.   134–5

24. William Mudge, Edward Williams and Isaac Dalby, 'An Account of the Trigonometrical Survey Carried on in the Years 1791, 1792, 1793, and 1794 . . .', *Philosophical Transactions*, 85, pp. 412–512, 1795: 'Plan of the Principal Triangles in the Trigonometrical Survey, 1795, 1796, 1799, showing the Directions of the Meridians at Black Down, Butterton, and St. Agnes Beacon'. Courtesy of Senate House, University of London, Special Collections.   138–9

25. 'Conventions used in the engraving of Mudge's Kent (1801), the Ordnance Survey maps of Essex (1805), and Kent (1816–19)', in J.B. Harley and Yolande O'Donoghue (eds), *The Old Series Ordnance Survey Maps of England and Wales*, Kent: Harry Margary, 1975, I, p. xxviii. Reproduced by permission of Harry Margary and Giles De Margary.   165

26. 'Observatory on the Cross of St. Paul's Cathedral' in 'Ordnance Survey of London and the Environs', *The Illustrated London News*, p. 414, 24 June 1848. Reproduced by permission of the City of London, London Metropolitan Archives.   173

27. William Brockedon, *Thomas Colby*, 1837. Chalk. © National Portrait Gallery, London.   178

28. William Blake, 'A Corrected and Revised Map of the Country of Allestone', in Benjamin Heath Malkin, *A Father's Memoirs of His Child*, London: 1806. Reproduced by kind permission of the Syndics of Cambridge University Library. Shelfmark: Keynes.U.4.10.   208

29. Samuel Taylor Coleridge, *The Notebooks of Samuel Taylor Coleridge*, ed. Kathleen Coburn, London: Routledge & Kegan Paul, 1957, Volume 1: *1794–1804, Text*, entry 1206 2.2. Reproduced by permission of Taylor & Francis Books UK.   210

Index

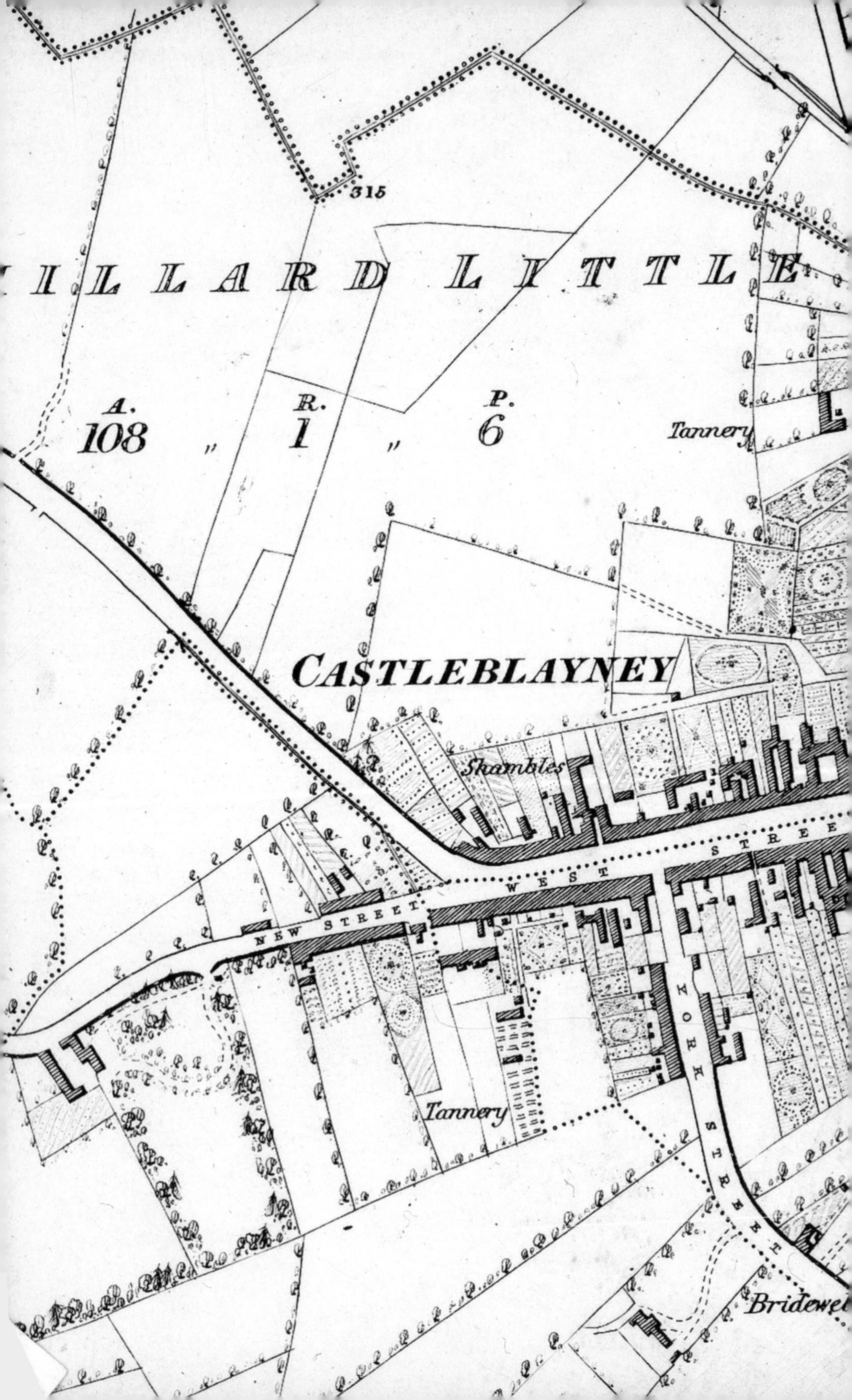

315

TILLARD LITTLE

| A. | R. | P. |
| 108 | „ 1 | „ 6 |

Tannery

CASTLEBLAYNEY

Shambles

WEST STREET

NEW STREET

YORK STREET

Tannery

Bridewell